# INVASION STRIPES

# INVASION STRIPES

*The Wartime Diary of Captain Robert Uhrig, USAAF
and the Dawn of American Military Airlift*

Brian J. Duddy
Lieutenant Colonel, U.S. Air Force (Retired)

Copyright © 2012 by Brian J. Duddy
Published April 2012
Second Edition


All rights reserved. No part of this book may be copied, reproduced or used in any way without written permission from the author.


Printed in the United States of America


ISBN: 978-0-615-62814-1

To the Uhrig and Duddy families and all the other members of
The Greatest Generation who ensured that our Nation remained
"The land of the free and the home of the brave"


Dedication from Bob Uhrig's Original Diary:

"To Toots"

Contents

Foreword by Colonel Arthur "Gus" Julian — ix

Preface — xi

Acknowledgements — xiii

Special Introduction — xv

Author's Introduction — xvii

| Chapter 1 | Day of Days | 1 |
| Chapter 2 | Join The Air Corps! | 3 |
| Chapter 3 | War Is Declared | 19 |
| Chapter 4 | "Airborne - All The Way" | 27 |
| Chapter 5 | By Air To Battle | 37 |
| Chapter 6 | Chasing the Afrika Korps | 45 |
| Chapter 7 | Traning and Tragedy | 75 |
| Chapter 8 | On to Sicily | 115 |
| Chapter 9 | Crossroads | 147 |
| Chapter 10 | Invasion | 161 |
| Chapter 11 | Liberation | 179 |
| Chapter 12 | Victory | 201 |

Epilogue - The Years That Followed — 219

Appendices — 221

Notes — 225

Bibliography/Sources — 227

# Foreword

Bob Uhrig and I were friends for many years. We first met at Wright Patterson AFB in 1948 when we were both working in Flight Test. He was the Chief of Maintenance and one day I returned from my first flight in an F-82 "Twin Mustang" with another pilot, Lt Breene. We taxied up to Captain Uhrig's maintenance hangar and instead of shutting down the engines and waiting for a tug to turn us around, Breene elected to swing the aircraft around using one engine. During this process, trash barrels were blown over and their contents scattered everywhere. Anything loose, such as engine cowlings and hats, went flying in the wind. I glanced at the open hangar door and saw this captain jumping up and down, yelling and waiving his arms. We shut the engines down and this captain (I later found out his name) proceeded to chew our rear ends off. That was Bob. After that we became great friends.

We spent a lot of time together on temporary duty at Muroc Field, later Edwards AFB, before I was sent to Korea. When I returned from Korea, I was stationed at Edwards with Bob and Toots and we worked together again in flight testing. There we had many great adventures and misadventures. Bob and I, Chuck Yeager and others used to like to go fishing up in Idaho. We had to get to some of our fishing spots on horseback since they were pretty remote. On one trip, one of our group, Tony Padavich, was having such a hard time with his horse, I recall Bob falling on the ground and laughing at him for twenty minutes. I also remember every time Bob caught a fish, he would scream bloody murder and you could hear him a mile away. He also snored so loud in his tent it kept all the rest of us awake.

On one fishing trip from Edwards to Twin Falls, Idaho, five of us - Bob, Yeager and I, Tony Padavich and our boss - had taken a C-47. On the flight back Chuck made the takeoff and then went back in the cabin to sleep. I took over as pilot and Bob was in the right seat. Although Bob was not a pilot he knew the old Gooney Bird well enough to navigate us home. It was late at night and we had to be at work at Edwards the next day, so Bob and I agreed to spell each other after 30 minutes of flying so we could also get some sleep. After my shift I turned the airplane over to Bob. I woke up some time later and looked over at Bob. He was asleep too, in fact everyone in the airplane had been asleep for almost two hours! I looked out the window and it was such a clear night I could see Mt Whitney. To my surprise the old C-47 had been flying along on her own, basically still on course all that time! I took over and got us back to Edwards with no further excitement.

I was very fond of Bob and Toots. Besides our off duty adventures together, I knew Bob Uhrig as a first rate Air Force officer. He was a natural leader, well respected by everyone and he was the best maintenance officer I ever worked with. I miss my old friend and our times together, but I am glad his wartime story is now being told.

Austin A. "Gus" Julian Jr.,
Colonel, USAF (Retired)

# Preface

The material in this book is drawn from two primary sources – Captain (later Lt Col) Robert Uhrig's 'war diary' which he faithfully kept and updated while in the combat zone from November 1942, through April 1945; and his letters home to his wife Ivea, who was nicknamed "Toots." The two primary sources have been integrated into a single story along with some amplification and background material where necessary for readers to understand the 'big picture' of the war and the scope of operations occurring in North Africa or Europe. Bob Uhrig kept a remarkably accurate and detailed diary and wrote hundreds of letters home to Toots during the war. Thankfully she kept almost every one and her family has preserved them along with the original envelopes. Reading them was like opening a time capsule from the 1940s. I was privileged to be able to open up and draw from this amazing and detailed grouping of wartime memoirs. Bob was careful to document every single important thing that he saw or did. It was also easy to see how devoted Bob was to his wife. He spoke of his love for her in every single letter and despite the long months of separation, his devotion to her never wavered. Their shared bond and faithfulness is truly inspirational.

Although some passages from the diary may seem routine and unexciting, they paint an accurate picture of life in a combat zone, what Greg "Pappy" Boyington called "hours and hours of dull monotony sprinkled with a few moments of stark horror." I have included as much material in Bob's original prose and style as possible, the thought being that someday these stories and recollections will be as important as the first-person narratives of men who had been present at Valley Forge, Waterloo or Gettysburg.

Bob's story was especially appealing to me for many reasons – it's a great WWII story, a great D-Day story and a great Air Force story. Bob's Air Corps career field was "engineering officer" which the Air Force now calls aircraft maintenance. As a former engineering officer and son of an aircraft mechanic myself, I have a special place in my heart for that career field and so I was especially excited to be able to bring a story like that to life. There are plenty of books about combat flying but almost none about what it was like to "Keep 'Em Flying" day after day, in many deployed locations and under harsh conditions. So his view of the air war in the Middle East and Europe is particularly important.

In every case that I could, I have included the sections of Bob's letters that mention his other comrades, squadron mates or friends from his previous assignment at Patterson Field. My hope is that the family members of these men might find additional vignettes about their loved ones described in Bob's own words. It would be wonderful to know that after 70-plus years they could find out something new about their service or adventures. As Bob himself said in the pages of his diary, "I would not take anything for the fond memories that I have. No matter how rich or poor one may become, these memories of things and friends will remain in our hearts forever."

Probably the most poignant part of moving through Bob's memoirs was reading the stories of his friends and comrades who would, later in the war, ultimately give their lives in combat or aircraft accidents. It is not a normal thing to see death on a daily basis, but as the war went on, it became a regular occurrence. Bob saw these men as they were during brief moments in time in their young lives and was no doubt deeply saddened by their loss. The significant tragedy of the enormous friendly fire casualties over Sicily affected Bob profoundly as shown in his diary. He was also conscious of the effect the long months overseas and family separations had on the morale of his men.

Bob's story is in many ways unique among Americans in WWII – he was a member of the pre-war Air Corps and was not part of the flood of volunteers and draftees that came into the service after Pearl Harbor. He is also unique in that he was deployed overseas for almost the entire duration of the Middle East and European wars – November 1942 through May 1945. During that time he took leave and passes in the local area but was never able to return home to the U.S. or see his family. Although airmen on flying status could rotate home after a defined number of hours or missions, support troops in the Army Air Forces were expected to stay deployed "for the duration of the war." Even the USAAF noted this disparity in treatment and its effect on morale in postwar studies.

Readers may notice some overlap between the letters and diaries. This was intentionally retained for the purpose of showing the difference between what the men in a combat zone could discuss or record among themselves, and what they were allowed to share with the folks back home. Much of this was due to security concerns.

A final note about language. Throughout this work, Robert Uhrig's words are his own. Other than very slight modifications or corrections to maintain readability, I have not changed the language or essence of his memoirs. They are what they are – including some words which may seem harsh to modern audiences. I leave revisionist history to someone else. But in all of Bob's letters and diary entries, it is significant to note that he never wrote a harsh or critical word about any of the men he served with. Perhaps it was one of the many positive lessons we can learn from the Greatest Generation – you're all in this together, so you had better learn how to get along and look after each other. Also note that in most cases, time is expressed according to the military "24 hour clock" method. In transcribing hundreds of hand-written letters I will also have undoubtedly made some mistakes with names or locations and those errors are entirely my own.

# Acknowledgements

This book would not have been possible without the perseverance and dedication of Robert Uhrig's daughter, Jan Uhrig Wiseley. Jan not only carefully kept and maintained her father's copious wartime archives, but lobbied without fatigue to have someone take up her father's story and make it known to the world. Jan's love for her dad is matched only by her excitement about his career and service to the nation. All of us who consider the Greatest Generation also the Greatest Heroes owe a huge debt to people like Jan who preserve our history for future generations. This is first and foremost Bob Uhrig's story and I am facilitating the telling of it with Jan's assistance. Jan and her sister Susie dug up all of Bob's records from the family archives for this book. In recent years, Jan's daughter Heidi also preserved some of her grandfather's memories in notes of her own.

Bob's original diary was typed up from handwritten pages by a woman named Kathleen Wright in the UK. These typed diary pages were discovered and preserved by her son-in-law, Alan J. Turner in 2008. Their family ran a pub in Nottingham during the war, close to Bob's base at Cottesmore. Alan transcribed the typed pages into a more readable version and it was through his diligent efforts that Jan received her father's diary.

Bob's great friend, legendary airman and test pilot Arthur "Gus" Julian graciously agreed to write the Foreword to give me another picture of Bob in happier times, when the two of them were stationed at Wright-Patterson and Edwards Air Force Bases. Gus' stories of their adventures after the war were every bit as exciting as Bob's wartime career.

A special thanks is also due to Mike and Nancy Ingrisano, the "Keepers of the Flame" for the 316th Troop Carrier Group. Mike's detailed chronicle of the 316th history: *Valor Without Arms*, was an invaluable reference to help tie together the events and personages in Bob Uhrig's story. I highly recommend it, along with Colonel Charles Young's *Into the Valley*, if readers wish to know more about Troop Carrier aviation. Of particular significance is Mike's fastidious roster of Group personnel, which aided in my identification of a number of individuals and the correct spelling of their names! As this book was going to press, we were saddened by the loss of Mike, but his contributions to Air Force history will never be dimmed. Nancy kindly allowed me access to Mike's scrapbook of wartime photos from the 316th TCG to help illustrate the book. Mike, as one airman to another, I salute you also for your service and devotion.

I was also fortunate to be able to interview some WWII vets for their slice-of-life view of Troop Carrier and Airborne operations – CBI "Hump" pilot Mahlon "Ham" Hamilton and 101st Airborne veteran, neighbor, military strategist and kindred spirit Richard "Dick" Ladd. I am always honored to spend time with you.

36th TCS veteran Col Bob Shawn provided some wartime photos of the 36th C-47 aircraft including the cover photo. Special thanks also to Alan Moore for his assistance in finding many wartime official photographs and proofreading the manuscript.

Unless otherwise noted, all the photographs in this book are from Bob Uhrig's collection.

# Special Introduction

Jan Uhrig Wiseley

Sue Uhrig Taylor

Thank you so much to our wonderful parents, who gave us the opportunity to travel and see the world. They were truly loved by us and many others around the world. Their adventures in life were akin to that of a movie. Letters that our father wrote to our mother during his 2 ½ year absence during WWII were that of a great love affair. We know it would mean a great deal to both of them that his WWII dairy that he kept for "Toots" his beloved wife was being published. During his later life he would often make the statement that "someday someone would want to read and share my exploits." Thank you so much, Brian, for making our father's wish come true.

# Author's Introduction

Airlift is airpower. Although airlift does not get much attention at war colleges because it is not as glamorous or exciting as strategic bombing or air-to-air combat, airlift is nonetheless a key instrument of airpower. It is not a nation's total force that the enemy fears, but only that portion that can be deployed and supported on the battlefield. In many parts of the world, airlift is the only means available to get combat forces to the fight in a timely manner. This work is about Bob Uhrig's experiences in WWII, but also the emergence and maturity of airlift as a vital component of combat power.

The Second World War began and ended with airpower. In Europe it began with German Stukas over Poland and for the U.S. it began with Japanese "Kates" and "Vals" over Pearl Harbor. The end came with two atomic weapons dropped from American B-29s. The emergence of airpower as co-equal partner with landpower and seapower drove all the warring nations to adapt their tactics, doctrine and equipment to new realities. Airpower compressed time schedules and forced nations to think in three dimensions. The pace of technology was unprecedented. When war began in 1939, all the major powers still had biplane fighters in their inventory. A mere six years later they had the first jet aircraft. For the first time, airlift forces and their airborne counterparts became significant elements in offensive operations.

Airlift is also frequently part of joint operations. Air transport forces are joined with ground forces to execute assault missions. With a surge of public interest in airborne operations after the release of *Band of Brothers* (both in book and film) there were some who came to criticize the actions of Troop Carrier crews in both the Sicily and Normandy invasions. They attributed the significant losses in those operations to aircrew incompetence or cowardice or both. In Bob Uhrig's account of the war there was no evidence that Troop Carrier men were any less brave, dedicated or steadfast than their paratrooper counterparts. Air Corps men were drawn from the same stock and population as the rest of the Second World War U.S. Army; with the same corresponding strengths, weaknesses, virtues and faults they all had. As President Roosevelt said in his D-Day address to the nation, "For these men are lately drawn from the ways of peace. They fight not for the lust of conquest. They fight to end conquest." In Bob's diary there are accounts of paratroopers refusing to jump. There are no stories of aircrew refusing to fly. As you will see from Bob's diary, even ground personnel such as he lobbied hard to participate in combat missions. No one wanted to be left behind.

Indeed, the men of the 316th Troop Carrier Group pioneered combat zone airlift, flying alone through hostile skies into flak and fighters to deliver their cargo to desert airfields still sown with mines and munitions. They were flying what was essentially a civilian airliner directly into the front lines. They demonstrated their devotion to duty in the Desert War. It's illogical to believe they would only summon courage for some missions and not others. If there is one criticism of their performance, it is probably fair to say that the crews did not have sufficient practice in night formation flying to be able to deliver thousands of men precisely on the drop zones while flying through cloud cover and flak. Or, that concentrated friendly ground fire was hazardous to tight formations flying at low altitude. The explosion in the size of airlift and transport fleets meant that training and preparations were not as thorough as they might have been in peacetime.

Once the Troop Carrier crews had better technology and sufficient practice, precision airdrops with light losses were achieved – in both Italy and Holland. Precision airdrops were also executed in the VARSITY operation but with significant losses in aircraft due to ground fire. Perhaps the fault lies in trying to adapt a civilian passenger transport – the Douglas DC-3 - to combat paratroop delivery. The concept of airborne warfare was still new and not thoroughly tested, and no purpose-built military transports with armor, armament and sophisticated electronics then existed. As will be seen in Bob Uhrig's narrative, doctrine evolved right along with technology.

The title of this book, "Invasion Stripes," has two meanings of significance. The first is the obvious: the broad black and white stripes painted on Allied aircraft just prior to the invasion of Normandy. As a squadron engineering officer, Bob was one of the first airmen to be let in on the secret plan for applying these "invasion stripe" markings as recognition insignia to avoid friendly-fire casualties over the beachhead.

He and his men worked non-stop to get them applied before the airborne forces loaded up for their epic nighttime drops over France. Even those with casual interest in WWII aviation know the significance of these now-iconic markings.

The second meaning is more broad. Bob Uhrig and his outfit were pioneers in the new military science of airlift and airborne warfare. They justly earned their "Invasion Stripes" with missions over North Africa, Sicily, Italy, France, Holland and Germany, while carrying the spearhead of invasion forces. They were pathfinders for the rest of the Air Force to follow. Their work and victories paved the way for other significant Air Force victories, starting immediately after the war with the enormously successful Berlin Airlift. The airpower concepts tried and tested by the Troop Carrier forces paved the way for that tremendous Cold War victory.

# Chapter 1

# Day of Days

*Almighty God: Our sons, pride of our Nation, this day have set upon a mighty endeavor, a struggle to preserve our Republic, our religion and our civilization, and to set free a suffering humanity.*
President Franklin Roosevelt, June 6, 1944

*Monday, May 29, 1944, RAF Cottesmore Airfield, United Kingdom*

"All we are doing is training and getting prepared for the invasion which I think is getting very close. We are just about finished and have our airplanes modified for the big thing. Everyone is getting nervous and jumpy. It is just like before the Sicilian deal. That is why I think it is pretty close because we are just finishing the same training as we did then. I am trying to go over the first night. I had Col Fleet, the Group Commander, just about talked into it, but now since he was killed on 12 May, I don't know how I will come out."

*Sunday, June 4, 1944*

"I had a 24 hour pass to go to London yesterday. Sgt Fenimore and I got all the way to the front gate of the base before we found out that there had just been a restriction put on, so we had to return. Our Squadron had their picture taken yesterday morning.

"About 1000 yesterday morning, Major McMenamin called all the Engineering Officers together and told us that we had to get our ships ready for a mission. We had to paint every ship in the squadron with 5 black and white stripes on the rear of the cabin and 5 on each wing. This had to be done besides doing all the maintenance and putting the equipment on. We finished at 0545 this morning, so we were all about dead. It has been kept secret about the painting so the Germans could not duplicate it. The Engineering Officers are the only ones in the squadrons that knew about it before hand. It is to try and keep our own men from shooting our ships down like they did in Sicily. If they can't see the markings on them now they should have their eyes examined.

"The paratroopers put their equipment on this morning. The mission is to be tonight and everyone is worked up. At 1300 they told us it was called off for 24 hours so we now have to sweat it all out again. The big show is about to start and we are all on edge. We transferred all but six of our navigators some time ago, so I asked the Squadron Commander if I could go along on the mission as navigator but as yet he has not received an answer. I want to be in on it and see what it is all about."

*Monday, June 5, 1944*

"It has rained all day and been very cloudy. We finished work early this afternoon. Sullivan, Welter and I talked about who was going to get what if someone did not come back. Most of our fellows were kidding each other about the mission, but you could tell that everyone was on edge. I thought most of it was impossible. We painted our war cry, *Tondeleyo* on No. 3 ship. Welter is flying No.3 and leading the second 9-ship formation of our Squadron. Sullivan is flying on his right wing. Tonight at 2300 our ships started taking off. All the ships in the Group got off OK. Each squadron has 19 ships flying.

"The place we are to take tonight has been bombed all day and 30 minutes before we dropped our paratroopers. The pathfinder ships are going over the drop zone with a flight of bombers to drop the men who will set up the radar equipment for our ships. The ground troops are not to start the invasion until 6 hours after we drop the paratroopers. Just now we are all sitting around Engineering, waiting. It is terrible to have to wait until they return. They are to drop at 0159 and be back here by 0405.

"We have a radio here in Engineering to listen to the news. About 1800 tonight the 'Berlin Bitch' said over the radio, 'Don't forget, we have a date with you tomorrow.' It is almost the same thing she said a year ago when we invaded Sicily. How they always get the information is beyond me."

*Tuesday, June 6, 1944*: THE INVASION OF NORMANDY (D-DAY)

"Our ships on the mission last night ran into some bad weather and had to break formation. We were all glad to see they all came back. They started coming in at 0400. As yet we have not received any information on the radio. On their way back, our ships passed over hundreds of boats so the navy should be giving them hell by now. I hope and pray that it is a success. If not, it is going to be a long war and it has been long enough already for me. We all feel better again because we feel we are doing some good once again."

# Chapter 2

# Join the Air Corps!

*Off we go, into the wild blue yonder!*

Half a world away from Cottesmore, England, and 30 years before D-Day, Robert Abraham Uhrig was born in November 1914 and raised on a farm in New Carlisle, Ohio. His father Harry died when he was just a boy and he was raised by his mother and grandparents. The family had modest means and the home life was very basic. Bob did all his homework by kerosene lamps and the family did not have electricity in the home until he was about ten years old. Before going to bed at night the family would heat up bricks in the wood stove and put them in the beds to keep them warm. Bob had his tonsils removed by the local doctor on the dining room table in the family house.

Bob worked helping his grandfather on the family farm for many years and learned a lot about handling animals and machinery. Like most country boys, he spent his free time hunting, fishing and swimming. He graduated from Olive Branch High School in 1932 and started "going steady" with a local girl, Ivea Nelle Fultz, known as "Toots" to her friends and family because she so disliked her given name.

Work after high school was tough to find in those days – the depth of the Great Depression. Bob was able to find work as a window trimmer in the department store where his mother worked. But like many young men of his generation, Bob was captivated by the romance and adventure of aviation. Since he had two uncles working at Patterson Field he tried but was unsuccessful at passing the Civil Service exam for a job there. Determined to succeed another way, he decided to enlist in the U.S. Army Air Corps.

The massive Air Corps installations of Wright and Patterson Fields were close to his family home in Fairfield (Dayton) Ohio, so in August 1936 at the advanced age of 21 Bob enlisted in the 1st Transport Squadron at Patterson Field. Initially, Bob only wanted to stay in the Air Corps long enough to get the necessary experience to pass the Civil Service exam. He had no idea he would remain in the Air Force for a full 30 years. During the interwar years, before the formalized basic training program developed in response to hostilities, enlistees were given their basic training at the unit where they enlisted. So, Bob's introduction to military life was all done near his hometown.

*Left*: Air Corps Recruiting Poster. *Right*: Contemporary postcard photo of Patterson Field, Ohio. (Author's Collection)

The pre-war Army Air Corps was a mere shadow of the organization that would grow to the most powerful Air Force in the world by 1945. In 1936 there were only about 100 Air Corps enlisted men in total stationed at Wright and Patterson Fields. The Wright Field side of town was almost exclusively dedicated to aeronautical research, development and aircraft production. Patterson Field was assigned to the Air Corps Materiel Division. Patterson was the home of the Fairfield Air Depot (FAD) that provided depot maintenance and logistics support to regional Air Corps units. The depot serviced 28 of the 50 Air Corps installations in the U.S. The transport squadron assigned to Patterson ferried high priority equipment and supplies to these bases as well as other Air Corps depots in San Antonio, Texas; Middletown, Pennsylvania; San Diego and Sacramento California.

The Patterson Field organization had grown out of a fairly recent concept for the Air Corps to perform its own supply function. In 1931, the Air Corps Materiel Division proposed creation of a new air cargo system. The four geographic depots would receive one cargo plane (initially a Fokker Y1C-14) and one enlisted pilot. The goal was that the Air Corps units could be independent of rail or truck transportation if they could accomplish all their transportation by air. The first of these air transport units would be at Wright and Patterson Fields. The airplane type was later upgraded. The Y1C-14s were replaced with the Bellanca C-27 Airbus.

Bellanca C-27 undergoing maintenance at Patterson Field.

Since Patterson Field was primarily a logistics hub, its subordinate military organizations were mostly supply and transport units. The initial concept of one plane and one pilot had now grown into a full Transport Group. The 10th Transport Group, activated May 20, 1937 at Patterson Field, was a consolidation of the 1st Transport Group and the 10th Observation Group. The 10th Transport Group subsequently was reassigned to Wright Field on June 20, 1938, but returned to Patterson on January 16, 1941. The Group then transferred to General Billy Mitchell Field, Wisconsin, on May 25, 1942.

The 10th Transport Group operated at Patterson and Wright Fields with C-27 and C-33 aircraft and consisted of five subordinate squadrons: the 1st (1937-l943), 2nd (1937-1943), 3rd (1937-1940), 4th (1937-1940) and 5th (1937-l944). The 2nd, 3rd and 4th Transport Squadrons were assigned to the San Antonio, Middletown and San Diego depots, respectively. The 1st Provisional Transport Squadron was activated July 15, 1935 at Fairfield Air Depot and assigned to the 10th Transport Group on May 20, 1937. While at Patterson, the squadron flew C-27s and Douglas C-33 and C-39 aircraft, as well as various civilian and military modifications of the DC-3. The 5th Transport Squadron was activated at Patterson Field on October 14, 1933, and operated a mixture of C-33 and C-39 aircraft.

When Bob joined the 1st Transport Squadron, it was barely a year old. Originally constituted as the 1st Provisional Air Transport Squadron on March 1, 1935, it was changed to the 1st Transport Squadron in June 1935 with a small cadre of enlisted pilots and mechanics. Bob joined the squadron as one of a handful of low ranking privates, responsible for general clean up and gopher work at the pay of $21.50 per month. He remembered that, "after they took out money for laundry and the Old Soldiers' Home, you had $18.75." Bob then began to work his way up the ladder as an apprentice aircraft mechanic. Although the squadron

was manned by military personnel, the depot activities were done primarily by Army civilians. The civilian workforce was very knowledgeable in aircraft maintenance and repair, and the 1st TS leaned heavily on them for support in maintaining their aircraft. Bob learned a lot by working on several different types of aircraft during those early years.[1]

*Left*: Bob Uhrig's identification card photo from Patterson Field.

*Right*: Some of Bob's buddies at Patterson. Note the Air Corps "wing and prop" insignia on their stripes.

The Fairfield Air Depot installation consisted of modern masonry and steel buildings and temporary living quarters. There was a central supply building and aircraft overhaul hangars. All of these structures were located on the high ground of the base that was known as "The Hill." The flying activities took place at the lower level of the field. The flight line had several old wooden hangars except the flight test building which was a steel building with metal siding. The Flight Operations building was little more than a small wooden hut manned by one person and a teletype machine.

Patterson Field hangars with Douglas transport.

After three months in the unit Bob was made a Crew Chief, responsible for one of the three-man maintenance crews assigned to each aircraft. Later this position was known as a Flight Engineer, since the lead man was expected to fly along with the aircraft when it went off-station. Bob was a quick learner and worked in every department in the Depot to get as much experience as he could on all types of aircraft.

The relaxed military environment of the Air Corps depot allowed the men plenty of free time between preparing airplanes for flight and helping out with visiting aircraft. There were very few officers assigned to the depot and most of the pilots were enlisted men. The enlisted crew chiefs also took turns going with the aircraft on cross-country flights to Selfridge Field, MI, Chanute Field, IL, Maxwell Field AL, El Paso TX and many other Air Corps bases. The pilots in the group taught some of the ground crew to fly during their off hours and Bob got a lot of stick time while assigned to Patterson. He had a desire to go to pilot training, but could not pass the physical due to color blindness. So instead he opted for a civilian flying license. Eventually Toots also got the flying bug and took lessons herself to get her pilot's license. Bob and a friend in his unit, Charles Lagana, bought a civilian plane and kept it in nearby Springfield Ohio.

The 10th Transport Group and its associated squadrons, located at various locations throughout the U.S., constituted the bulk of the Air Corps airlift capability in the pre-war years. This airlift operation was exclusively a peacetime logistics and supply mission. There was not much thought given to what the role of the Group might be in wartime.

By 1939 the 1st Squadron at Patterson only had about 100 enlisted personnel. The only officer assigned was the Squadron Commander and he had so many duties that the only time Bob and the other men saw him was when they mustered for pay day. This situation would soon change by 1940, when the Squadron expanded significantly. It was during this critical pre-war time that the Air Corps was experimenting and practicing with equipment and techniques that would be essential to the conduct of air operations WWII. As war clouds gathered, Bob and his comrades would soon be getting swift and immediate "on the job training" on how to operate an air logistics operation. In the years he was at Patterson Field, Bob quickly rose through all the ranks from Private to Master Sergeant.

As the 10th Group approached wartime status, additional Transport Squadrons were being formed. Patterson Field started receiving a large influx of new recruits to fill out the rosters of these new units. Unfortunately Bob felt many of the new men did not have the same level of mechanical knowledge as he and the other "farm boys" did and they required additional training. Bob took to recruiting additional local boys that he knew could handle the complex equipment.

In 1938 Bob and Toots decided to get married. To avoid the lengthy marriage approval process in their native Ohio, they traveled to nearby Mount Olive, Kentucky to take advantage of the single-day wedding system. They were married by a country preacher and had the two gravediggers from the church cemetery as witnesses. Toots was teaching school in North Hampton Ohio at that time and in those days only single females were allowed to teach. So their marriage had to be kept secret, both from the school and later from the Army. Two years after their marriage, the Army was discharging men who had married without permission. This time Bob had to go through the formal approval process and the couple was remarried in Mount Olive.

In these pre-war years, Patterson Field and in particular the neighboring Wright Field became the center of activity for new warplane development in the U.S. It was here that almost all the new aircraft designs were tested and evaluated. Bob Uhrig, in the center of this activity, was able to witness the operation of all manner of new ships of all types – bombers, fighters (known as "pursuit" aircraft in those days), ground attack aircraft and transports. Many of these aircraft were unique, one-of-a-kind prototypes. Some would never make it much past the drawing board, but others like the B-17 and P-40 would become the mainstay of the Army Air Forces in WWII. Bob had a chance to see them all in their original prototype form.

The Transport Groups were also receiving new aircraft on a regular basis. Eventually they traded their old C-33s and C-39s for newer Douglas C-53s and C-47s which were fitted out with equipment for military cargo transport and airborne (paratrooper) operations.

For a short period of time a third transport group was located at Patterson Field. The 316th Transport Group was constituted February 2, 1942 and activated February 14 also at Patterson where it was equipped with C-47 aircraft. On June 17 the organization, with its subordinate transport squadrons (the 36th, 37th, 44th, and 45th), was reassigned and relocated to Bowman Field, Kentucky (now Louisville Regional Airport.)

The day of the Pearl Harbor attack all the men assigned at Patterson Field were ordered back to base immediately. Bob and his maintenance crews spent most of the day parking aircraft that were instructed to land at Patterson and turn their aircraft over to the military. The same day the Alaskan Command asked for transport aircraft to be sent to Alaska to provide transport until they received their own aircraft. The first wartime mission for Bob Uhrig was close at hand.

# Bob Uhrig's Patterson Field Scrapbook
## and
## Images of Pre-War Air Corps Aircraft

## Aircraft Accidents

*Left:* Consolidated or Fleet biplane trainer crashed amidst the Patterson Field buildings.

*Above:* Early model Martin B-26 bomber that overshot the runway and went through the airfield fence.

*Above:* Wreckage of a pursuit aircraft.

*Above Left and Right:* Early model Boeing B-17 (B or C) bomber with collapsed main landing gear.

*Right:* Consolidated P-30/PB-2 attack aircraft after a gear-up landing.

## Pursuit (Fighter) Aircraft

*Left:* Boeing P-26 of the 1st Pursuit Group.
*Right:* Curtiss P-36

*Above and Left:* Curtiss YP-37 prototype/forerunner of the P-40.

*Below Left:* YP-37 and YA-14
*Below Right:* Seversky P-35

# Bomber Aircraft

*This page:* Boeing Y1B-17 "Flying Fortress"

# Bomber Aircraft

*Above and Below:* Douglas B-18 "Bolo"

*Above and Right:* The Boeing XB-15 on the ramp at Patterson Field. Only one of these prototypes was built. First flown in 1937, it was the largest landplane ever built in the U.S.

## Observation, Attack and Other Special Mission Aircraft

*Left*: Grumman OA-9 "Goose" amphibian. Army version of the aicraft widely used by the Navy and Coast Guard for anti-submarine patrol.

*Left*: Douglas O-46 observation aircraft.

*Left*: North American O-47 "Owl" observation aircraft.

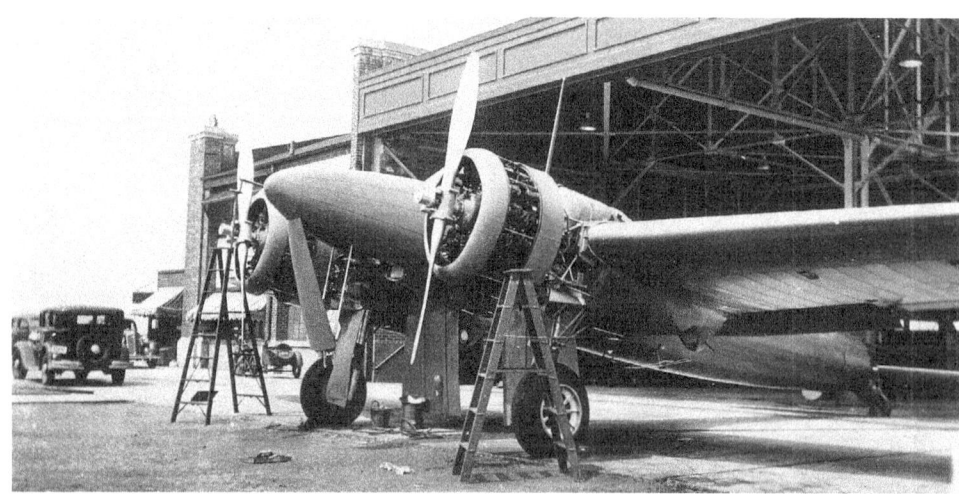

*Left*: Curtiss YA-14 attack aircraft undergoing maintenance at Patterson Field.

## Other Scenes from Patterson Field

*Right:* Northrop YC-19 "Alpha"

*Right:* Ryan YO-51 "Dragonfly" short takeoff and landing aircraft.

*Right:* Northrop BT-9 trainer. Fixed landing gear forerunner of the famous T-6 "Texan."

*Right:* Civilian monoplane with early B-17s in the background.

# Transport Aircraft - Douglas C-33s based at Patterson Field

There were 18 C-33 aircraft produced by Douglas. They were assigned Army serial numbers 36-70 to 36-87.

*Left:* Serial number 36-71.

*Above and Below:* Serial number 36-75. The F.A.D arrowhead insignia on the rear fuselage stands for Fairfield Air Depot.

# Transport Aircraft - Douglas C-33s based at Patterson Field

*Right*: C-33 at Camp Skeel, Michigan in July 1938.

*Above:* C-33 serial number 36-75 that taxied into the hangar door damaging both the aircraft and the building.

*Left:* C-33 marked below the pilot's window with Staff Sargeant Bob Uhrig's name as the Crew Chief.

## Transport Aircraft - Douglas C-33s Based at Patterson Field

*This page:* C-33 maintenance at Patterson Field.

# Transport Aircraft - Douglas C-39s Based at Patterson Field

The Army contracted with Douglas to produce 38 C-39 aircraft. They were assigned Army serial numbers 38-499 to 38-537. The vertical tails were marked with the last two digits of the serial number.

*Above and Below*: C-39 serial number 38-507 at Patterson, April 1940. The aircraft were delivered in natural metal finish with the colorful pre-war Army Air Corps insignia.

*Above*: Sparse interior of the C-39s.

*Above*: Two classics of the late 1930's.

*Above:* C-39 serial number 38-507 in the markings of the 10th Transport Group. The rear portion of the cargo door has been removed to permit easier exit for paratroopers in training.

*Above:* Bob Uhrig (kneeling, center) with the other crew chiefs of the C-39s.

# Chapter 3

## War is Declared

*As Commander In Chief of the Army and Navy,
I have directed that all measures be taken for our defense.*
President Franklin Roosevelt, December 8, 1941

In short order after the Pearl Harbor attack and the December 8th declaration of war, the 10th Transport Group was alerted for wartime action. Less than a week after formal hostilities were declared, the 5th Transport Squadron received their first deployment orders. The men and planes of the 5th were to support the movement of a squadron of Curtiss P-40E fighter aircraft from Spokane, WA to Alaska. The P-40s were from the 11th Fighter Squadron, a unit commanded by Major John "Jack" Chennault, son of Brigadier General Claire Chennault of "Flying Tiger" fame. The P-40s would need the assistance of the transports to navigate over the uncharted wilderness and to ferry the necessary supplies to support the fighter squadron in Alaska. The Alaska Defense Command also requested that transport aircraft be sent to support a buildup of Army forces in Alaska. This would be the first wartime test for the nascent USAAF airlift forces.

Bob and his comrades left Patterson field around December 10. Their first stop was Wichita Kansas. Here Bob penned the first of hundreds of wartime letters to his wife, a practice he would sustain throughout the war, regardless of the location or conditions. "Well, here we are still in Wichita. You should see our flying equipment they gave us. We have electrically heated underwear, shoes and gloves. Then we have fur lined coats that reach clear down to you knees. Fur lined pants, fur lined boots that are about 20 inches long. We have fur lined helmets and gloves and also a face mask and goggles."

Bob was able to send another letter home from Albuquerque on December 13. "If we keep on at this rate we will never get to Alaska. This is supposed to be a ten day jaunt but if the weather keeps on like this it will be more like 20. We at least found out where we are going -- we are supposed to go to Anchorage, Alaska. We can't fly at night anymore out here because they have blackouts on the coast and shut off all the radio beams, therefore you would not know where you were, and that is bad. A fellow was just telling us here that they shipped 1800 fellows from here to the Philippines and they got word today that twenty-some of them have been killed already. Send your letters to Elmendorf Field, Anchorage, Alaska."

From Wichita the group made its way to McClellan Field in Sacramento California on December 14. "Well at least we arrived at Sacramento." Bob wrote home, "Half of Patterson Field is here. The fellows that came out with the ten transports last week are still here. They are keeping them here for maintenance and they don't know when they get to come home. They are sending all the transports to the west coast. One good thing the orders have not changed for us and we are still to go to Alaska."

Sacramento was a place the Patterson Field men already knew well, as it was one of the large Air Corps depots in the Materiel Division. Here the C-53 aircraft were refitted for cold weather operation, and awaited the arrival of the P-40 unit, but there were complications with their deployment. Bob relayed this to his wife, along with the feelings of some of the rest of the men on their first deployment and perhaps their first Christmas away from home. At least he was able to room with his flying buddy, Charlie Lagana. "Here it is another day and we are still here. The men that we are supposed to take up there have not arrived yet. How this place knocks your spirits. The rest of the fellows are just stuck here and they don't know for how long. Lagana is lying on his bunk practicing drawing his pistol. All the way out here we would get out of the ships and see who could draw their pistol first. Two more ships came in from Patterson today so that leaves only two ships at home. The whole patch of ships will be out here in another week. We finished work on our ships today and now all we can do is wait."

But another day in Sacramento only brought more waiting. On December 17, Bob wrote again: "Tomorrow we are supposed to go to Ogden Utah, and we think we will have to stay there until January 6 before we even start to Alaska. The P-40s that we are supposed to go to Alaska with will not be ready until then. They will

have to be winterized yet and it will take that long to do it. All we can hope for is that they cancel the trip all together."

The next day did bring movement to Utah, although there was no room for them at Ogden, so they landed at nearby Fort Douglas in Salt Lake City. But their final destination was still uncertain. "We sure belong to the Lost Legion, it looks like by now. The pilots sent a wire to Wright Field tonight to see if they wanted us to stay here until we go to Alaska or not. So we will know tomorrow if we stay here or go to Ogden and then I will tell you where to send your letters. I should have letter scattered all over the country by this time."

The stay in Utah was brief, as the group was re-directed to Geiger Field in Spokane Washington. Crisscrossing the U.S. again, they arrive in Spokane in late December, just before Christmas. They would be stuck there in limbo for almost a month – unable to proceed with their original mission, and not permitted to return home for the holidays. Although he missed his family very much, a local family hosted Bob and two other members of his unit for Christmas dinner. It took some of the edge off of being deployed, but could never replace being home.

*Above:* Bob was treated to some spectacular scenery on his trip across the U.S., including the new Hoover Dam.

Since there was no flying mission for a while, Bob and the other members of his outfit had time to hike and explore the countryside around Spokane. "Spokane is a pretty nice town with very wide streets and also nice stores. It sits in a beautiful valley with a rapid river flowing through it. They say the fishing here is wonderful, but right now everything is frozen." Unfortunately, the mail had not yet caught up with the men and they were all a bit frustrated. "If I don't get some mail from home I think I will start walking to find out what the holdup is."

Bob's cross country trip also formed some of his ideas about activities for after the war. "When this shit is over, I want to take plenty of time off and see this old country of ours. You can hardly believe all there is to see and how beautiful it is. If anything happens to me, I want you to see all of the United States." Sadly it would be almost four years before he would get that time off. The war was now in full swing and experienced men like him were in great demand to form the ranks of the new transport and troop carrier squadrons.

*Above*: On their way through Los Angeles, the group stopped at the North American Aviation plant. Lined up on the ramp are new trainer aircraft awaiting delivery to the Army Air Forces.

Bob's first phone call home was right after New Years, 1942 and it certainly helped with morale, but their mission was still held up. "Two of our ships left today. And the rest of us are supposed to leave in a few days. But no one knows for sure. Yesterday they gave us more Arctic equipment. Listen sister, if I can ever get home with half the equipment that I have we won't have to buy any more warm clothes for 20 years. You cannot imagine what swell clothes that we have. Krebs said we would wear out the engines hauling clothes home. At White Horse where we will make a stop it was 60 below zero this morning. At Fairbanks it was 50 below. Also only two hours of daylight at Fairbanks this time of year. Some fun, eh?"

The days at Geiger field dragged on. "I have spent a lot of money so far on this trip. Everything that I have has gone on the blink. Both radios have had to be fixed. The rest is for chow. You know by now that soldiers cannot buy anything to drink only between six and ten in the evening. We spend most of our time in the evening going to the show here on the field. Last night we saw *Flying Cadets* and it was just piss poor. If *Swamp Water* ever comes to town you sure do want to see it because it is the best picture that I have seen in many a day. They say they are going to flood a field in back of the barracks so that we can ice skate."

On January 5 work on the aircraft continued but still no movement in sight, probably due to adverse weather along the route. "Today again we had the ships up on the line ready to go and then as per usual at about one o'clock they told us we were not going. It looks like we are never going to get started on this trip. Lagana bought a new pair of ice skates yesterday. He paid $15 for them. He is out trying them out now and I am listening to Jack Benny's program."

*Above*: On the trip the transport group passed through a field with a P-40 air defense unit. The aircraft are parked together in the middle of the field to guard against sabotage which was a major concern after December 7, 1941.

January 9th came and they were still stuck at Geiger Field. "We almost did get started today, but they found out that the weather was bad so now we are supposed to leave tomorrow. The first day out I will write and tell you where to send my mail. Lagana has to stay behind a few days so he will gather up all the mail and fruit cake and bring it up when he comes. We are going in town tonight and get some chow to take along with us. We are afraid eating will be a problem on the way up there. We are going up with about half of the original P-40s. The rest of them are cracked up all over the country. That is what Lagana is going to do, wait here and see how many of them can be fixed up to send on their way."

The weather cleared on January 14 and the transport squadron began their long-awaited trek northward. Their first stop was Calgary, Alberta. It was a welcome diversion from the lengthy stay at Geiger Field. "Well at least we are on our way. It just does not seem right. Now maybe I will get home soon. We are staying here in town tonight and expect to leave early in the morning." It was here Bob had his first exposure to the Allies he would fight with in the coming months. "We landed at a Royal Air Force field and you can hardly understand them. They seem very nice and could not seem to do enough for us. You can talk to men from all of the English possessions right on this one field."

Bob soon discovered something that soldiers from many nations have discovered over the centuries; that is, they have more in common with other soldiers than with some of their own countrymen. "We have had some long talks already with the English fellows or chaps. The one I liked best is from Dublin, Ireland and was a pilot over here training other pilots. We exchanged addresses and had long talk about London, autos, money and rank. It is sure fun to compare notes. Really, old bean, it is jolly good fun. The one from Dublin has his wife coming over in about six weeks. His eyes just about popped out when he saw an American silver dollar. He gave me an English two cent piece and I gave him a penny and a nickel."

Bob's first priority as a crew chief was the condition of the aircraft for the long trip ahead. "My right engine about jumped out twice today and I could not find out what was wrong. Then it smoothed out and has run all right ever since. I don't know how far we will go tomorrow. I hope we get plenty far because the more time we make the sooner I will get home."

The next day the transport group resumed the flight north, but they were diverted to search for the crews of three B-26 bombers that were lost in the Yukon Territories. The new and unproven B-26s were part of the 77th Bombardment Squadron, stationed in Alaska. One of Bob's squadron mates, Cecil Petty, picks up the story around January 15. "I flew co-pilot for Frank Krebs on this trip, during which we saw some beautiful country. Not long after our arrival in Ft. St. John, we heard about a lost flight of three Martin B-26s. We had met the crews earlier in Edmonton, and remembered that they had not been very well trained in instrument flying… We had some information they were in a large, wide snow covered valley somewhere, but nobody knew where….The next morning we took off in two C-53s each with five P-40s on our wings and began our search. Fortunately we figured right and we found them that same day, everybody okay."[2] Bob was also concerned about the fate of the men, and in a short note to his wife from Nelson Lake he wrote: "The ship just came in for the mail. They have not come in with the B-26 crews yet."

P-40E fighters on the wing of the C-53 transports as they made their way to Alaska.

On January 16 Bob noted that the temperature was "real mild here compared to only ten days ago when it was –23. Of course we expect some real cold weather farther north with some to run as low as –60. Sure hope we can finish our mission soon and get back to our house."

Most of the airfields they used on the journey north were not finished – only care-takers were available to assist them. By January 17 the transport unit was at St. John, British Columbia and Bob described the area

and their situation. "The field here had one runway marked by cedar trees which they cut and stick in the snow along each side and the ends. The snow is about 18 inches deep, too much to keep off the runways so they pack it with heavy rollers. One ship a week passes through here which is the Yukon Southern. All the fields were started about last July. They cut the trees and then pulled out the roots. Ft. St. John is the last place close to a railroad which is 60 miles from the town which has 200 people, one hotel which can sleep nine people, one beer hall and two cafes. Even the highway ends here.

"From here is it nothing but wilderness. You have to service your ships from 50 gallon drums. Be sure you have passed over the end of the runway before you land! The town is about two miles from the field. The water sells for 40 cents a barrel because they have to haul it 16 miles over a rough trail from the river. At the airport they have an oil house and one radio shack which operates only during the daytime which is about seven hours. This time of year you can land on any river or lake, but you must watch on landing because the snow is deep and you have a very good chance of nosing over. If you are ever forced down you want to stay with the ship because you would never have a chance by yourself. One more thing, they would never see you from the air and you could never find your way out unless you were very lucky."

But at least there was some good news on the B-26 aircraft and crews. "The B-26s that were ahead of us, only two got through. We have been in contact via radio with another two that have cracked up about two miles apart. They are going to call us again at three o'clock so we can take a reading on their direction and then we are going to take supplies and look for them and drop them out of the airplane. They will have to bring out the men by dog team."

As for their primary mission of supporting the fighter aircraft: "We are having real good luck with our airplanes. One of the P-40s ran into a Canadian primary trainer at Edmonton and cut both the upper and lower wings off about two feet. It took only a half an hour to repair the P-40, but it will take quite a while to repair the trainer. As yet we have not lost any of our ships. The first group lost three of theirs but no one hurt."

The P-40s still together as a group at one of the intermediate stopping points on the way to Alaska. All the airplanes had to refuel by hand using the drums stacked around the parking area.

The next day the group had made their way to Ft. Nelson, BC. "The field here also has just the one runway marked with cedar trees. They also pack the snow down here and service the ships from 50 gallon drums. The post consists of a construction camp of about 50 men. You can get a bunk in one of the bunk houses and use your sleeping bag. The windows are of cloth. You can buy a meal for 50 cents which is cheap. There are about ten buildings which are built out of logs and the snow banked around them. They are heated with old oil drums used as stoves. The Ft. Nelson Trading Post is about three miles south of the field. The town there consists of about eight houses. They have only one store and pool room, one trading post, the British Columbia Police and of course the Hudson Bay Company, of which every town has one. The rest are log cabin homes.

"Everyone had a dog team and a toboggan. The town is right on the bank of a river. They use dog teams for transportation in the winter and boats in the summer. The timber here is mostly pine trees which are 70-80 feet tall and straight and very close together. From the field to town it is all down hill and only a dog team trail. All fuel and supplies are hauled in by air – no other connections with the outside except by radio. You must run your gasoline through a piece of felt or chamois skin to filter the water out of it. All of this territory is a new air route which is now being used for the first time so you might say we are pioneering it. All the Alaskan flights have been next to the coast where the climate is mild and there is more civilization.

"The air bases and fields were being constructed at the same time as the Alaska Highway. The B-26s had

managed to land next to a small stream. There was deep snow in the area, but no fatalities in the mishap. The crews had plenty of supplies and food but the area was so rough that the bush pilots could haul only a few men out at a time."

By January 19 the 5th TS arrived in Watson Lake, Yukon, where Bob dashed off his last letter of the trip, confirming that the bomber crews had been located. "The Junkers brought out four of the B-26 crew members this afternoon. They had to leave the police behind so that they could get off the ground with the injured. They were not hurt terribly bad. All their injuries were taken care of here. They are going to send them to Whitehorse tomorrow. There was an army doctor sent down from Fairbanks to take care of them. It will take three more days for the rest."

*Above and Left:* The wrecked B-26s in Alaska. Repairing them without heavy equipment or facilities in this environment would have been next to impossible.

Bob was now experiencing his first taste of life in the frozen north. "To get heavy machinery in here can only be done in the summer. From Vancouver to Wrangell Alaska it goes by coast steamer. From Wrangell to Telegraph Creek by river boat 160 miles. From Telegraph Creek to Dease Lake 72 miles by Caterpillar Train. Across Dease Lake from Dease River to Lower Post on the Liard River is 200 miles. Lower Post to Watson Lake 23 miles by Caterpillar Train.

"Potatoes are $27 for 100 pounds and the rest of the chow is just as high. The 20 men in this construction camp are all that exist within 20 miles and at the end of 20 miles is a metropolis of 30 people and a Hudson Bay Company. When they build something up here they must first have a camp. This field has a large building with running water, something plenty new in this part of the country and a mess hall and sleeping quarters all under the same roof. The men here have made room for us to sleep using our own sleeping bags."

Ordinary items had great value in such a remote location. "If you ever come up in this country, please bring shit house paper and whiskey. You can get about anything for that. All outside toilets with no paper. Of course if there were news boys we would have plenty."

On January 23 the C-53 aircraft took off to go to Whitehorse to get more gasoline for the Junkers transport that rescued the B-26 crews. But half way there the ice on the wings of the aircraft was so thick that they had to turn back. Bob's last letter on this journey was posted back to Ohio via a Yukon Southern Steamship.

The 5th TS finally completed their mission to deliver the P-40 unit to Alaska where they later took part in repelling the Japanese invasion of the Aleutians just prior to the Battle of Midway in June 1942. Bob and his unit stayed in Alaska for four months, supporting the newly-formed 11th Air Force by shuttling supplies around the northern defense perimeter of the U.S. and Canada. The majority of flights were to the Aleutian and Kodiak Islands.

During his time in Alaska, Bob labored to keep the aircraft flying in austere conditions. All maintenance was done on the flight line – there were no hangars – and portable heaters were used to warm up the equipment just enough to keep the tools from freezing to the men's hands. Wing covers had to be put on the aircraft every night as there was no deicing fluid available. Bob even became proficient at driving a dog sled after a few lessons from a local trapper.

*Above Left:* The extent of the fueling and maintenance facilities in Alaska.

*Above Right:* Dog sled team.

*Left:* After a long deployment, a welcome rest on the way back to Patterson Field.

Upon their return to Patterson, Bob discovered that some of the senior enlisted personnel had been offered commissions as officers. Bob also volunteered to accept a commission and was sent to Warrant Officer School. He passed the Warrant Officer exam but later he came to question his decision to accept the promotion. As the squadron settled back in to Patterson field they prepared for their next challenge – developing airborne troop carrier tactics with the newly-formed parachute infantry regiments. The squadron would soon be on the move again.

Bob Uhrig's Alaska Scrapbook
January-April 1942

# Chapter 4

## "Airborne -- All The Way"

*Where is the prince who can so afford to cover his country with troops for its defense so that ten thousand men descending from the clouds might not, in many places, do an infinite deal of mischief before a force could be brought together to repel them?*
Benjamin Franklin, 1784

**While** Bob and his squadron were delivering the P-40 unit to Alaska, major organizational changes were underway for all U.S. air transport units. Douglas C-47 aircraft were now pouring off the assembly lines. These updated versions of the C-33 and C-39 aircraft would be used to fill the squadrons of the new transport groups. The 316th Transport Group was activated in February 1942 at Patterson Field while Bob and his squadron were running cargo in Alaska. The units formerly stationed at Patterson now became the nucleus of the 36th and 37th Transport Squadrons. Most of the enlisted men such as Bob were taken from the 10th Transport Group. Bob was assigned as a crew chief in the 36th Squadron and was one of the experienced cadre of men who would form the backbone of the wartime units. In June 1942, two more squadrons (the 44th and 45th) were added to the unit which was then re-designated as the 316th Troop Carrier Group (TCG). That same month the Group was moved to Bowman Field, Kentucky, and Bob was transferred along with it.

The stay in Kentucky was brief and the 316th was moved again to Lawson Field-Ft. Benning, Georgia on August 9, 1942. Ft. Benning was the new home and training field for the U.S. Army airborne force. This force included paratroopers and glider-borne infantry, artillery and other combat units that would be delivered to battle by air. It was a new concept in American military doctrine, and Ft. Benning was the place where the concepts were developed and the airborne soldiers and entire units were trained and prepared for war. Lawson Field was the Army Air Forces (AAF) facility at Ft. Benning.

The framework for American "Air Infantry" was built beginning only in 1940 by then-Major William Lee. Lee would later be known as the "Father of the American Airborne." He spent the first half of 1940 working with the Air Corps to bring together the aircraft and aviation expertise necessary to support parachute and glider forces. This concept included new parachute equipment for both the paratroopers themselves and the transport aircraft that would carry them into battle.

The first Parachute Test Platoon was formed out of the Infantry School at Ft. Benning in the summer of 1940. These volunteers would practice jumping with new military troop parachutes from transport aircraft like the C-39, C-53 and C-47. These specially designated "troop carrier" aircraft had to be configured with heavy cables running the length of the passenger compartment of the fuselage. The paratroopers would clip the "static lines" (ripcord) of their parachutes to these cables so that as soon as they leapt from the rear door of the aircraft, the parachute would be pulled out of its backpack and open almost immediately. This was necessary in order to get the maximum concentration of paratroopers out the door and on the drop zone in the least amount of time. The aircraft would normally be flying at a low altitude, so there would not be much time between parachute deployment and the jumper touching down on the drop zone.

The concept of airborne infantry was new on many levels. The paratroopers had to be trained in parachute packing, deployment and landings. The aircrews had to be trained and practice precision navigation to find the drop zone, and close formation flying to deposit thousands of men on a drop zone in enemy territory with speed, accuracy and a minimum of losses. And finally, the aircraft themselves had to have suitable parachute equipment including large rear cargo doors, cables, precision navigation equipment and air to ground communications. Aircraft like the C-47 which were originally designed as civilian airliners had to be modified for these specific requirements. All these tactics and equipment had to be in place and well-rehearsed in order for the airborne units to be effective in combat.

None of these things had even been tried – much less perfected-- in 1940. Once hostilities began the airborne training began in earnest. The first U.S. parachute battalion officially came into being in September

1940 but the first parachute group was not formed until March 1941, and then only as a training school cadre for the creation of future airborne regiments and ultimately, airborne divisions.[3] At that point, the new "Jump School" training went into high gear to prepare volunteer soldiers for these airborne divisions. In addition to infantry soldiers; artillerymen, engineers, signal troops and medical detachments also had to be integrated together in airborne units. All of their weapons and equipment had to be "re-packaged" so that it could be transported and delivered by air, rather than by truck or horse (the Army was not yet fully mechanized) and then quickly reassembled when the paratroopers hit the ground. A tremendous amount of work had to be done to prepare even a small unit to be ready to drop into combat and survive.

Jump School training was rigorous and intense. A paratrooper of the 101st remembered it this way: "At that time, entire battalions would go through jump school as a unit. About 800 men started the jump school class, but only 600 to 700 made it to the actual jumps. We took off from Lawson Field and headed to the nearby dropzone next to the Chattahoochee River. At Benning, the C-47s carried 24 jumpers – two 'sticks' of 12. All the training jumps were made at 1200 feet above ground level. The aircraft flew as single ships, trailing each other about five minutes apart. When we reached the drop zone, the first stick of 12 jumpers would exit, then the aircraft would circle around and the second stick of 12 would jump. We made a total of six training jumps at Benning, five in daylight and one night jump before we received our wings as qualified paratroopers."[4]

In parallel with parachute troops, another innovation in airborne warfare was being explored in the U.S. The Germans had had some success with troops delivered to battle in gliders so the U.S. Army began experiments with glider soldiers. The glider concept called for an unpowered aircraft, capable of carrying a squad of troops, and towed behind a transport aircraft. These gliders would be "cut loose" from their tow-planes a short distance from their intended landing zones and the gliders would all land in a designated area and discharge their soldiers who could immediately go into battle. The U.S. glider concept eventually formed around the WACO CG-4 troop glider. This aircraft could hold up to 15 troops, plus two pilots; or a jeep or a light artillery piece. The glider delivery plan as it later developed was to tow 2 CG-4 gliders behind a C-47. As with parachute drops, this maneuver also required well-developed flying skills to keep the gliders and tow-planes in formation and deliver them over long distances. In its final form an airborne division then would include both parachute and glider infantry and artillery units.

None of these concepts and techniques had been perfected when the 316th TCG arrived at Ft. Benning. The first large-scale airborne practice exercises were held as part of the Army Louisiana Maneuvers in September 1941 and the Carolina Maneuvers in October-November 1941. Bob and his unit participated in the Carolina exercises, operating out of Shaw Field along with the airborne troops. "I am writing this letter at 8000 feet, on our way to San Antonio to pick up infantry troops. There are 43 airplanes up here this morning and are they every pretty. We are so close to the ship next to us that I cannot get the whole ship in the lens of my camera. It is about 15 feet from the wing tip. We will leave San Antonio tomorrow at 0500 to come back to Sumter (Shaw Field). The field is a new one; in fact it is far from being finished. We have a room for every four men and is that ever swell. We have hot water and plenty of heat. The field will be used for British cadets starting in December."

*Above:* Tent City at Shaw Field (US Army)
*Left:* Aircraft from the San Antonio Air Depot with temporary exercise markings (red cross).

Once the troops were all in place, the full scale exercise began. As Bob described it, "We cannot leave the post for six days because the war started at midnight last night between the Red army and the Blue army. We are fighting for the Red army this week. You should see all the troops in the woods along creeks everywhere you look. One of our generals got captured already. Don't worry if I don't write because we are busy 24 hours a day."

Unfortunately the small number of Troop Carrier aircraft in the USAAF did not permit the large-scale deployment of airborne forces thought to be representative of future operations. "In November 1941 for instance, the wing participated in the Army's famous Louisiana (sic) maneuvers in which ill-equipped troops used broom handles to emulate machine guns and jeeps to simulate tanks. Airdrops had been planned, but the 50th Troop Carrier Wing had barely improved its readiness in this area. The wing had difficulty providing 39 airplanes for the airborne segment. With these airplanes, however, the Army did make its first airdrop of more than one company of paratroopers at one time."[5]

In spite of the limited numbers of aircraft available, the Carolina Maneuvers were good practice for the mass parachute jumps that would come later in the war. Bob Uhrig saw these maneuvers from the cabin of a Troop Carrier aircraft. "Yesterday we dropped out 400 parachute troops at one time. Then we rushed back and got 500 infantrymen and landed them in the field where we had dropped the parachute troops. If you ever want to see a beautiful sight you should see 400 parachutes in the air all at once. If you were in the maneuver area you would think that you were actually in a war. You can't believe how much it is like pictures from England that you see."

*Above and Below*: Mass drops of paratroopers over Lawson Field, Georgia, 1942.

The Troop Carrier crews were busy each day of the exercise, throughout most of November. "We dropped 300 troops yesterday. We dropped them in a practice field just at dusk. They were supposed to capture a bridge from the enemy. We have not heard whether they did or not. This afternoon we go to Ft. Benning to take part of the troops home. The engines on ship 07 just about jump out of the airplane every time we take off. The sand down here has ruined about half our engines. I will have to change the ones on my ship when I get home and they only have 50 hours on them. But they are supposed to go for 550 hours. The fellows on the ground say it looks like we are sky-writing there is so much burned oil coming out of them. Koczan parks his ship beside mine and his is very nearly as bad."

By this time in the war there was an urgent need for qualified men to become officers and pilots. Bob had a strong desire to do both but unfortunately was held back by his minor medical condition: color blindness. Color blind men would not be admitted to pilot training. While he was at Shaw Field, "They sent me to Ft. Jackson [SC] to take the exam for officer candidate school and of course they caught my red and green eyes. But I am going to take the next one at Patterson and Krebs says we will fix it then. They take married men now." Bob was hopeful he could somehow beat the system and finally get into pilot training.

A lack of sufficient transport aircraft (either C-53 or C-47) slowed the pace of airborne development. The aircraft were needed for both jump school training missions as well as practicing mass jumps as part of these maneuvers. All these missions were in addition to their regular transport and logistics duties. As it was now impractical for the large airborne units to "borrow" transports when they needed them, a permanent party of troop carrier aircraft had to be stationed at Lawson Field side of Ft. Benning. The first two U.S. airborne divisions – the new 101st and the original 82nd – were declared "operational" on almost the same day that the 316th arrived at Benning.

All of this points up another change in the Air Corps thinking that was now required. Air transport units were traditionally viewed as non-combat logistics support organizations. Now the newly-formed troop carrier units were expected to become fighting units. This meant not just combat support, but front-line units that would drop the airborne forces in the midst of heavily defended enemy positions. Their unarmed and unarmored aircraft would be particularly vulnerable to fighter aircraft and antiaircraft artillery. But that was the concept for airborne troops and it required a dramatic revision of thinking in Air Corps circles. It would also require significant training if a combined airborne-troop carrier team was to be successful.

It was in these untested waters that Bob Uhrig and his comrades in the 36th TCS found themselves in August 1942. Once settled in, his first letter home from Lawson Field was on August 10. "Here we are down in the Deep South and have we ever been busy. It is just about twice as hot here as it was at Bowman, so from that you can form your own conclusions. It will be about two or three days before I can get down to serious letter writing." Bob found an on-base apartment and wrote home to Toots on August 12, "Sister is it ever hot here! Bowman Field is like being up north compared to down here. This post is so large that it takes about a half an hour to drive off of it and we don't even have any way to drive off so here we are. As yet I have not been off the post even once. You should see the exercises that the parachute troops go through. It is enough to kill any ordinary man. And what! I mean they are tough. After they go through their course I would even be afraid to look at them twice. We are now about ready to start operations again after so long a time."

Two weeks later, Bob wrote home again and he was already in the midst of the frantic pace of the airborne training regimen. At Lawson Field the 316th received 52 brand new C-47 aircraft and a dramatically increased crew complement. The Group then began intensive training in formation flying, carrying paratroopers and solving problems involving airfield capture.[6] Bob and his mates now had a massive workload to contend with. "I pity myself so terribly much. We are working something terrible. They try to get 12 hours a day on these airplanes and the just won't take it. Since I came back I have never been to bed before one thirty in the morning. Everyone on the line is just about ready to drop because we have very few that know enough to fix them yet. I am taking time to write this while we are waiting on some parts. I just can't find time to write every day. When we do get finished, we are even too tired to wash so we just go to bed dirty."

In the weeks that followed, the pace did not relax and on top of everything it was the middle of hurricane season in the South. "They are still doing as much flying as ever and we are doing as much work as ever. But there is nothing else to do so one does not mind it so very much. We are now staking down all of our airplanes because there is a storm warning from the Gulf. If the wind gets too bad we have to take them off and fly them out of the storm area."

*Left:* Bob in front of his barracks at Lawson Field.

*Right:* Contemporary postcard of Lawson Field hangars. (Author's collection)

By September 1942 the 316th was training to a high standard and the men sensed that another movement was near. They expected it might be overseas, but no one knew for sure. "I am just about dead all of the time. The only time that I can write is in between events. There is something going wrong about ever half hour around here. Boy, if it stays like this very long I will have to throw in the sponge. I think I will be with you in the near future so just keep your fingers crossed until it happens."

Bob and his mates continued work non-stop to support the airborne training and exercises, but with so many new aircraft and so few trained maintenance men, he was overwhelmed. "I have to tell you again about my writing. It just seems impossible for me to find time to write. You remember how much I wrote when I went to Alaska, but then I was not so busy. I just go around in a spin half of the time. I don't think I know what I am doing. There seems like something else coming up every two or three minutes."

The something else came soon enough but not what the men were expecting. The next move for the 316th came at the end of September. Bob wrote again on September 11 just before their departure; believing at first that they were going to Michigan and he was excited because it was close to home. "We are now about packed up to move. It is supposed to happen Monday or there about. As it stands we are going to Selfridge Field, Michigan. I knew it two weeks ago, but could not tell you because I told the captain that I would not say anything about it. The ground echelon goes to the Port of Embarkation and we go to Selfridge. The shipping tickets in the supply depot are made out for Selfridge. The sooner we leave this hole the better for me."

Four days later, plans had changed. "I think the Selfridge Field deal is out for the present. I think we are going to remain here for another month or more. So I will just have to continue sweating out coming home on week ends just like I have in the past."

And then a few days later the plans changed for a second time. This time Bob was being promoted from Master Sergeant to Warrant Officer and given new duties. He had already attended Warrant Officer School at Patterson Field and successfully passed the exam so the promotion came quickly. It was both a blessing and a curse; it was a significant step up, but as a result he was being reassigned to another unit. "I am being transferred to Bowman Field Kentucky in the next day or so. Colonel McCaulley and Captain Garland did all they could but just the same I am still being transferred. I am just sick, thinking of leaving this group. At least we will be much closer to each other and we can always get one day off. We pitched a good one here in the room last night and I felt pretty good for a while but then later on I began wishing that I had never heard of a warrant officer."

At the end of September the 316th did finally move, but this time without Bob Uhrig. It was not overseas as was expected, but to Del Valle Air Base, Texas. A mass flight of aircraft departed Benning for the base near

Austin on September 29. For the next six weeks the Group trained with a traditional infantry division, rather than the paratroops they had been living with. The Group moved troops and supplies from San Antonio to the Mexican border.[7] This massive training operation foreshadowed the first combat action the Group would see in North Africa.

Bob was left behind at Lawson Field/Benning as part of the permanent party of Troop Carrier crews supporting the parachute school. Bob was despondent; both because he could not remain with the unit he had been a part of for so long, and because he could not return to his home and Patterson Field. He wrote home to Toots on September 24. "I wish I had never heard of a Warrant Officer. And if I would have known then what I know now they never would have gotten me to take the examination. I think I would rather be a Master Sergeant any time. Today will probably be my last day here before I have to move up to Bowman. It just about kills me to think of leaving the group. But maybe in the long run I will be better off."

The only positive side to the new promotion was the chance to have Toots join him at Ft. Benning. They were both thrilled. By early October the couple was planning to find an apartment off the base and be permanently reunited. "Last night I asked the laundry man if he knew of any furnished apartments. He said he would have one on the eleventh. So I went out to see it. The fellow that is living there now showed me through. The people living their now say that it is just like having your own home because you hardly ever see the people that own it. It rents for $50 a month. It is by far the best setup I have ever seen, because places down here are sure hard to get."

The prospect of being with Toots again was still not enough to make Lawson Field more attractive. "I would sure like to be back at Patterson because these Troop Carriers are not what they used to be. And they probably won't get any better in the near future. This dear old Fort Benning has not changed a bit and just the same old hell hole as always and with plenty of sand to go with it. We are not flying as much as when I was here with the 316th but we are still doing our share of it. The only good part is that I am now working with the squadron instead of the group and that makes things a little better."

The couple continued their plans for staying at Ft. Benning up until the end of October, and then another radical change presented itself. One day at lunch time there was a loudspeaker announcement on post that Bob had been summoned by his former commander and would be collected in two and a half hours to re-join the 36th Troop Carrier Squadron which was being deployed to Egypt. Bob was recalled to the 316th Group in advance of their next big move. The call had come on November 10, when the 316th was alerted for overseas movement. The Air Echelon consisting of planes, aircrew and a few select support troops would depart directly for the Middle East while the rest of the ground echelon would go by ship from Virginia. After five days of preparation the Air Echelon took off as scheduled. They would pick Bob up on their way overseas.

Toots had already arrived at Ft. Benning to stay for good. But almost as soon as that happened, Bob was hurriedly deployed overseas. There was little time for preparations. Toots was left to clear Ft. Benning by herself. On top of that the engine was out of their family car and was being overhauled. When the car was fixed Toots would be returning to Dayton less than a month after she had arrived in Georgia.

Among Bob's papers was a curious letter dated November 13, 1942, to the "Commander of the 316th TCG," requesting that, "in the matter of Warrant Officer Robert Uhrig," recently transferred to his command: "It is requested that the amount shown be collected and remittance mailed to the custodian, Lawson Field Officer's Mess." As anyone knows who has ever been in the Air Force, you're not allowed to sign out of your last duty station until you've paid your Officer's Club bill, even if you're being sent to war!

One of Bob's last letters to Toots before he departed the U.S. was on November 13, 1942. "Always keep your fingers crossed and maybe in the near future I will get to see you again. It was terrible having to leave you in the condition that everything was in. You can write right away to the above address (APO, New York) because I don't know if I will get to see you before we leave. Keep the home fires burning for me and I hope to see you soon." It was a poignant letter, because Bob and Toots would not see each other again for almost three years.

Training to Go to War:
Bob Uhrig's Scrapbook from Lawson Field/Fort Benning, Georgia, 1942

*Left:* Regular maintenance was required to keep the airplanes operational with the hectic flying schedule.

*Below:* Every available transport was needed to support the jump school and airborne training exercises.

## Paratroopers 'Chute Up for their Practice Jumps

*Left* Aircrew/crew chiefs or jumpmasters with the new T-4 parachute developed especially for paratroops

## Formation Flying for Exercises and Jump School

# Formation Flying for Exercises and Jump School

# Chapter 5

## By Air To Battle

*After the war when a man is asked what he did, it will be quite sufficient for him to say,*
*'I marched and fought with the Desert Army.'*
Winston Churchill

Operation TORCH, the Anglo-American invasion of North Africa, had commenced on November 8 so the urgency of supplying additional American air power to the Middle East was considerably increased. Besides transport groups like the 316th, American fighter and bomber units were hurriedly deployed to reinforce the British Royal Air Force (RAF) units, who were now fighting two major campaigns in the region.

Seven days after leaving the U.S. the 316th arrived in the Middle East. Newly-promoted Warrant Officer Bob Uhrig was lucky enough to deploy with the Air Echelon of the 36th TCS, so he was spared the long sea voyage from the U.S. to Egypt. But the aircrews and their C-47 aircraft still had to fly the long southern route across the Atlantic. At this period all Troop Carrier units moving overseas went by the same route, whether their final destination was the Middle East or the Far East (the China-Burma-India Theater). The continental U.S. departure point was Miami, followed by stops in Puerto Rico, Brazil, Ascension Island and then Africa. The Group aircraft would make many stops at colorful and remote areas of the globe, places a country boy like Bob never dreamed he would see. Most of the fields where they would land had been established and developed by Pan American Airways before the war for their own routes in South America and Africa. This meant that sufficient support existed along the way for aircraft spare parts and repairs, which became necessary to accomplish the long deployment in the minimum amount of time.

From Miami, Bob wrote home on November 14. "You should see my nice set of hair. It is sure short now just like I always wanted to get it cut. Almost everyone in the squadron has a short hair cut. How would you like to spend the winter in Florida with us?" One week later, and they had arrived in the war zone. "We are still on the move, so I do not have a home as of now. We have sure been having our share of work on the ships. We fly all day and then work most of the night on them. Crook and I have a good start on a beard, and I think I am going to keep mine. We have been sleeping out almost every night since we left home. Tell everyone I will write in the near future. Just don't worry about me because I am just as safe as if I were back at Patterson Field."

At this point, Bob Uhrig began a project which he would continue almost every day for the remainder of the war in Europe. He began to keep a diary – a detailed record of his wartime activities, impressions and experiences. Along with his letters home to his wife and family, Bob would document his days and thoughts, particularly those events he could not share in letters due to security concerns or personal sensitivities.

Bob Uhrig as a newly-promoted Warrant Officer and now Engineering Officer for the 36th Troop Carrier Squadron, 1942.
He would not remain in this grade for very long once the 36th arrived in North Africa.

Bob's first diary entry was November 16, right after departure from West Palm Beach Florida, headed to Puerto Rico and then Port of Spain Trinidad. "Well here I am back in the good old 36th Troop Carrier Squadron and we are just ready to start for Egypt. Some little jaunt, eh? We have 13 ships in our squadron. We left West Palm Beach this morning at 0215 and arrived at Boriquen Field, Puerto Rico at 0940. During the night we ran into thunderstorms and had to buck up our formation, and from there on it was every man for himself. But just the same, all of our aircraft came through OK. The field at Puerto Rico is camouflaged plenty good. From about 3000 feet you can hardly see the field at all. It is sure beautiful with all types of tropical vegetation. All ships are dispersed and parked in back of the mounds of dirt built like a horseshoe [known as revetments]. This is to protect them from bombs. The 37th squadron took off 15 minutes behind us this morning and when we left Puerto Rico one of the ships had not arrived yet.

"We took off from Puerto Rico at 1115 and arrived at Waller Field, Port of Spain Trinidad at 1635. We had three ships with rough engines and one with a bad oil leak. So we had to work this evening until about 2100 to get them finished. We had been talking to many of the fellows around here and most of them have been here about ten months. Last week they killed a 23 foot Boa Constrictor which weighed about 250 pounds. They killed it in one of the barracks which had been empty for a while. How would you like to wake up and find one in bed with you! The guard killed three coral and one bushmaster snake just outside the Guardhouse. Plenty of snakes. They say that this island is just alive with snakes. About 50 feet from the field the jungle starts, and I do mean jungle. It is so thick that you have to cut your way through. The island is completely covered with tropical vegetation. In other words, it's just plain jungle. The barracks are built about four feet off the ground to keep the snakes and sand fleas out of them.

"I am the Engineering Officer of the 36th Troop Carrier Squadron and tonight they gave me a jeep to take care of my ships. I just about cracked it up because they drive on the left side of the road and that is sure hard to get used to. We finished work on our ships at about 0100. No one could go into town because we were restricted."

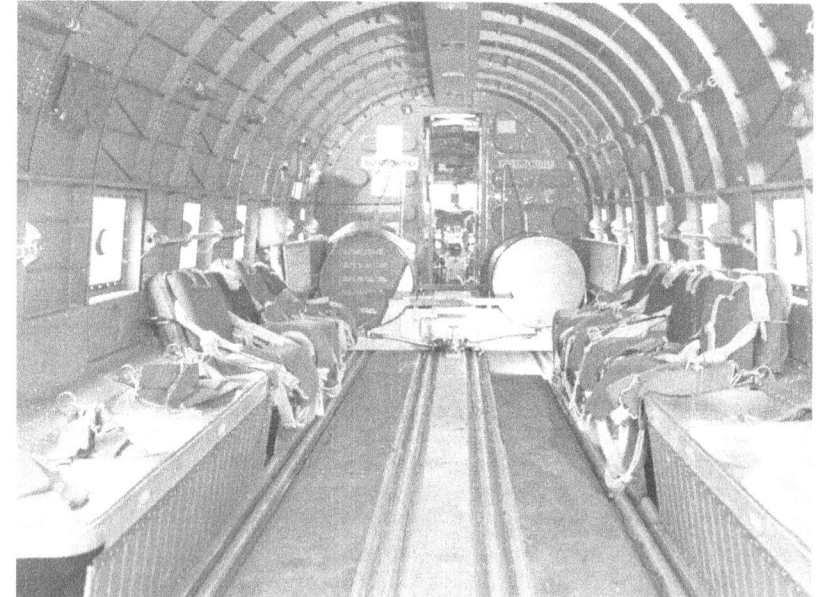

Interior (looking forward) of the cabin of a wartime C-47 fitted for long-range flights. The temporary fuel tanks are mounted in the front. Note the parachutes for the passengers and the metal bench seats which could be folded down to expand the cargo area. Note also the absence of any cabin insulation, armor or soundproofing.
(U.S. Air Force photo)

The next day, November 17, the formation was on its way to Belem, Brazil. Many of the fields the Troop Carriers would "stage" through were used by Pan American Airways to run their fleet of DC-3 aircraft, almost identical to the 316th C-47s. "We were up this morning at 0400. We were supposed to take off at 0530 but just as we were getting ready to leave we had a flat tire on the ship I was on. Pan American had one so we traded them our old for their good one. So that made it 0735 before we got off the ground. The 37th Squadron was just coming into Trinidad as we left. We arrived at Belem, Brazil at 1435. This, too, is a Pan American Field. The mouth of the Amazon River is just about 100 miles from here and we flew over it on our way. Today's flying was almost all over water.

"Most of this field was built by the Germans. And there is still a German hangar here on the field but last year Brazil took it away from them. The women here are terrific, so they say. They tell us that they will proposition you right on the street. And I guess they have everything there is to have. The people here are very short and look like little kids. The food here was nothing to brag about, but it was food. We went to the Post Exchange (PX) and had some beer at 25 cents a bottle but they are one-fifth gallon bottles so it is not too expensive. The beer is just fair, nothing to brag about. The people all get around here mostly by boats because they all live in these small inlets and rivers of which there are thousands. You can see one grass hut after another on the rivers and no way to reach them but by boat. They say they have spotted as many as six German submarines at the mouth of the river. Almost all of our ships have 25 hour inspections due on them tonight. Also three of them have rough engines but nothing serious as yet and we are living in hopes we don't have any trouble that will make us leave any of our ships behind. I would sure like to get all 13 of our ships across the Atlantic. That would be some record. The 37th Squadron got in late this evening and we are sure trying to beat them across. They have only eleven aircraft left, one had engine trouble and one had to turn back with hydraulic trouble. There sure are plenty of machine guns around here and from what I hear they know how to use them. Brazilian soldiers guard the field at all times and the funny part about it is they wear the old German uniforms."

Bob's diary entries continue on November 18 with a story of the hop across Brazil to Natal, their final departure point for the Atlantic crossing. "We took off from Belem this morning at 0835 and arrived at Natal, Brazil at about 1600. The field at Natal was mostly sand and was very white. The town has about 65,000 people, but as usual we could not go into town. We had the food here since we stayed on the field. All of the barracks are dispersed so in case of a bombing attack they will not destroy too many of them. Around here the jungle is just as thick as the rest of the places that we have passed. The maintenance kept us busy until 2330. I ran into an old friend of mine down here who is now a lieutenant and a Service Pilot ferrying ships to Africa. The last I saw of him he was a civilian in Arizona and we used to run around together when I went on cross-countries. We shot the bull out in the airplane until all the maintenance was finished. He is getting seven days off when he goes home before he has to make another trip and he is going to call Toots for me and try to get my radio and camera to me through Ferry Command."

C-47 cockpit. Note the instrumentation. These simple instruments plus the skill of the navigator were all the crews had to get them halfway around the world to Egypt.

On November 19 the 316th departed the Western Hemisphere for their long overwater flight to Africa. The only intermediate stop was at Ascension Island, a small rocky island in the middle of the South Atlantic, barely larger than an aircraft carrier. "We left Natal at 0835 and arrived 1745 at Ascension Island which is 1255 miles. All of the flight was over water and is just as long a flight as I ever made. This is the only stop between South America and Africa. If the Germans were to take this island it would be just too bad for America in ferrying small ships across. The island is just six miles square and is nothing but volcanic rock.

Therefore there is nothing on it in the way of vegetation. The British gave it up and said it was impossible to build an airfield on it but American engineers moved a whole mountain and built one. There are about 200 British soldiers here and the rest are Americans. After you land, you taxi up to what looks like a highway around the mountain and park on a high plateau. It is the strangest flying field that I have ever seen.

"The men here have everything rationed, including beer and water. In fact, there is nothing at all on this island. After you are here for a few hours you are covered in dust which makes you a copper color. There are plenty of large guns protecting the island and also some bombers and pursuit ships; in fact it is the lifeline for small ships going to Africa. All the gasoline is in drums scattered all over the island. They drive around and fill their gas trucks from these drums. There is only one mess hall where both enlisted men and officers eat. It is just about time it got like that for when it comes to a man's stomach, there should be no difference in food. The Colonel here told me that they are always under German U-Boat observation, but as yet have never been bothered. The closest German base is at Dakar, Senegal, Africa. Most of us are sleeping in the ships tonight because they have no place here for us to sleep. The nights are very cool and the days are hot. Ferry Command has a base here and so does Pan American. They both have pretty fair equipment to work with and try to help you out all they can. Tonight we have the right engine rough on ship number six, solenoid out on number eight, radio trouble on number eleven, collector ring broken on number twelve and left engine rough on number thirteen, so you can see where most of our evenings go."

With all the maintenance done and the ships repaired, the Group set off again on November 20 for Africa. This leg of the flight was to Accra in Ghana. "We were up and had breakfast this morning at 0430. We took off at 0700 for the dark continent of Africa. Today's flight was over water with no land in sight but was our last hop across the Atlantic. One ship had to turn back with a rough engine so that put and end to our near perfect record. That was the first ship in our squadron that had to turn back or that we had to leave behind. We arrived in Accra at 1615. A flight of [British] Hurricanes flew out to escort us to the field. This place is surrounded by Vichy French territory and no one trusts them too much. They have native guards here on the field and they cannot speak English. All they know how to do is ask for your pass. It was dark before we got back to the ships to work on them and the guards ran eight of us into the guardhouse. We did not know about the pass business and we soon found out. It sure is some feeling to have a gun pointed at you and the man on the business end not able to understand English. Just try it some time!

"During supper we had an air raid warning but nothing happened. About 2100 we had another and the anti-aircraft guns about ten miles from the field began cutting loose. As of now we do not know if it was a sham battle or the real thing. These people over here are playing for keeps. Some welcome on our first night, 700 miles from the fighting front. The field is alive with airplanes that are being ferried through. The combat planes in this area are stationed just 25 miles away. Pan American has a base here and all through this trip they have done everything possible for us. We had our first bath in two days tonight and it sure did feel good. We have also broken into our emergency rations because it is hard to get enough food. We work so late at night that the men get hungry and there is no place to eat. It is now 0130 and it will be another two hours before we are finished with the airplanes. You can see how much sleep we have been getting. The only salvation is that we get to sleep during the day while we are flying."

Sunday, November 21 was a "down day," as the 36th Squadron awaited their comrades from their sister squadron (37th) who had been a day behind them. "This morning we had to wait until the 37th took off because they were parked in front of us. When we had found all our pilots a storm was blowing so our flight was cancelled. In the afternoon we played cards. In the evening the Colonel's ship came in and we had to get a couple of men to work all night changing the carburetor. His engine had been rough through the whole trip from Ascension Island. The only way he could keep it running was to keep it in automatic-rich, since it would not run at all in automatic-lean. Love, Tucker, White, Cull and I slept in the aircraft as usual. We each took our turn on the carburetor change."

On November 22 the group was on the move again, east across Africa to Nigeria. Their plan was to skirt around below the Sahara to avoid the fighting in the desert. "We were up this morning at 0330 and went to eat breakfast. We took off at 0550 and landed at Kano Nigeria at 1135. We serviced some of the planes and

40

then went to lunch. There are thousands of native workers on this field who are building runways. They carry all of the material in baskets on their heads and there are endless lines of them. They have enough different kinds of dress to stop you in your tracks. Most of them look like they just got a piece of cloth and threw it around their shoulders but they all wear loin cloths. All this territory is under cultivation and is divided into plots of ground like small gardens. All of the people and their buildings are inside a brush and stick fence. Instead of having their livestock fenced out in the field they have them fenced in with them. Each family has ever so many buildings built close together and fenced in. There are quite a few trees in this part of the country. They do not grow very tall but are very large at the base. It is terribly dry and dusty here with visibility very poor, only about half a mile.

"One native stands at the edge of the runway and whenever an airplane is coming in he blows a horn which is about eight feet long, just like I have seen in pictures but never expected to see for myself. On the field here there are about ten Spitfires and ten Hurricanes. Also there are a few American ships on their way through.

"When we went for food I saw my first vulture sitting on the mess hall roof. There are quite a few flying around in the air. In fact, we almost hit some of them on our way in to land. In the middle of the morning there were three herds of giraffe sighted, but as usual I was asleep and did not see them. We are now getting into the part of Africa where no one has to tell you that it is hot. You can tell yourself. You must keep your sun helmet on all the time. We are having quite a lot of trouble with our engines heating up. We have to fly with the carburetors in automatic-rich and the cowling flaps open most of the time."

That afternoon they continued the trek eastward across Nigeria. "We took off from Kano at 1422 and landed at Maiduguri at 1625. This part of the trip we flew at 11,000 feet to stay on top of the overcast so there was not much that we could see. When we landed at Maiduguri they took us to camp to eat and even if it is deep in Africa, it was just as good a meal as I have ever eaten. There was one waiter to every two people at the table. They brought us one course after another until I thought every man was going to bust.

"The drive to camp was about five miles and the roads here are terrific. It was dark but there is a full moon tonight and we were able to see quite a few native huts and kids. And I mean they don't know what birth control is by the gang of kids around every hut. Close to the field there is a native village and they were having a dance. It was my first time to hear the well known tom-toms. They must be artists because they were beating the hell out of them and hollering at the same time so you might know what that sounds like. We are going to fly to Khartoum, Sudan tonight which is eight hours away. So that will mean we have been on the go for way over 24 hours.

"While we were waiting around to take off I had a long look at the bright moon and thought of Toots and what she was doing. At home it is four hours earlier than it is here. I hope she was thinking of me at the same time. You can always say that we have the same moon to look at. I only wish we were together again. I wrote all of today's part while we are flying at 12,000 feet. White is sewing on his Master Sergeant chevrons. Cull is asleep beside me. The first pilot and radio operator are flying (we take turns at the flying). The crew chief is asleep between the two extra gas tanks. The extra passenger is in the latrine smoking. The co-pilot is asleep on the floor and me, well; I am just sitting here giving the diary hell. We have a pretty good set up on our aircraft. We have four extra gas tanks in the cabin with a walkway in between them and we have two sleeping bags there. Then on the floor in the back we have two more sleeping bags and one on the five seats that we have set up, so you see five of us can sleep at the same time. The high altitude makes this or any other fountain pen run all the time, which you can see by the blots. We just passed about three large jungle fires and each one was five miles long. I wonder how or if they are put out. I will close for today. I am now going back and get some instructions from the navigator. By the way, I did not mention him above, but he is busy most of the time."

By November 23 the flight of C-47s was at their last stop before they reached the combat zone. It was one more leg north from the Sudan to Egypt. "We arrived at Khartoum this morning and went to the Pan American base for breakfast which was very good. Then we went back to the ships to try and get them in

flying condition which is getting to be quite a problem. It got very late so the captain took twelve ships on ahead so they could get to Cairo before dark, to keep from getting shot down. One ship was too late in getting repaired, so I stayed behind to help finish it up. Therefore, we had to spend the night at Khartoum at the Pan American base. They have the mess hall air-conditioned and the barracks are beautiful with a large patio in the middle of each one with green grass. They flood this part about once a week to keep the grass from dying. There are about three men to a room with beds made in the States with Simmons mattresses and also a Room Boy who waits on them hand and foot. I met two fellows here that used to be in my squadron at Patterson Field. They were Puscuss and Schafer. Both are now stationed here with Ferry Command. The PX [Post Exchange] is located on the roof with the best breeze, if any, blowing across it. The climate here is the best that I have ever been in. The air is as soft as silk and the temperature is just about right, but in the day time it gets pretty hot. I saw the Nile River today for the first time and it is a beautiful thing with plenty of green vegetation growing on both sides. It seems odd to see nothing but sand, then right along the edge of it, vegetation. But of course you know that everything has to be irrigated all the time. In the evening we saw a film of Sonja Henie in Iceland. They showed it on one of the patios and we just went and lay down on the nice green grass with a full moon overhead and enjoyed the show."

On November 24 the long ferry flight was over. Bob described his first exposure to the desert: "We took off this morning and headed for Cairo, Egypt. We have a ferry pilot in the lead ship so all we have to do is follow him. There is a 37th Squadron plane with us. They made a belly landing yesterday out in the desert when both engines quit. All they had to do was put two new propellers on and fly it out. We landed at Luxor because the ferry pilot had to drop some passengers off. We are now beginning to see many camels and we are now in part of the Sahara Desert. We flew over 900 miles and saw nothing but sand. You can hardly believe that you can fly for so long a time without ever seeing anything but sand.

"We arrived at Cairo about noon. We flew over the Pyramids on our way to Cairo and they looked just like the pictures you see of them. Every place you land on the way you have to shoot a flare. Each part of the day you have a different colored flare and this is to show ground crews and other aircraft that you are friendly aircraft. They took us to the English Officers' Mess at Cairo. The food was pretty good but as usual you have to go through all the English formality. It was a beautiful place with fan-tailed pigeons all around and they were sure strutting their stuff. Our Squadron (the other 12 ships) had just left here this morning. We took off after dinner and circled Cairo a few times and it certainly is a beautiful city.

"We caught up with our Squadron at Deversoir, Egypt and this is where we are going to be stationed for a while. The 36th and 37th Squadrons are going to be stationed here and the 44th and 45th are to be stationed at Ismailiya, Egypt, which is about 17 miles from here. We are the only squadron which arrived here with all 13 of their aircraft. One of ours as you know was half a day late. The officers are living in long, low-roofed buildings and the enlisted men have huts like the ones in Iceland. The captain told us to take the rest of the day off and get settled. We had a few beers and then went to bed."

Deversoir, Egypt: *Above*: Barracks. *Right*: RAF Control Tower. (US Army Signal Corps via National Archives and Records Administration - NARA)

Eventually, the entire 316th TCG – 52 aircraft - arrived in Egypt. Their initial mission was to evacuate American personnel and citizens from Alexandria in the event the Germans threatened to push all the way to the Suez Canal. In late 1942 that was a real possibility. However, by the time the 316th got settled in Egypt, the fortunes of war had reversed and after the Battle of El Alamein in November 1942, the threat to Alexandria and the rest of Egypt had receded. British General Bernard Montgomery and the 8th Army now had the Germans on the run, back towards the west and into Libya.

Regardless of the tactical situation, the 316th had accomplished a historic first for American airpower. The Group had deployed straight from the U.S. and directly into the battle area. While this type of movement would soon become commonplace in the war, it was a significant achievement for American airlift forces.

The Allied strategy in North Africa was now to squeeze the Germans from both sides of the desert – the Eighth Army would drive them from the east and the Anglo-American forces of Operation TORCH would drive them from the west. Eventually the two campaigns would intersect in Tunisia. Before that happened, there would be several hundred miles of pursuit across the deserts of North Africa as the British attempted to destroy the German forces before they could escape to Tunisia. To keep that pursuit in motion required massive resupply by air and that became the new mission of the 316th.

American C-47 over the Pyramids, ca. 1943.
(US Air Force)

## "Toots"

The photos of his wife that Bob Uhrig carried with him through three years of war.

# Chapter 6

## Chasing the Afrika Korps

*The first essential of an army to be able to stand the strain of battle
is an adequate supply of weapons, petrol and ammunition.
In fact, the battle is fought and decided by the Quartermasters before the shooting begins.*
Field Marshall Erwin Rommel

The 36th and 37th TCS were now based at Deversoir, Egypt, an airfield built by the British before the war. It was located just adjacent to the Suez Canal with a number of other British installations that were constructed to guard this vital waterway. The rest of the 316th TCG – the 44th and 45th Squadrons were located a short distance away at Ismailiya. Their mission had shifted from evacuating U.S. citizens to fully supporting the Eighth Army drive to expel the Afrika Korps from Egypt and push them across Libya. The British forces were pursuing the German troops as they fell back along the Libyan coast road, the only hard surfaced highway in the entire North African Region.

Army Air Force fighter and bomber units had already been operating from Egypt since August 1942. They joined units of the British Royal Air Force (RAF) in support of British and Commonwealth troops in the eastern part of North Africa and the Middle East. By October 1942 the US Army Middle East Air Force had 63 heavy bombers, 46 medium bombers and 56 fighters based in the area.[8] The 316th was charged with supporting those units as well as RAF squadrons and Allied ground forces.

The desert battlefield had a complexion all its own. It ran for some 1,500 miles from the Nile River to the border of Tunisia. It was also a battlefield with two "open" flanks – the Mediterranean Sea to the north and to the south the Great Sand Sea and the salt marshes of the Qatarra Depression. There was nothing to sustain a large army in this region. It was one of the most arid places on earth with very few towns and areas where fresh water could be obtained. Every bit of supply had to be hauled in – particularly food and fuel. There were few large ports in North Africa and the Mediterranean was a dangerous place. Supply ships were always under the threat of air or submarine attack. Resupply of the armies by air was the only viable alternative, especially to sustain a mobile campaign.

Bob was first able to write to Toots on November 28 that he was, "Somewhere in Egypt" and happy that regular mail service had just begun. "Our mail service is supposed to be very good. They say it only takes about 14 days but we will wait and see. They take all the mail to Florida by air and from there it goes regular mail unless you send it air mail. This country has not changed a bit unless they have a little less sand because I think I have eaten about half of it. The natives are just as filthy as ever and are pretty apt to stay that way. They have flies over here that about drive you nuts. You just can't scare them a damn bit. They sit on you and you make a pass at them and they jump about six inches and are right back again. Some of the fellows have made trips to Palestine and Jerusalem and say that they are beautiful cities and very clean and that is more than I can say about this hole. They say that there is rich black oil with plenty of fruits and vegetables in each city. They brought us some of the oranges and they are just like California navel oranges."

The next day, Sunday November 29, the Squadron was still recovering from their transatlantic trip. "We are off today. The captain said that the English could take Sundays off so we can too. In the morning about everyone went to town, the name of which is Ismailiya. We had lunch, such as it was. You can only get one order because everything is rationed. There are only a few places in town in which we are allowed to eat. The fellows rented bicycles and we went riding about town. Eight men together on bicycles, you might know what would happen. During the day we went to the canal that goes through town from the Suez Canal and watched the boats go through the locks. They are 90% sail boats (dhows), most of which are one mast jobs and very crude. There is a line of boats a mile long waiting to get through the canal. There are beggars and peddlers everywhere you go. They just hang right on you and you can hardly get away from them without

knocking them down. Men lay around the streets with sores all over them, and flies circling them like a piece of meat. In the evening we went to a part where the English have a lovely tea garden. It is very clean and they have good food, but they feed you like you were on a diet. Americans must eat twice as much as other nations. We then took a taxi to the airfield which cost about six dollars in American money. When I walked into the Officer's Club I found that Cull, Tucker, Davis and Smith had got into trouble in town, had got beaten up, and were in jail. They had already sent a truck to town to have them brought home."

The next day the Squadron went right to work. They had to prepare their C-47s to operate with the RAF; with different communications equipment and even RAF-style "fin flash" insignia on the tail. "We all left for Cairo this morning to have some radio equipment changed and the Allied insignia changed on the tail. We all had permission to go to town so that is where everyone headed as soon as we landed. We caught the tramway to Cairo which took about 20 minutes. The ride to Cairo from [RAF Base] Heliopolis where we landed was really something. You pass some of the most modern apartment buildings that I have ever seen. You also pass some places where you can hardly believe some people live. They have a common well from which women carry water in five gallon jugs or cans. Most of the women have veils. The filth is just beyond description. We went to a Greek restaurant to eat and got a pretty fair dinner of fish. It being a meatless day we were unable to get any. We then looked around the main part of town for about three hours and did a bit of shopping. I bought some groceries and a swagger stick. Everyone bought a few small items. There are many horses and carriages to ride in and quite a few taxis. All the street cars are open because there is very little rain here. We just returned to Deversoir where we are stationed, just at dark. There are no lights on the field so you just have to know where to come in."

The squadron settled in and set about establishing normal procedures for flying and servicing the C-47s as well as normal housekeeping chores. On December 1 Bob anxiously awaited the mail. "Up at 0630 this morning so I could tell Crook that his ship had the mail run for the next three days. We have a ship that goes to Cairo, then up to the front and back each day. They gather up mail, passengers and freight. We have one ship stuck in the sand and one wing just clears the sand by two feet. The Colonel, Captain Garland and Captain Frank McMenamin went to Ismailiya to find out about the trouble the boys got into Sunday night. They returned late in the evening and said the fellows were not so much in the wrong as had first been expected. So Captain Garland is going to let them go."

By December 3 full scale flying operations had already begun and housekeeping was established. "We have most of our ships cross country today, so there is not much to do. Pappy Street is changing a set of spark plugs in his ship. Love is building a table for the hut and I am building a wash stand to put a can of water on, with glass on top so that it will warm the water enough to shave. I also made a clothes rack for our good clothes. There are four of us in a room and we use out steel helmets for a wash basin."

Through early December, Bob was still trying to adjust to the harsh desert life. His next letter said in the upper left corner: "The stuff in the envelope is what I eat and sleep with." He marked his location as, "Where?" and tried to describe their living conditions. "Still no rain and plenty of heat with nice cool evenings. I am now like a chicken, I just open my mouth and let it fill up with sand and let it digest my food for me. You should see the way we live it is a joke, but we do have quite a bit of fun. We fight over old wooden boxes and crates so that we can get the nails out of the boards. Then we make the most beautiful furniture that you have ever seen. I am living in luxury. I built a wash stand with a soap, towel and toilet article holder. In the middle I have it fixed so that it holds my wash basin which is my steel helmet. They make pretty good basins and flower pots. Then I built another stand a little bit higher and put a can of water on it then I have a piece of glass on top and the sun warms the water enough to shave. I have a piece of rubber hose in the can and when I want hot water all I have to do is lower one end of the hose and the water runs into my helmet."

The men had little free time at that point and sleeping provisions were primitive. "In the evenings we play cards and chess. Last night I played black jack and won some of this cheap rag money that they have over here. It gets dark very early over here and we are always in bed by 9:30. There are four of us in one room. I am getting my share of sleep, but I don't get to sleep late in the morning and that is getting me down. Our beds are set about a foot off the floor and are made out of wicker. The first night here some of the fellows did

not have any sleeping bags and they looked like a waffle in the morning."

On December 5 Bob was able to ride along on a hop to Cairo. "We took off for Cairo this morning at 0730. After arriving we ate at the St. James Restaurant and hired a guide to take us to the Pyramids. We rode camels all around the grounds. We saw the Sphinx, most of the tombs and all the Pyramids. After we went back to town, I went to the American Legation to find Major Graft, but he was not in so I got a jeep to take me back to the airport at Heliopolis. We came home at about 1630."

Bob finally received mail from Toots on December 6 and replied to her: "The last one was marked November 22 and that is good service in anybody's language. The English sure do gripe because it takes their mail at least three months. Some of our fellows received letters that were post marked just eight days ago." He responded to Toots that the men were finally able to take a break and travel a bit and further described his visit to Cairo. "Yesterday about ten of us got a trip to Cairo so we had the whole day off. We all went to visit the Pyramids and all of the Tombs. It is sure some sight. How they ever built them all is just beyond your imagination. We went up in the middle of the large pyramid where the king's tomb was, and then just below the king's is the queen's tomb. The caskets are made out of solid granite. All of the stones are fitted together with no mortar used at all. The temple of the Sphinx has granite blocks that weight 25 tons. We also had a long ride on camels. It is not such hot riding. I am just as well and healthy as any one could be, but these damn flies are about to drive me nuts. We have plenty of good food now and hope that it stays that way."

That day also Bob described some of the environs around the base in his dairy. "Most of the airplanes are gone today, so the Captain said that I could let most of the men off. We have one jeep and an English truck, so we took the truck and took everybody for a ride around the far side of the lake. There is some good farm land close to the lake where they irrigate it. Where we live is just about a city block from the entrance of the Suez Canal. The Great Bitter Lake that we live on is just part of the Red Sea. The water is beautiful. Freighters anchor out a bout a mile from shore and they unload them with small boats which bring the freight into shore and load it onto trucks. Some large sail boats take part of the freight up the Suez Canal to Ismailiya, and from there they take another canal which takes them to Cairo. Our barracks is about a half a mile from the Red Sea and faces it, so every morning we can get up and look out on the Red Sea."

The next day was December 7 and Bob reflected on the significance of that day to Americans. "Just a year ago she started and I wish she was over now. All our ships are gone today, so there is very little for me to do."

The following day, Bob got his first glimpse of the ravages of the desert war. "I rode up to the front today with the Captain in one of our ships. You can never believe the equipment left and destroyed in the desert. You see tanks everywhere you look and trucks by the dozen. Some places they caught as many as 25 German planes on the field and destroyed them. Tonight we heard that we are going to move up to the front tomorrow."

**The Squadron Moves to Libya**

As the Eight Army pursuit of the German and Italian forces developed, it became immediately obvious that the only way to keep the enemy off balance was to stay on the offensive without pause. This would require a significant logistics operation and one that could only be implemented through the use of airlift -- in this case both British and U.S aircraft. General Montgomery himself recognized the magnitude of this commitment:

> "To get the full value from having established the air forces in the Cyrenaica bulge about Martuba, they must be able to operate at full blast against Rommel's supply routes by sea across the Mediterranean, the port of Tripoli and the enemy communications between Tripoli and Agheila. The air force daily requirements for these tasks was given to me as follows:

By 28 November – 400 tons
By 2nd December – 800 tons
By 9th December – 1050 tons
By 16th December – 1400 tons (1000 at Tobruk and 400 at Benghazi)

These were big tonnages for the air forces alone. But if Rommel intended to stand and fight at Agheila, we should have to build up army recourse of supplies, petrol and ammunition before we could attack."[9]

To sustain this operational tempo, the full resources of the 316th and other Troop Carrier groups were essential. All four squadrons were now equipped with 13 C-47s apiece, 12 primary mission aircraft and one "spare" for a total of 52 aircraft in the Group. Although the Air Echelon of the 316th arrived in Egypt in November, the Ground Echelon would not arrive until February 1943. That meant that Bob and the few other support personnel that had arrived with the C-47s were the only squadron members available for maintaining the aircraft, and so they had to move with them. In many cases the moves were to advanced landing grounds that had recently been occupied by the enemy and hotly contested. The 316th went into action almost as soon as their initial bed-down in Egypt was complete.

As the Allied forces pushed the Axis troops westward the fighter units were also on the move. Their ground personnel were formed into "A" and "B" parties so one could move to the next landing ground while the other stayed behind to service the aircraft. Then when the next section of aircraft moved forward, the other party went with them or "leapfrogged" ahead to the next forward field.[10]

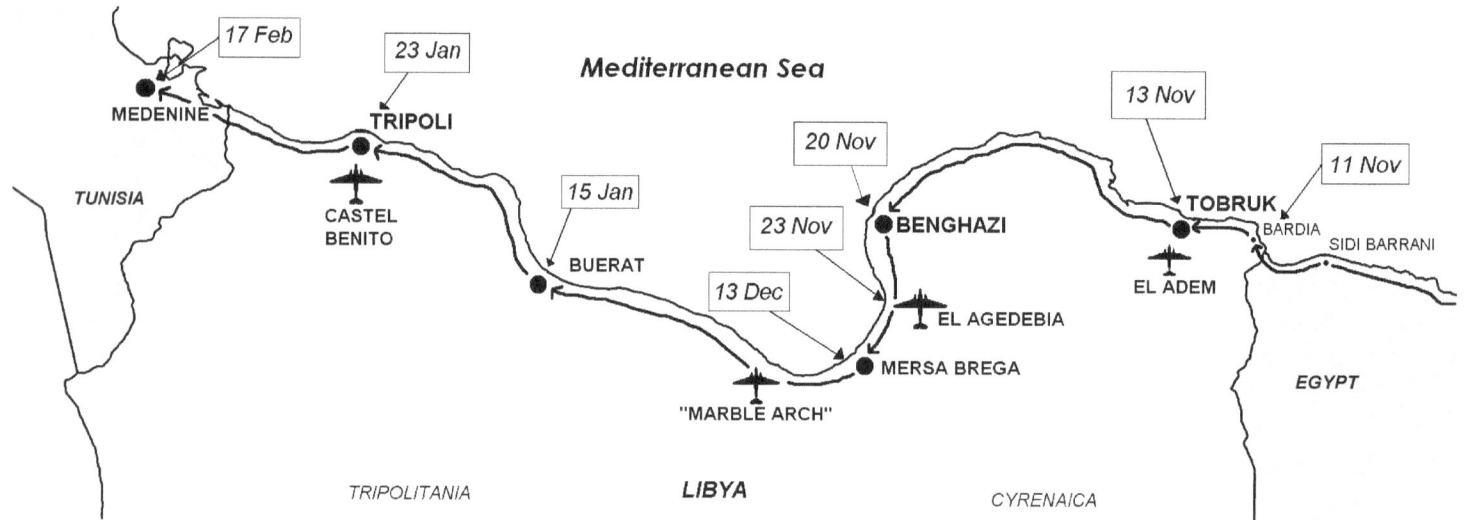

Map of the locations of the 36th TCS bases in North Africa during the advance into Tunisia, 1942-1943.

On December 9th, the 36th TCS decamped from Deversoir, Egypt and redeployed forward with the advancing forces. "Well we finally got ready to move about 1030. We left about 1200 and headed to Tobruk. We are now stationed about 12 miles from Tobruk at a place called El Adem, Libya. There are about 20 German planes around the field that have been destroyed. There are also plenty of tanks around too. We are living in an old Italian barracks. There are three men to a room, and about four rooms to a barracks. They are built plenty strong to stand plenty of abuse. They are shot up something terrible, and most of them have had the roof blown off. We use canvas for roofs and you should see some of our inventions to make life comfortable, they are really good."

The next day, December 10, airlift operations began in earnest from the new landing ground. "This afternoon we hauled our first load of gasoline up close to the front. This is what we are going to do, haul gasoline from Tobruk, where the oil tankers come in, up to the advance bases for the fighter planes. Up here you fly about 20 feet above the ground as protection from enemy planes."

In one mission December 11-12, six of the 316th's C-47s flew over 11,000 gallons of aviation fuel to desert landing grounds near Tobruk and returned with 77 wounded men.[11] Bob described his activities for those days. December 11: "We sent one ship to Cairo today, the rest hauled gas. In the afternoon we went for a walk in the desert and got some guns and ammunition. Boy, you can find just anything that you want. You must be careful because there are land mines around. All of our ships make two trips a day, so we have to meet them every day at noon, then again in the evening."

December 12: "We sure had our share of bad luck today. First thing at noon, No. 12 (Crook's aircraft) ran into No. 7 (Winnie's aircraft.) Crook's hydraulic system went out so they didn't have any brakes or control over the ship. Then No. 10's hydraulic system went out and then they ran into a 37th Squadron ship and tore their rudder up. Crook's ship has a large hole in the wing and the de-icer boot torn. Winnie's aileron and wing tip were damaged. This evening we were working on the ship when we heard them yell 'Lights Out!' By the time I got out of the ship, the Germans were sure giving us hell at Tobruk. The whole sky was lit up from explosions and bursting of our 'ack-ack.' The shells were going in every direction and looked like Roman candles. There were quite a few large explosions and the Germans must have dropped 100 flares. Even here, 15 miles away, it was as light as day. They boys all know what a funny feeling it gives you. It is no so hot, that's all there is to it." The next day was quiet. "All of our ships are flying but No. 13. They are still flying gas to the front, and to think of it, all of them are still in commission. None of them has been shot at yet."

His next letter to Toots told of life in the desert and the dangers they were facing. "We have moved again and living conditions are just like being up in Michigan roughing it. I have plenty to eat and a good warm sleeping bag, so don't worry about me. All of our fellows are still OK, but two of them have been burnt by benzene. We take a bottle and put benzene in it then we put a rag in the bottle and let part of it stick out of the top and we use these for lights. Don't worry if I don't write to you for a while because some times we are rushed to death and whenever we move it takes time to get the mail going again.

"You should see this desert where all the fighting has been, it is just impossible to believe what equipment has been left and destroyed. We all now have British battle dress uniforms because the natives up here don't know of any uniforms but British and if they see any other kind they knock their can off. They shoot first and then walk over to see what they got. I am here to tell you the desert is sure one cold place at night. It does not even get too warm in the day time up here. We are all trying to find a good motorcycle but as yet all we have found are ones that are all shot up. We take walks out in the desert and you can find anything that you are looking for. But we have to watch for Booby Mines. The Germans have them on everything. They even put them on their dead so when you go to bury them they will blow up. They put acid in fountain pens so it will spray you when you take hold of them. They have a thousand and one different ways they use them. We are living in some old Italian barracks which have been plenty shot up and all the roofs have been blown off. There are four rooms to a building and they are made from cement. We put tents over the top for a roof. You should see all the inventions for living out here. I am not a lieutenant yet but the Colonel has sent my name in for me so now all I have to do is wait about a year for them to say no."

**Change of Command**

The 316th now became part of the newly-formed U.S. Ninth Air Force which was the headquarters for all American air units in the eastern part of North Africa. The Ninth was commanded by Lt. General Brereton and he immediately recognized the significant part that American military airlift was playing in the drive to eject Rommel from North Africa. "On one day, forty nine C-47s carried 48,510 gallons of gas from El Adem to Agedabia, a distance of 425 miles by truck and 250 miles by aircraft. The flight was completed in an hour and five minutes. It would have been a three-day trip for fifty-nine trucks."[12]

The Eighth Army was continuing offensive operations, and the 316th was right along side. By December 14th Mersa Brega and Agedabia were in Allied hands and the C-47s would soon have to move as well. The aircraft were experiencing their first set of troubles with the desert environment. "We were up this morning as usual at 0515. Tucker and I went to the front with No. 6 which has been cutting out at take off for a long time. And this morning as usual she cut out at about 50 feet from the ground. We unloaded the gas in about 15minutes and started back. Another 'big push' is starting today and there is about all the excitement and activity that you would care to see. Where we landed, the Germans had just a week ago, and they are still clearing the land of mines. You can hear the big bombs and guns from the field, which is about 30 to 40 miles from the fighting. I got to take her off coming home and flew it all the way back at 50 feet. We held her out of the afternoon mission and changed the carburetor to try and make her stop cutting out. Love went up to the front in one of the ships this afternoon and has not come back yet."

Bob's letter to Toots on December 14 described necessary precautions for the ever-present threat of air raid. "The Germans gave us a very pretty side show here last night, but they were about 15 miles away. It looked like the Fourth of July with all the trimmings. We are sure working plenty hard and also plenty of flying going on. But we have to be very careful how we work at night. There are supposed to be no lights visible at any time, but you just can't work on these crates without some light showing. You should see our Engineering Office. We have about twelve 50 gallon gas drums filled with sand and then a canvas thrown over the top of it. And we have the best bomb-proof shelter in back of it with all the modern conveniences. We also have a gallon gas can that we fill with gasoline. Then we hook the air hose from our compressor to it and we have a burner coming out of the top of the can also, then when we turn on a little air and light the burner it is just like a blow torch. We can make coffee on it in about three minutes. We now have eaten up about all of the emergency rations out of the ships for our snacks."

In addition to new locations and equipment, Bob had another challenge to deal with. His recent promotion from enlisted to warrant officer meant he was no longer just "one of the guys"; a crew chief and mechanic. Now he was a supervisor and had several men working for him, including some who were his close friends. He would now be called upon frequently to exercise his leadership skills along with his extensive technical knowledge of the aircraft.

Their moves closer to the front brought increased danger as well. On December 15: "No flying today because they are pushing the Germans back. It would do no good to haul gas to where the fighters are because they will be moving further up in a day or two. We will then haul it to their new location. Love's flight came back this morning. They got lost yesterday and flew clear over the front into enemy territory. They arrived over their destination so late that they couldn't return yesterday and had to remain overnight. Love saw one Messerschmitt shot down and two trucks blow up by mines. The Germans also strafed the field where they were just before daylight."

The Americans in the Troop Carrier units were getting their first exposure to combat. On the other side of North Africa, other American units struggled to push the Germans into Tunisia. On December 16, Bob recorded that, "Only three ships had to go up to the front this morning: Nos. 2, 3, and 11. The wounded are now starting to come out from the push. We haul gas up and the wounded, those that are in the worst condition, back. Some of the fellows are pretty bad off. After we get them here, they still have to be hauled over the worst road possible to Tobruk. Yesterday we had the jeep out in the desert to a hill where a large battle had taken place and you could find everything from hand grenades to bibles. Most everyone here carries a compass because it is so easy to get lost in the desert."

**Ruins of the Desert War**

By December 17, Bob was seeing first-hand the results of three years of desert war and described it to Toots. "Well, I was where things are buzzing day before yesterday. We were just south of Benghazi about 25 miles. Saw two trucks get blown up by land mines. These land mines are the worst thing the boys have to contend with, the Germans have them everywhere. If you could see this desert you would think you were

dreaming. Just around this field I knew that there are about 20 tanks, about 40 trucks and about 20 airplanes destroyed. The boys that have been fighting back and forth out here sure do deserve a lot of credit. I have never in my life seen so much equipment destroyed and laying around. You can fly for three hours and you can't look out once without seeing old trucks, tanks, guns or some war material lying around. Parts of everything are scattered all over the desert. Some are burnt out and others just broke down but most of them have been blown up."

Bob fully appreciated the lot of the Eighth Army men. "When some poor foot soldier says he fought in the Libyan Desert you buy him the drinks for the rest of the night. Just put yourself in his shoes, flat country just as far as you can see with nothing at all to hide behind and then some one shooting at you from everywhere. You can see where they piled up little piles of stones for a little bit of cover. Where most of the major battles took place, the Germans are holed up and dug in. There are also places where there are some small hills out here and they were really fortified. We flew over where one of the largest tank battles took place and over 300 tanks were destroyed. I just can't tell you how terrible it looks. Boy it sure takes a lot of pots and pans to make all of the equipment." Here Bob was probably making reference to the "scrap drives" going on in the U.S. to collect metal and household objects to be recycled into munitions.

Along the way, he also saw the handiwork of the Desert Air Force. "Lots and lots of trucks and tanks were destroyed by airplanes because they had no place to hide. It looks just like they stood toe-to-toe and the best man won. Flew over one field where 25 German planes were destroyed on the ground."

Also on December 17, he saw the ruins of Tobruk, which had changed hands several times during the Desert War. "Today we went to Tobruk in a truck to get some water. It is (or was) a town about the size of Urbana [Ohio]. It just makes you sick. Most of the buildings are made out of cement or cement bricks. Just make a bunch of houses out of cement blocks and then shake them up good and you have Tobruk. What is left of the buildings you can see are full of holes and bullet marks from where they fought in the streets. It must have been horrible. You can hardly lay your hand on any building in town without touching a bullet mark both big and small. The harbor has over a hundred boats sunk in it and that is no small amount of boats. The Germans let the harbor have it about every four or five nights. Crete is only 165 miles away so it does not take them long to fly over."

On December 18, operations continued, but Bob's unit and airfield at El Adem came under air attack. "We were pretty busy today. Most of the ships hauled gas to the front and hauled injured back. Tonight we were bombed for the first time. They first bombed Tobruk and it must have gotten too hot up there for them because we could see the stuff [anti-aircraft fire] going up and bursting. There were three enemy ships that came down here. They circled for about half an hour overhead. The anti-aircraft could not fire on them because an English night fighter had signaled them that he was up after them. He finally chased two of them off but, oh boy, the third one really let us have it. It is a terrible feeling. He dived down through the clouds to about 200 feet with his machine guns blazing all the time. Then he pulled out to level flight and came across the field wide-open and dropped the 100lb bombs on his way. We were in a trench and you could see and hear the first one and then they kept getting louder and closer and you wonder if they are going to stop dropping them before they get to you. You also pray a little. Some of the fellows were on their way back from the flight line so they jumped into the first trench that they could find and they had gravel thrown all over them. After it was over I went out and got a piece of shrapnel to keep from my first bombing."

**Operations from Mussolini's Marble Arch**

The next day, December 19, the C-47s resumed operations to a curious location on the Libyan Front known as the Marble Arch. It was an airfield sited near a large decorative stone arch in the middle of the desert, astride the main highway that ran along the coast. The arch was built at the direction of Mussolini to commemorate Italian conquests and mark the boundary between the two major regions of Libya, Tripolitania in the west and Cyrenaica in the East. *L'arco dei Fileni de Mussolini* in Italian was sardonically re-christened "The Marble Arch" by British Tommies in honor of a famous landmark near Hyde Park in London. It was

the only major landmark in Libya for hundreds of miles of desert. The airfield area had been captured by the New Zealand Division on December 17. On the 18th, RAF P-40 Kittyhawk fighters of No. 239 Wing began operations there. The 316th flew in supplies to them at this remote location.

On the 19th Bob went along on the flight to the new landing ground. "We went to Marble Arch today to change a tire on No. 7 which had a flat yesterday. While we were there, three flights of 12 each P-40s took off with bombs. They go out and drop the bombs on the Germans and then strafe them until all of their ammunition is gone, then they come home. The Germans held this field until three days ago. All around the field are mines. They took 500 mines off the flying field alone. The crew of No. 7 saw three men killed and four injured by a landmine about 100 yards from the plane." As it turns out, the wounded men were ground crew from No. 3 Squadron, Royal Australian Air Force. One had set off a German "S-Mine." It was a particularly nasty type antipersonnel mine filled with steel balls that shot up in the air to chest height before exploding. The troops called them "Bouncing Betties." Fortunately, there were American C-47s nearby and the wounded were loaded on the aircraft and transported to Benghazi for treatment.[13]

After a few harrowing days at El Adem, the squadrons began to take more safety precautions starting on December 20. "After being bombed the other night, the Colonel got worried and moved us all out of the old buildings into tents. The 36th Squadron went about a mile from the field to the south and the 37th Squadron moved into tents about a half mile from the old buildings."

Since the fighting had now become more static, some of the 316th was re-deployed back to Egypt, but the 36th was left in Libya to carry on. In spite of the extra precautions, dangers remained. Bob recorded on December 24: "This, my friends, is the day before Christmas. The 37th and 44th Squadrons moved back to Ismailiya and Deversoir today so we had to break camp and move up to where the 37th Squadron lived because the Group is staying and they won't move down with us. We finally got moved back and the tent we got was something you read about it books. In other words, it is about shot. I have a pound bet on tonight that we get bombed, because it is Christmas Eve and the Germans have not been over for about three nights. One of the ships ran over a land mine today which did not do any damage to the ship. Crook's ship is now on flying status again, if you want to call it that. We have a patch on the leading edge of the wing, made out of old metal and sheet metal screws and we have an aileron on it from one of the 45th Squadron ships that was all shot up. They turned a 37th Squadron crew into the status of 'Missing In Action.' We have all looked all over the desert for them and have not found any part of the aircraft or crew. They were over German lines with the 45th Squadron ships and they think the 37th Squadron ship was shot down. Tobruk sent up some flak while we were eating supper so we had to put out all the lights. None of the enemy ships came over here and we are not sure if any were over Tobruk so I don't know if I have won my pound yet or not. Some Christmas Eve. MERRY CHRISTMAS you all."

Bob's Christmas letter to Toots described some of the "scrounging" efforts of the AAF men. "I now have myself a German motorcycle to ride to the ships on and she runs pretty good too. You don't look for roads out here you just take a compass and start out. Only one thing, there are many sharp rocks that cut the heck out of tires. I don't see how any equipment holds up in the desert. It takes a lot of pots and pans to keep things going here. We had steak for dinner today and it was very good and we were lucky to get that. Supplies are hard to get in this neck of the woods."

His diary for Christmas day described the few comforts of home available to the men in the desert on this holiday, including a crude ersatz stove known as a "Benghazi Burner," which was an invention borrowed from the British. "Up this morning at 0800 to get two crew chiefs. We only have two ships flying today. We build a fire out of gasoline and sand. We take a can, fill it with sand and mix gasoline with it. Then we touch it off. This is the way we heat our water. This being Christmas I think it is time I take a bath as I have not had one in two weeks. I took one towel and soaped it up good and splashed it all over me. Then I rinsed off the same way. For dinner we had the best steak and all the trimmings. I slept off a good headache in the afternoon. It was from last night's drinking. I don't believe there was one sober man in the camp last night. Men were drunk that I never seen ever drink before. Then we all got in a big circle and sang Christmas carols. It was poor music but everybody had the spirit. We even went in different tents and read from the Bible. But

in the back of my head I was thinking of Toots all the time. It was a beautiful night with a full moon. One thing, you don't enjoy a full moon over here because the Germans can see to bomb too well. We could hear large guns all day but don't know where they are.

"I now have a German motorcycle that I got from an English salvage crew. It sure does run good and has a drive shaft instead of a chain drive. This afternoon it got cloudy and when it gets cloudy it gets cold. Stracke came in today from salvaging his wrecked airplane. One of our ships stayed out in the desert last night because they could not get in before dark. You can land most any place around here, but if you don't get in before dark you run a chance of being shot down. In this country you have different colored flares you shoot from your aircraft for recognition and the colors change twice a day. There are only four of us in the tent now since the 37th Squadron moved out."

Although Bob's diary and letters speak of English or British troops, the Allied forces in that part of North Africa were a mixture of British, Australian, New Zealand and even Indian divisions, all fighting as part of the Eighth Army. In fact, Bob states that the American units attached to the British troops had to wear British battle dress uniforms because the Arabs despised the Germans and Italians and would shoot anyone not wearing a British uniform.

December 30 brought a "foraging mission" for Bob in the local area, a time-honored task for armies on the move. "I went with the Mess Sergeant today to get some eggs. We went about 30 miles from camp to some native camps. We traveled on the highway about half the distance and the rest of the way we just cut through the desert and sweat out landmines all the way. These natives live in long low-roofed wool tents. They are all different colors. They live in small settlements of about five or six tents. They didn't want to sell anything but they wanted to trade for sugar, rice, tea and coffee. They are miles from nowhere so money does not do them much good. The natives raise small patches of wheat, goats, sheep and camels. Most of the goats are black and white and the baby ones are very cute. They even look ornery out of their eyes. One today got on the roof of one of the tents and they tried for about 15 minutes to get him down. He would just jump up and down and run back and forth. They were churning butter in a goat skin at one tent. We traded a beer can of sugar for three eggs. Then we had to go from one settlement to another to get enough eggs. We were never able to get any guns. One thing most of them have is some piece of military clothing on. All types of military equipment is around their tents. They even water their chickens from mess kits. We were told that the Italians drove off all of their livestock. The main thing I wanted to get was a very small automatic pistol that the Mess Sergeant saw last week. I wanted to get it for Toots but we were never able to get or find one. Tonight I feel like I had been beat on all day long because the ride was something terrible. Even the road has been bombed so often that you might just as well cut cross country."

**Desert Conditions Impact Operations**

The New Year 1943 arrived and brought with it a horrible sand storm to El Adem that lasted for several days. It was a small taste of what Allied and German soldiers alike had dealt with for several years. Bob's diary starting on January 2 gives some sense of the severity of it. "Today we had our first real sand storm and they are nothing to laugh about. The wind blows about 35 miles per hour and you can only see about five or six feet in front of you. The sand even cuts into your skin and every place that is bare. You have to wear goggles and respirators on you or you cannot be out in it at all. The tent just blows away and the sand filters into it and covers everything with it. Tonight I got lost on the motorcycle and spent an hour coming from the ship to the tent. I even lost the flying field and was out in the desert. I also ran into a shell hole and just about didn't get out again."

One of Bob's mates preparing for one of their many moves to new fields.

The next day, the wind continued but flying operations had to continue and the aircraft had to be serviced in spite of the weather conditions. "Up this morning at the usual 0515 and helped get the ships going. Last night we had three flat tires and this morning one more so that makes a grand total of four and right now we have no inner tubes and only one casing. The supplies out here are terrible. We fix the ships the best we can and let them go at that. People at home would not even get in one like these, let alone fly it. We are now patching our inner tubes with Leicer patching. We have No. 3 ship out in the desert somewhere and have not heard of him since yesterday morning. We hope they landed because of the storm and are OK, but as yet we don't know. One of our other ships got lost and had to land in the desert at a native camp and ask where Tobruk was, and then they came on home. Today it is raining and the wind is blowing just as hard as yesterday, but the sand is not near as bad because of the rain. We are changing two cylinders on Crook's ship which have been leaking oil. One engine uses about four gallons of oil an hour."

His January 4 letter to Toots spoke of coping with the intense desert winds, known as *ghiblis* to the locals. "The third day and the wind is still going strong. We have sand in, on and around everything now. Just think of the worst blizzard and then replace the snow with sand and put up with it for three days and you will know what we are putting up with. It is about the worst thing I have ever been in yet. Our mess tent blew over last night and we spent all morning putting it back up in this sand storm. Everybody's face is olive drab from being out in the sand. I have been sleeping with my respirator on because every time I take it off I start coughing. I sure do miss my radio and I would love to have my camera because I have sure seen plenty to take pictures of."

**Permanent Move to Marble Arch**

The next week was hectic for Bob's outfit as they prepared for another move, this time deeper into Libya to the remote Marble Arch airfield. His diary does not pick up again until January 10: "Well today we left good old El Adem and moved to Marble Arch. We had to leave Sgt Boggs and Sgt Middendorf behind with their ships Nos. 47 and 6 because the engines had to be changed. We helped them set up a tent and left them ten days rations. We then came on to Marble Arch and found out that No. 2, Sgt Johnson's ship, also had an engine burned out. Oh yes, and we also took all four tires off of the ships back at El Adem so that we can keep the rest of them going. We now have taken both tires off No. 2 also. You see, the fields are so rough up here that they just break the tires one after another and as yet we have been unable to get any. In fact, we have been operating about two months without any supplies at all except what each crew chief had in the back of their ships. The supply situation here is terrible."

On January 12, Bob recorded his own impressions of Mussolini's Arch. "They do not have all of the land mines cleared out of here yet so we have to be very careful. The Marble Arch is located at the very south west corner of the flying field here. It was built by Mussolini for his soldiers to march through after they had conquered Africa, but the last time they ran through it. It is built of marble and is about 100 feet high. It is located out here in the middle of the desert with no towns at all close by. It is built over the only highway out

here. We are about half a mile from the Mediterranean and some of the fellows have been swimming in it. At about noon it gets warm enough that you don't quite freeze. But they say the water is plenty cold. And it gets plenty cold at night here even if it is the desert. The 37th and 45th Squadrons moved in today and between the two of them, they only have 13 out of 26 ships flying. And we have 9 out of 13 so we are not doing too badly. Today from noon on we had one of those good old sand storms. The damn sand gets in everything and about drives you nuts. Our tent is about 100 yards from the highway and it is bad."

Mussolini's "Marble Arch." When compared to the size of the vehicles in the photo, it is easy to see why this was one of the major landmarks in all of North Africa.
(Courtesy of Paul Plumley collection)

During January as the Eighth Army moved up to Buerat, the ground situation became static for a time. So after several weeks of desert living, Bob got a brief leave on January 14. "Lt Farris, Lt Coursen and I left on the mail plane this morning for Alexandria to spend a three-day leave, but it takes a day to fly there and a day to fly back, so you then have one day and two nights in Alexandria. We arrived at Landing Ground (LG) 174 at 1530. This field is about 32 miles from Alexandria. We then caught a ride into town in a command car that one of the pursuit [fighter] squadrons was sending into town. After arriving in town we found that the Cecil Hotel was full, so they sent us to their annex, the Dorchester. By the time we got the desert filth off of us, and cleaned up it was about 2100. We had dinner at the Dorchester and then went to the Metropolitan Night Club. It is very nice and they have a very good orchestra. The floor show was also good. At 2230 they stop selling drinks so after that we had a steak, so that made two meals in one evening. A Major close by had a couple of girls that Lt Farris and Lt Coursen danced with, but I cannot dance so hot, so I just sat and listened to the music. Alexandria is completely 'blacked-out' at night, so it makes it tough to find your way around at night. Taxis are very had to get at night also. We walked to the Night Club along the sea wall and there was a three-quarter moon and it made everything beautiful. All of the buildings along the sea wall are very modernistic and silhouetted against the sky and with no lights at all in any of them made them look like great monsters. Everything is so very quiet except for the sea."

The next day the men had a few hours to become regular tourists, now far away from the war. "We got up this morning at 1000 and went to the Cecil Hotel and got a room for the three of us. Then we went shopping. We engaged Jimmy Hassan who has always shown tourists and movie stars around. I was going to send 200 dollars home to Toots but I saw a dinner ring that was a beauty. I had to pay 40 pounds ($160). I thought a long time before I bought it but Toots likes dinner rings so very much that I thought this one would hold her for a while. It is Egyptian style with a large diamond on each side, and then it has a raised center with chipped diamonds in it. Then around the large diamonds there are blue sapphires. We ate lunch, roast lamb, at the Union Bar which is a very good place to eat. I also bought a watch because I am tired of not knowing the time. It is an Alma watch which is both water and sand or dust-proof. It cost 5 pounds and 75 piasters or $22.50. In the evening we ate in the Union Bar again and we had the best steak I have ever had anywhere and best of all she was a good big one. We had French Fries, new peas and fried onions. For dessert we had

strawberries with whipped cream. Then we went to the Carleton Night Club. It is just about the same as the Metropolitan. We went to bed at 0100."

Bobs' letter to Toots on January 16 described his current situation and his recent leave in Alexandria. "You want to know what I am. Well I am still a Warrant Officer and am Engineering Officer of the 36th Troop Carrier Sqd., and a damn good one too. I don't think I am going to Officer Candidate School. Since we are a way up here your mail gets here very late. We only get mail once a week. Lt Farris, Lt Coursen and I went to Alexandria for three days. It was sure good to get under a shower again and see the dirt run off of you. I think I about washed myself away. We have been in the desert for five weeks now so you can tell just how we felt. I bought myself a waterproof and dustproof watch so that I can now tell time again. Most everyone buys them because their good watches just can't take it out here. The first night in town we ate until we could hardly walk and then we took in the Metropolitan Night Club, one of the best and it sure is beautiful inside. The next day we slept until 1000 and then went shopping. In the evening we went to the Carleton Night Club and it is just about like the other one. Alexandria is a very beautiful city with all the good apartments, homes and hotels facing the [Mediterranean] sea. They have a total blackout at night and it is sure something picking your way around in the dark. You will open a door and step into a place full of lights and also very beautiful, it is almost unbelievable."

Earlier that day they had left Alexandria to return to their "home" and desert base. "We were up this morning and 0600 and had bacon and eggs for breakfast. We then hired a taxi to take us out to the field and this cost us 2.50 pounds ($10). We were all so tired that we slept all the way home on the mail plane. When we arrived here today we found that they had had some excitement. Major Dekker, 45th Squadron Commander had run over a land mine with one of their ships and it had blown off the right wheel and propeller, and tore a large amount of holes in the wing. These mines are powerful devils. Then he was lying in his bed and a truck ran over a land mine just outside his tent and tore 57 holes in the tent just over his head. Don't you think that is enough trouble for him to have in one day? When he ran over the landmine, his co-pilot had a piece of shrapnel go through his flying jacket, through his leather jacket, and put a black and blue spot on his stomach. The radio operator had a piece of shrapnel cut the sleeve out of his jacket. The Colonel had us take the ships to a field about ten miles from here because they have been expecting the Germans every night. The Marble Arch and the highway are such good landmarks that they could hardly miss finding the field as bright as the moon is."

The following day, January 17, Bob was back in the thick of the action again and supporting the RAF fighter squadrons of the Desert Air Force. "We loaded all the ships this morning with gasoline, bombs, rations and oil. We are hauling gas to a field that the Germans left yesterday. We will be the first to land there and the pursuits [fighters] will land one hour later and start operations. The pursuits carry one 500lb bomb. They fly over enemy lines, drop the bomb then strafe until they run out of ammunition, then they come home for another load. Everything went off as per schedule. As yet we have lost no ships and we have been flying to within 15 to 20 miles of the front."

The pursuit ships Bob speaks of were likely P-40 fighters operated by USAAF or RAF Squadrons. As the front moved steadily westward, the Allied fighter units followed closely behind the ground troops in order to keep their aircraft in range of the front line. By the end of the North African campaign (Nov 42-May 43), one USAAF P-40 Group had moved 34 times to new airfields from Egypt all the way to Cape Bon, Tunisia.[14]

The relentless pursuit across the desert continued as well as maintenance difficulties with the C-47s. On January 18: "We have Nos. 1,3,4,7, 8 and 11 still flying and Nos. 2,5,6,9, 12 and 13 each with a burnt out engine. This sand plays hell with our engines and also us. Nos. 12 and 13 both burnt out their engines this morning just after taking off. The few men that we have are working our heads off. This afternoon I went along to the front where we are hauling gas because they said that No.8 had returned after takeoff from the front with engine trouble. On our way we flew over Sirte, the town that the Allies captured on Christmas day. It is not much of a town, just about 60 buildings, but it is a very beautiful place with palm trees all around and built right on the sea. The new field is very hard to find because it is in the desert with no roads at all, just what they call tracks through the desert. No. 8 had a burnt out spark plug lead which is why they returned.

The crew chief had already repaired it but the engaging solenoid was burnt out as well so it kept blowing fuses and he could not get the engine started. Sgt White and I started it and they flew it on home, so we will have to work on it tomorrow.

"All the way back I was flying and we went along the coast and I know we saw over 100 mines washed ashore. We also passed over thousands of trucks on the main highway and tanks and all sorts of war equipment. We always fly just about 50 feet above the ground and the fellows on the ground are always waving to us. I got to take her off and land her today. That is the first time for about a week now – no, I take it back – I got to fly one over from the dispersal field the other morning. The trucks are still going by our tent by the hundreds every day and night. Where they put them all is beyond me because you see very few of them coming back. Oh yes, two more landmines let go yesterday. One blew up a large crane and the other a truck but no one was killed. They now have the sappers [engineers] back here looking for mines again. They found three of the large ones [anti-tank mines] which blew up one of the transports. It makes an old man out of you, walking around here, because you never know if your next step is going to be your last one or not. We all call them 'iron tulips'."

The following day, Bob was finally able to see Mussolini's famed Marble Arch up close and witness the delicate business of explosive ordnance disposal. "This afternoon we caught up with most of our work so I took about six fellows down to the end of the field and we climbed up the ladder inside of Marble Arch to the top. It is very high and on the landing next to the top there is a bronze soldier lying on each side facing the highway. On the landing below this we all wrote out names. The walls are just full of names so Cull and I put ours on the north side on the ceiling. No one else has their name there yet. While we were up on top we saw some sappers looking for mines and they found one so we climbed down to watch them dig it up. They dig all around it to see if the Germans have hooked a booby trap to it. Then if they do not find a booby trap they pull the mine out of the ground and then they put the detonator on 'safe' and screw it out of the mine. The mine is then harmless. We were very interested so they explained different mines to us and how they worked. They say their own mines are more dangerous to take up then the German ones are."

The next day, January 21, the rest was short lived and the flying continued. By now the Eighth Army had pushed the Axis forces all the way across Libya and almost to Tripoli. The fighting was tougher for the Allies now since the Axis supply lines were shorter and less vulnerable to air attack. In contrast, the Allies were still being supplied through Egypt until the Libyan ports could be cleared of wreckage, mines and bobby traps. "Our six ships tried to take off this morning and No. 8 came back with two rough engines and No. 7 was one hour late taking off because they could not get the left engine started. There was a vapor lock in the carburetor. About noon we got the new engine finished in No. 2 and then this afternoon we flew it for one and a half hours and everything worked out OK. Then we all went swimming in the Mediterranean Sea. It was my first time. The first time in you think you are going to freeze to death but after a while you are numb and you don't mind it at all then. The air was pretty warm so that helped.

"When the sun is out here it gets very warm but once it gets behind a cloud it turns cold. And the nights here are still very cold. Lt Farris, Doc Hamilton and I are all writing letters this evening. Yesterday I received two from Toots, two from Bessie and one from Mom and Pop. Not bad for one day, but it had been about a week since I had received one. The Colonel flew up and landed close to where we found the missing 37th Squadron ship; the one that got behind enemy lines. Well, it was crashed and burned up. It must have been shot down by anti-aircraft fire. The Germans had buried the crew and hung their dog tags on the crosses and wrote in German that 'here were buried the crew from the American Transport plane.' So they must have all been killed in the crash." [This may have been the aircraft piloted by SSgt Howard Robeson of the 37th that was shot down on December 19, 1942. Robeson was the only one killed in the crash. He is buried at the American Cemetery in North Africa. The other members of the crew got out safely but were taken prisoner.[15]]

In his next letter to Toots, Bob recounted some of the more colorful aspects of desert living. "It is sure good to receive mail. That is the first that we have received for the past ten days. Just now our chow is not so hot because it is pretty hard for them to get it to us. And I have never seen fresh milk since we left the

States; every bit of it is canned. We are also back to the rationing of water because we are 60 miles from any at all. Here the squadron doctor, operations officer and I share one tent. There are only the three of us so we have plenty of room. We also have a field telephone in our tent so that they can get a hold of us whenever they want to. Every time there is a ship going to the rear for anything we all send for cheese and all types of canned goods. Only one thing you can't buy the first cracker over here because they just don't make them. In our tent we always keep quite a bit of food.

"Yesterday all the ships took off so I did my laundry. I also aired out my sleeping bag and shook the sand out of most of my clothes. We made a new Desert Lilly for our tent yesterday. Oh, you want to know what a 'Desert Lilly' is? Well it is a latrine for you to go #1 in, but not #2. You dig a hole then fill it with rocks, then you knock both ends out of a five-gallon can and set it on the stones then you pack sand all around it, then you take another can, knock the top out of it and punch holes in just half of the bottom, then you put it in the other can at an angle and there she is, all set for the trial run. We are able to buy quite a bit of whiskey and beer whenever a ship gets out of here because the British have stores which sell it to just squadrons. You can buy it by the case or by the bottle. If we were not kept so busy up here I believe a man would go nuts, but we work most all of the time, because the sand gives our engines fits.

"I won some money playing cards which I will send home to you when the post office man flies up here again. They are moving forward so fast that the engineers can only clear the mines from the camp area and flying field so when we walk anywhere we find a nice truck track and stay in it. The damn things are a nuisance, they hinder your fun. But here they have brought another lot of engineers back and they are clearing larger areas. I watched them dig one up yesterday and I don't believe they have a brain in their heads because then bent right down over it and kept digging the sand away from it until it was ready to come out. The Germans even put booby traps on the mines to catch the men that dig them out so they have to be very careful to look for these while they are digging one out."

In his diary, Bob documented the difficulties in keeping up with the fast-moving Allied ground forces. The C-47s continued to fly missions loaded with 5-gallon gasoline cans, dubbed Jerry Cans since they were copies of the German design. These were easier to load and unload from the aircraft, directly into supply trucks or armored vehicles. "Only four ships were sent to the front this morning and the rest of ours were in commission so I went as crew chief in No.2. We hauled gasoline as usual. They told us up there that during the night one fighter squadron had flown to Tunis and bombed and strafed the Germans. Then today another fighter squadron left to do the same. When they do this they must fly across Rommel's army. The Germans are retreating so fast that they do not have time to plant mines on the flying field so they plow great furrows all over the field. One airfield that one of the American fighter groups are going to move into was being plowed up like this so they sent a flight over and killed the Germans doing the work and they are going to patrol the field from the air until the ground forces can move on to it. Some stuff, patrol a field that is still in enemy hands. They expect to take Tripoli today (January 22). We passed by a large convoy of ships. There were 14 of them, escorted by three cruisers and plenty of aircraft. They are waiting until they can take care of Tripoli and then they will move into the harbor. I got to take off and fly to the front and also land the aircraft. We flew over the 37th plane [Robeson's] and there is not much left of it. We hauled a load of wounded back as we do every day and some of the boys look pretty bad off."

Rommel's divisions had been driven out of Libya and started their withdrawal to the so-called "Mareth Line" defenses on the border of Libya and Tunisia. On January 19 he had ordered the demolition of Tripoli's harbor installations to prevent their use by the Allies. On January 23rd, Montgomery had taken the formal surrender of the city as the Axis troops fully retreated into Tunisia.[16] The Allies were now closing the ring on the Axis forces. They were being forced back into Tunisia from both the east and west. Now that the front had been significantly contracted and more ports were available, there was less need for airlift assets to supply the Allied forces.

# The Return to Egypt

As a result of the Allied success on the ground, Bob's outfit was ordered back to Egypt from their temporary base at Marble Arch on January 24. "Yesterday we received orders to move back to Ismailiya so we spent most of the night packing and loading the ships. It will sure feel good to get back to where we can live half way like human beings and every time you take a step you don't have to sweat it out as being your last one from some damn landmine. Also we can sleep again without having to go out to the slit trench during an air raid. We got off the ground at about 1000. On the way White, Tucker, Love, Best and a few other fellows played blackjack until the air became so rough that we all got half sick and had to stop playing. We arrived at Ismailiya at 1635 and then we had to put up our tents so it was well past dark when we finished. We left Crook, Singer, Milliman, Robinson and Babbitt behind to change engines in Nos. 12 and 13."

The next day the men settled into yet another new airfield – not knowing how long the stay at this base would last. "Today we went to Ismailiya and took the town in. We are stationed much closer to town than we were the last time we were here. I am having a British battle dress jacket made out of camel hair cloth and a gabardine shirt and pants. It is going to cost me about 15 pounds or over $62, but I think they are worth it. Major McMenamin is having an entire gabardine uniform made and it is a beautiful thing. Lt Farris, Capt Hamilton and I ate all we could hold and then we went to have our photographs taken together. Then in the evening we went to a picture show just off the edge of the field. It was a very old film but something to look at anyway. It doesn't seem like the same life when you come out of the desert and can go places again, even if they are not so hot."

Once he was settled, Bob penned another letter to Toots and enclosed some photographs of his squadron mates. "Well here we are back from the front and still OK and I expect to stay that way for a long time. It sure makes you feel good to be able to sleep in peace again and walk around in peace. But we still have the good old sand with us, but in the distance you can see a few palm trees and that is something. Yesterday we went to a small town close by and we ate until I thought we would burst. Then we had our pictures taken together. The ones in front are Lt Farris and I. He is the operations officer, and in the back seat is Capt Hamilton, our squadron doctor. We all sleep in the same tent. Since I left the States I have seen only one American flag, so how about you sending me a small one. I am sending you a Palestine Pound ($4.15) so that you won't go broke - now don't go and give it to church just because it is from the Holy Land. Major McMenamin has the best looking summer uniform that I have ever seen, so I am having pants and a shirt made just like it. He should look like a 1-star job[general] when he goes out in it." The next day he sent off another letter with some additional photographs of his men. "The picture I am sending you, you can put away for me. It is, left to right, MSgt White (Line Chief), MSgt Love (Flight Chief), and TSgt Tucker (Flight Chief)."

Although they now were closer to civilization – such as it was -- the Squadron still struggled with a lack of spare parts to keep the aircraft flying. "Slept pretty late this morning (January 26) and Major Mac came over and said that I could have two new engines and four new tires. Then at about 1000 we received a message saying that we had No 2. down at a field with the right engine burnt out. So I don't know what we will do. I do know it will make No.2 with two brand new engines when we get the right one changed. We were in the desert for six weeks and four days without any relief and no equipment except what each crew chief had in the back of his ship and if you don't think that is tough to try to keep them flying like that, just try it sometime. It is like coming into a new life to come out of the desert and see life, something green and most of all good equipment to work with."

*Left:* C-47 engine maintenance in the harsh North African environment. (US Air Force via Air Force Historical Research Agency - AFHRA)

But things were starting to look up for the 316th by the beginning of February. After a long sea voyage, the remainder of the Ground Echelon had finally arrived. Bob was sure glad to see them. "The ground troops arrived on the boat from the States yesterday so this morning Major Garland and I went down to see them. They are stationed at Fayid, Egypt which is about 30 miles from here. Most of them looked like they needed a good feeding. The trip over took them 42 days, and it was a little more than some of them could stand."

Bob and his fellows were now having a quieter time being further from the front, which gave him a little more time for letter writing. On February 3, he wrote to Toots and told her about his current roommates. "There is not much news that I can send you from here. We are all taking it sort of easy just now but we don't know how soon they will put us to work again. I am sitting in my new wicker chair with a kerosene stove beside me, one electric light, two kerosene lamps and one kerosene lantern. Some light, eh? Well at home it would equal one good electric light bulb. Farris is cross-country and the Doc went to the show. We will have one more in the tent starting tomorrow. He is Captain Wannamaker, our squadron adjutant. He used to live with us until he was sent to the hospital for ulcers. He is back for a while now, and I hope he gets to stay because he sure is nice to live with."

In the weeks that followed, the group concentrated on repairing their desert-worn aircraft and integrating the newly-arrived ground echelon airmen into the units. Bob's next diary entry was February 10: "Ever since we moved here we have had very little to do. We have most of our ships grounded for engine change. Some of them could fly a while longer but if we do that they would want to send us back in the desert and then we would have the same old thing, airplanes scattered all over the desert. So we just keep them grounded until we can get some engines. In the day time we have been working on the ships to catch up with the maintenance. All of our ships are here now except No. 6, which is in the desert getting another engine. We are also working on our trailer which we took with us from Marble Arch. We have received some of our supplies that came over on the boat and are now using part of them on the trailer. We have an air compressor, light plant and booster pump mounted on it. We are going to build a work bench over the light plant and build boxes to store our drills, light cords and other equipment. When we pull up to a ship we will be about able to do anything to her. Then we still have about ten feet of platform to haul jacks, wheels, maintenance stands and other equipment. The trailer is an old bomb trailer which is built very low to the ground so it is just what we need. Last week we had two trips to Algiers, and they are going to start a mail run. One of our ships leaves in the morning for the first scheduled run. It goes over this route: Hilio, Benina, Castel Benito, Biskra and Maison Blanche. Tonight we went to the show which is located just off the post here. We have been going every night since we moved here. The price for a box seat is only seven piasters or 28 cents. Jim Farris is keeping us all up with his packing for the trip tomorrow. Every time someone goes on this new mail run, someone has to go along that has been on the route before. So, Farris is going tomorrow because he has been before."

The maintenance trailer built by Bob's crew in the desert. (AFHRA)

## A Break from Combat Operations

By mid-February 1943, Bob's letters to Toots were mostly about family and friends, since the frantic pace of activity had died down and there was less excitement now compared to his previous desert adventures. "I am now going to try to answer all of your letters that I have received. The news about Tunisia you can just skip for we are the ones making it. And we get all of our information first hand. At least from this side of it. I wish we were on the other side with the Americans. I want my radio and camera so bad I can taste it. I could have brought both of them along if I knew then what I know now.

"There just is not much to write anymore. Bones shaved off his long side burns and part of his long mustache. We now have some good equipment so we are building everything we can to help take care of the ships in case we go back in the desert soon. You should see some of the things. If we don't have what we need we just make it. We are sure getting good chow now and I hope that it keeps on just like it has been. Everyone is getting along just fine and me, I am fat and sassy."

The Group continued rotational flying throughout North Africa, supplying both sides of the front as the Allies were near to victory. "This morning (Feb 13) we had three of our ships and three of the 37th Squadron ships leave for El Adem for a three week period. The 36th Squadron will be in charge of maintenance, so we had to send our mechanics. I sent Sgt White, Sgt Young, Sgt Fenimore, Sgt Babbitt, Sgt Wallace and ships No. 8 (Sgt Kawoll), No.9 (Sgt Ridgway) and No. 4 (Sgt Cooke). They are going to move troops from Tobruk to Tripoli from where the transport group in Algiers will pick them up and take them to the front on the other side of Tunisia.

A group of Bob's mechanics in the desert.

By the next day, most of the aircraft had departed and the field was quiet. "No. 11 left this morning for Tripoli. We were all off today. In the afternoon, Captain Wannamaker and I went to down to eat. Then we took a horse and carriage to the U.S. Club to take a bath and on our way we saw the usual sights along the Suez Canal. In the evening we all went to a banquet given for the Heads of Departments of the Group. It was the Group's first birthday, and we are now one year old. Col McCauley gave a speech and made us all feel good."

After the brief respite, the work continued. On February 15: "Yesterday morning Crook's ship froze when we went to run it up, so I told him to just quit until today. Well, we found out today that the generator had frozen and was holding the engine from starting. Ship No.1 left this morning for the front. No. 11 returned today with their load still on board because they could not land due to the muddy condition of the field. We now have our trailer about finished and she is going to be a big help. When Best returned today he told us that they were already having flat tires on the ships. They sent two tires back with Best so tomorrow we will have to get them ready to go back. The front is sure hell on tires and airplanes."

Although the 316th was now a little further back from the front, normal accidents and disease claimed the lives of some of the men. Bob records a somber event on February 18: "Last night at 2330, one of our navigators died from polio encephalitis. He was sick for less than 36 hours. He had been married less than six months, and his wife is only 18. His name is 1Lt Richard Hillegas. Most all of the Squadron went to the funeral, which is not very many right now. They put you in a box that is hand made by the natives. We had a firing squad and we were able to give him a military funeral but we had to bury him out in the desert, in a small graveyard. It sure would be one lonesome place to spend the rest of your time. It makes you feel pretty low to see one of your own men buried out in the desert in such a lonely spot. That morning the 44th Squadron buried an enlisted man that was killed when a truck ran over an embankment during a sand storm."

**Field Promotion**

Bob Uhrig had been a warrant officer for less than a year, but the wartime demands on him and his men and his new level of responsibility meant that he was ready to take on the new position of a commissioned officer. This required a request to move through Army channels and ultimately an interview with higher headquarters. On 19 February, he was sent to meet with the promotion board. "I caught the mail plane this morning at 0645 and went to Cairo for my interview with the Board to see if they are going to give me a commission or not. I found out that I did not have to appear before 1500 so I looked up Colonel Graft. He is Assistant Engineering Officer for the Middle East Air Force. He fixed me up with a room at the Continental Hotel. At 1500 I was there, but it took until 1700 before they got to me and then they only asked me a few questions. In the evening I had a few drinks with Col Graft and then went to a movie.

"The next day I spent most of the morning with Col Graft at his office. I told him about our ships and the spare parts we need or should have. In the afternoon I took a carriage ride through old Cairo and visited three mosques, of which there are over 600 in Cairo. They are the most beautiful things that I have ever been in. Each one must have cost a fortune. I went to the top of the tower on one and took pictures of Cairo. Old Cairo has the narrowest streets that exist anywhere. The shops are all open with everyone making things by hand. Just like in the books, they have everything - thieves (plenty of them), beggars (just as many) and peddlers. At 1530 I went out to the airport at Heliopolis, where the mail plane comes in, to go home. One of our ships landed on their way back from Algeria so I rode home with them. On their trip the Germans had bombed them at Algeria and had strafed the field at Biskra but had not hit the ship at all.

"When I arrived at Ismailiya, I found out that the 36th, 37th, and 45th Squadrons were all quarantined for the disease Lt Hillegas died from [polio encephalitis]. We went over to Group to call up and find out if we were all quarantined and they said that we were so we went over to the Squadron. We think it will last for about 21 days, which is a long time to stay in just one small area."

Now under quarantine, Bob had little to do but work and write to Toots. "It rained today (Feb 22) believe it or not so tonight it is pretty cold. We took some tea, sugar and cream from the mess hall and we are going to make our own tea tonight. We sent some fellows that are not quarantined into down after grobbies. They are small cakes of different kinds and are the best I have ever eaten and everyone also thinks the same. We have a small gasoline stove that we bought so that we can heat our water and cook if we want to. So tonight in our tent with the wind blowing outside we are going to have our tea. We gave the tent a good cleaning today. We

decided that if we were going to have to stay here for 21 days we might just as well live in style. The officers played the enlisted men in softball and the score was two to two when it rained us out. So tomorrow I think we will fight it out again. I thought I was going to enjoy this evening writing and drinking tea around a nice warm fire. Well it is now 2200 and I just returned to the tent. It has been the busiest night for over two weeks. I have spent the whole evening looking up crew chiefs and checking on airplanes. But now I am finished and they have cut our portable light plant off so I will have to finish this later by kerosene lamp and lantern."

On February 26, a serious mishap occurred in the 36th Squadron again and Bob was in the middle of it. "Today all of our Staff Sergeant pilots were sworn in and are now Flight Officers, which is about the same thing as a Warrant Officer. In the afternoon some gasoline in the Supply Room caught fire and burned the Supply Room to the ground. When it caught fire, Sgt Hershberg ran out with his clothes on fire and we threw sand on him and put the fire out. Then Sgt Jones and I ran in with fire extinguishers to try and put the fire out but we had only got a good start when the ammunition began going off from the terrific heat and we had to run out. All of the Armorer's equipment, flares, and ammunition was in the building so it was a hot time for about an hour with flares shooting up into the air and shells going off constantly. Most of our equipment had been moved to Fayid but we lost our welding equipment and a lot of spare parts. Also the personal belongings of Lt Hillegas were burned up. They had just finished packing them to send home to his wife. The medical quarantine was lifted today so most everybody hit town and this evening what few were left went to the show, but I stayed here in the tent and wrote letters. It is pretty cold tonight so I had to keep the kerosene stove burning pretty high. The lights went bad tonight so Sgt Miliman brought over his gasoline lantern and we both wrote letters."

His letter to Toots a few days later included a press story on the Troop Carriers. "Yesterday we worked until about noon then Lt Farris and I went to town and he got his new uniform. Then we went to a place called the French Club for dinner. In the afternoon we took a horse and carriage and went to the United Services Club and took a hot bath. Then we rented bicycles and went riding for most of the afternoon about the town seeing the sights. Most of the nice places here are for officers only, that is the English doings. Then in the evening we went to the show as usual. We had some pictures taken and I am going to start sending them to you. I am sending you another *Parade* magazine dated Feb 7, 1943. I want you to read about the Transport Ships. Then you can tell about the English. Only one of the ships in the picture is English. All of the rest are in our Squadron and not even the first word was said about Americans. You can even see the star under the wing on Pappy's ship that one at the top of the page. They mention about fighter escort and the only ones we have ever seen were around the fields where we were hauling freight and the rest of the time we are on our own. They also talk about the ground crews unloading them; well we pushed gasoline drums off of our ships until we were blue in the face. I am marking all of the ships that belong to us. Almost everyone in the squadron is burning up about how they grab all the glory."

Bob's earlier interview with the promotion board was history and he made no further mention of it as things were still busy at their base in Egypt. But on March 4, life had a surprise in store. "Came in tonight from the show at 2200 and the Charge of Quarters (CQ) said that Colonel McCauley had called and wanted to talk to me. So I called him and when he answered the phone I said, 'Warrant Officer Robert Uhrig speaking' and the Colonel said, 'You mean Lieutenant Uhrig, don't you?' and I said: 'You don't mean it Colonel!' and when he said he did I just about passed out. Then he had me come over to Group Headquarters to be sworn in as a Second Lieutenant. The he took me to the Officer's Club and bought me a few drinks and we talked about old times. This is one of the happiest days of my life and I would sure like to call Toots and tell her about it."

## Move to Fayid Egypt

On March 6, 1943, it was time for another move for the 36th TCS. This time it was to a more permanent base at Fayid Egypt. The C-47s were still flying rotational flights up to El Adem, but the bulk of the squadron moved from Ismailiya. Fayid had the advantage of more permanent facilities such as barracks vs. tents for

the men. In the next week, Bob got his men settled at their new location. "Today we tried to get the men back into Flights and also in Crews just like we had them back in the States. We have so many now that the crew chiefs are in their seventh heaven. We also put up an Engineering tent for Sgt White and me and for the Line Crew to come and get Technical Orders to read. There are about five mail runs each day now and I think that we get stuck with about half of them. Aircraft 12 and 13 are out for engine change. We have the engines out of them but we have no accessories so we might just as well not have any engines. We are living in barracks now with three men to a room. They are very large rooms and we have our grass rugs on the floor and a table in the middle of the room. We have a blanket on the table. We also have a stand with three drawers so that gives us one apiece. There is Lt Farris, Captain (Doc) Hamilton and myself in this room."

Mess hall building used by the 316th while at Fayid. (AFHRA)

Bob's next letter home was the first one where he was able to pass on the news of his recent promotion and some of his impressions of the recent desert campaign. "Received three letters from you today but I will wait until tomorrow to answer them because it is getting very late. In fact it is almost 2200. But you did talk about the Eighth Army. Even if you don't like the English you have to give them credit this time and besides, who do you think the 316th is working for? Well I will tell you, it is the Eighth Army. They have retreated and then charged so many times that this desert fighting is old stuff to them. And if you had been there when the big push was on you would have changed your mind. The push was something terrible; they never once let the Germans even stop for a minute. We know because every day we would move our landing ground where we hauled the equipment from 40 to 60 miles farther on. The whole thing was well planned and carried out. In fact they were wonderful and while you are talking about this, just how are the Americans doing on the other side [Tunisia]? If I read it right they were pushed back. If anybody deserves a rest it is the Eighth Army. Of course there are not very many British but lots of Australian and New Zealanders and South Africans. We have some men up at our old home site but plenty far from here and there was a German ship shot down the other night and the crew jumped out and they captured them in the night. The gunner was 14 years old. What do you think of that? We have moved again but this time it was only a few miles. We now have barracks and the officers have three men to a room. It is getting better and better. In fact this is like any field back home."

His letter on March 10 told of settling in to their new semi-permanent base. "I am writing this letter from our engineering tent that we have just set up here at our new home. We have a door and a couple of wooden horses for the Flight Chief's desk and they found a table and painted it for my desk. For a chair we have the bottom or seat out of a broken chair and it is bolted to a nail keg. We also have a storage tent set up for the men to keep their equipment in and to have a place to hang out in. Having so much help now the men don't know how to act. They can't believe it. The crew chiefs used to pull an inspection and I would tell him, well, maybe I can give you one man in about a half an hour so you will have to use your radio operator and now they have an inspection and I send 8 or 10 men out and they just about faint. We have a room in a building for the clerk and inspector so they can keep their files out of the weather. You should see our trailer, we are so proud of it. You know that we built it when there were very few of us to do the work and no material except what we could find ourselves. The group inspector gave us a letter of commendation on it. We hook on to her and pull up beside an aeroplane and do just about anything. This field is, as far as fields go, just like being at

any field in the States. It has everything that you could ask for as far as flying fields go. But after that is said, the rest is a blank because we are so far from the town that we will be lucky to go there once a week."

Bob was able to get some welcome leave at this time, and wrote of his experiences to Toots. "I have just returned from a five day leave but we could not get reservations on the train so we had to remain another day. We were to the Holy Land and saw Jerusalem and Bethlehem and all the points of interest. It sure was some trip. When I have more time I will tell you about it. I will make $317 per month with my new rating (2nd Lt). But many more trips like this one and I can't send money home. A pound ($4.15) is just about like a dollar back home. Now don't get mad because I thought that I should see the Holy Land. I spent over $80 for bibles but now the problem is how to get them home. I thought a bible from the Holy Land would make a good gift and would last a life time. The one that I got for you cost $32. I haven't bought a camera yet. They want as much in pounds over here as they want dollars for them back in the States, so I borrow Jim Farris's all the time."

This was a time of rest and reorganization for the Squadrons, and with his recent leave it was a full week before Bob had enough news to make a diary entry on March 15. "Number 47 came home last night with an engine burned out, so that makes three ships out now for engine changes. We were up this morning at 0530 because I am going along to Tripoli and on the way home stop in El Adem and see how the boys are getting along. We were off this morning at 0700 and flew to a small field down by the Canal to pick up 19 men and their equipment and take them to Tripoli. We had to stop for gas at Baninah which is just outside Benghazi, and then we flew on to Tripoli. The field here was used by the Germans and Italians as a repair shop, so when the British took over the Germans had to leave about 150 airplanes behind. I never saw so many destroyed airplanes in my life.

"The country around here is beautiful and if one did not know different you would think that you were in California. The dispersal areas are small roads are cut out through the gardens, fruit groves and vineyards. It looked very funny to see pursuit ships under trees and under grape arbors. Poppies grow wild here just about as thick as dandelions do back home. They are beautiful and there is field after field of them. Everything is so green that it is hard to believe that you are on the edge of the desert, and that the front is only 100 miles away. Most of the pursuits have been already moved out of here and are much closer to the front but with the harbor being here at Tripoli they can haul gasoline so close to them that it would not pay to have us up here yet. So we go on hauling mail, personnel and freight. Almost all of the civilian population is still living in Tripoli and most all are Italians and quite a few French."

While at Tripoli, they had to endure yet another air raid. "We all sat in the ship and listened to the radio until about 1900. Then Christman and I made up our beds under the wing of the airplane. The rest – Love, Tucker and Quisenberry slept inside. About 2300 the ground was shaking so badly it woke me up. When I looked towards the harbor the sky was lit up like day. The Germans were here and they already had dropped about 24 flares. The anti-aircraft was cutting loose and it looked like about 150 roman candles going off in different directions at the same time. The searchlights were also seeking out airplanes in the sky. Every once in a while you could see the flash of a large bomb and in a few minutes the whole earth would begin shaking. The anti-aircraft guns also make quite a lot of noise. In other words, when there is an air raid you have a war all your own. In a few minutes there were planes circling over our head at a very low altitude trying to find the field. Everything on the field was as quiet as could be and not a first glimmer of light shown. The guns around the field could not open up because there were three British Beaufighters [night fighters] that had taken off and were up there after the German planes. Every so often one of the ships would cut loose with a burst of machine gun fire but we could not tell who it was. The German ships did not drop any bombs on us and no ships were shot down close to the field, but one was shot down over the harbor. The way the Germans do it is to have a heavy escort of pursuit ships with the bombers at night to keep the British night fighters away from their bombers. They put four 100lb bombs on the fighter wing racks. Then when the bombers are working on the shipping the escorts go to the flying fields and let go with the small bombs. The British send up night fighters and they fly around the fields and just around the edge of the anti- aircraft fire and then they try to catch the German planes coming out of the flak. There are over 100 anti aircraft guns just around

the city of Tripoli itself."

The next day, the small flight of C-47s with Bob aboard headed back to their base. "We were up this morning at 0700 and walked about a mile and had breakfast. We had powdered eggs, bacon, coffee and bread. It was pretty good. We found out this morning that the air raid last night was the worst they had had yet. It lasted for over 45 minutes. All they had to send back with us was two bags of mail, so we were traveling very light. We had to stop for gas at Baninah because we could not make El Adem on the gas we had. The boys there are doing very well with their three ships but Richey in No. 9 was there with a burned out engine so that makes four ships out for engine changes. This one blew the intake pipe off and they had to feather the propeller and stop the engine. They had a load of oxygen bottles on board so they opened the door and threw them out so that they could keep the ship in the air. We have taken off from El Adem and are about one and a half hours from Cairo. We have to stop in Cairo to put off three wounded we picked up at Beninah. It is 1620 now so we won't be home before about 1800. It is about 35 minutes flying time from Cairo to Fayid. I have some poppies pressed in my notebook and I am going to send them home to Toots and my mother."

**The Stars and Stripes**

While Bob was up at the front, he received a very special package from Toots, one that had deep significance for him. She had sent him two American flags for the men to hoist above their installation. To someone so far from home for so long, the sight of an American flag is very welcome. As soon as possible, the men had the Stars and Stripes flying and Bob relayed his deep appreciation to the folks back home. "I just came back from the front. We were at Tripoli and it is just like being in California with groves of trees everywhere, gardens everywhere and vineyards everywhere. It is just beautiful country. But then you see all the destruction and it ruins everything. Planes destroyed everywhere and the beautiful gardens have paths cut through them where they can taxi planes away from the flying fields. Poppies grow wild just like greens do back home. So I am sending you two of them that I picked beside a destroyed Italian plane. We were only there one night so we had to sleep under the airplane wing. At about 11 o'clock the Germans came over and were giving the harbor hell and the sky was lit up by bomb bursts, antiaircraft fire and search lights. You should see antiaircraft fire some time because it is beautiful. One cannot explain how it looks. Every once in a while you could hear bursts of machine gun fire but whose side it was on you could not tell. But no bombs were dropped on the field, so here I am back home safe. On the way home I stopped in the desert to see how some of our men are getting along out there and everything was OK, so I came on home.

"The two flags are just wonderful and all of the fellows are so very proud of them. Today the men made a flag pole that runs up through the center of the Engineering tent and sticks out the top about three feet. They have a pulley at each end and a small rope runs around both of them. The flag is sewn to the rope and all you have to do is pull on the rope and up she goes. Then you pull on the other rope and down she comes inside the tent. In fact she is hanging in front of me right now. Tonight when we took her down MSgt White whistled retreat and the rest of us gave her a good salute. We are sure proud of her. So you better send another one because the wind gave her pretty much of a beating. The one that you hang on the wall I am going to put in my room. We can have lights on at night so here we are back to our old tricks of working nights. It is now 2045 and we won't be finished for another hour. I have my men split in half and we change every week. I have a nice collection of newspapers. You better tell John to brush up on his German because I have some German newspapers. I have found them in an old destroyed building and in dugouts in the desert."

**Witness to Tragedy**

After their return to Fayid, there were more tragic accidents on the field. "This morning while we were on the line, there was a 'Wimpy' [British Wellington bomber] that came in for a crash landing but no one was hurt. But while we were eating dinner, a B-26 cut out and crashed and burned about 100 yards from the Mess Hall. We all ran out but she was burning too badly to be able to get close to. When it crashed, two men were

close by and pulled the pilot and copilot out but the man on board just going for a ride was unable to get out and burned to death. After the fire died down you could see the body but there was very little left of him and his skull was pure white from the meat being burned off."

By the 21st of March the sand storms had resumed and there was not much to do but wait them out. "This is the third day of very high wind with the sand blowing. It is not as bad here as at El Adem because the sand there was much finer. It is all we can do to keep our tents standing on the line. Today Lt Chiros gave me rifle instruction and tomorrow night I am supposed to get some more. We are making coffee and shooting the bull. Lt Farris is getting cleaned up to go to Tel Aviv on a five-day pass. Tel Aviv is, as we have been told, like the American cities. We shall find out just how good when he gets back."

Two days later, the wind had let up just enough for maintenance work to continue, but unfortunately, so did the aircraft mishaps. "Received five letters from Toots this morning and have just finished answering one of them. Now we have five ships out for engine change. No. 8 yesterday had an engine freeze up and we don't know as yet what caused it, but we are going to change both engines. We also sent two engines and the equipment to change them to El Adem this morning to put in aircraft No. 9. It is now 1625 and we have been watching a P-40 crack up. He did an inside loop then went into another and all at once did a snap roll that was a snap roll. We saw something fly out of the ship and it looked round. Then the ship went into a dive from about 3000 feet straight into the ground. We saw it explode and burn. The ambulance rushed over but we didn't know the score until they got back. It was about a mile from the field. We see about as many killed here as we did at the front. Yesterday, two other P-40s collided in mid-air and blew up and there was not a piece of engine or airplane left larger than two feet square. One pilot was thrown clear and opened his parachute, but was beat up pretty bad. The other they only found a few pieces of. We swung the engines into Nos. 12 and 13 this afternoon and then put the propellers on. Tonight we are trying to get one ship ready to run."

The results of the attempt to rescue the pilot from the crashed P-40 were unsuccessful as they had feared. "The P-40 that cracked up yesterday went about ten feet into the ground. They were digging it out last night and all they found so far of the pilot was his billfold cut in two, his keys and the top part of his pants with his belt still buckled but not part of him. This morning Love and I took the motorcycle and drove over to the far side of the reservation looking for some lumber to build benches. We found some old bomb boxes and bomb fins so we are going to try to make tables out of them."

In his letter to Toots, March 26, Bob conveyed his continuing frustration with the difficult desert weather conditions. "This wind is just about to drive me nuts. It is the only time in my life that weather ever got on my nerves. It blows all day long and we have to fix the Engineering tent about three times a day and the sand gets in and on everything. We must keep the planes flying so it means that the men must work in it with no protection. Every ten or fifteen seconds there is a gust that about blows you over and so you can tell from that how the sand cuts your face. I stay out on the line most of the day and part of the night because they don't feel so bad about it if you are out there with them. The tents flap and bang until you could scream. The wind also changes very often so we have to push the ships around by hand to get them headed into the wind. Crook has new engines now and we test-hopped it yesterday and she was in fine shape. So he left this morning on a mission. We also tested another this morning and she will be ready to leave tomorrow. We still have all of our ships and they are now in pretty fair shape. We have such horrible inventions to use to help is in our work over here, but they all work. If you don't think it blows, just run your fingers across this paper. I have just taken it out of a closed box and am inside of the Engineering tent but we still get sand. It is still not half as bad as we had at Tobruk. We could not even see at all there for two or three days at a time.

"A P-40 was cutting up over the field day before yesterday and lost control and dove straight into the ground from 3000 feet. They never did find over two or three pieces of him. We have barracks to live in as I told you before and they are very nice. But we have an English latrine. They cut out a board and put it on a bucket and there you are. I don't like it because the wind blows up your back. In the desert it looks very funny to see men sitting on them right out in the open."

On April 1, the 37th Squadron lost another C-47. Bob's account was logged in his diary: "Both engines

cut out in the desert and they cracked up on the side of a hill." There was no mention of casualties as Bob was getting ready to travel to the desert himself to visit the small 36th Squadron detachment. "We left this morning for El Adem. I am going along to see what is holding up the engine change on No. 9. We are taking men and ships up to replace those that have been up there for the past month. We are taking Nos. 12 and 13 up and we are going to leave No.2 back here for now. Crook is going on a pass so he won't have to go up until he comes back. We arrived at El Adem at 1420 and Flight Officer (F/O) Cull greeted us with open arms. It was pretty cool at night so about eight of us got in one tent and had Cull play his accordion. He bought it in Tripoli and paid 43 pounds for it ($172.00). It sure is a beauty and he can play just about anything in the book. You just tell him what you want to hear and he cuts loose with it. Out here in the desert music like that just does things to you. I went outside and thought a long time about Toots. We did not get to bed until 0130."

Bob's letter to Toots that day again mentioned the difficult working conditions. "We are getting our heads worked off just now. In fact we have more than we can handle. I would sure like to come home and get some new ships to bring over here. It is getting plenty hot here now and I suppose it will continue to do so for some time to come. But as you know the heat I what I like but I don't know if I can stand as much as they put out over here or not."

The next day, "We were up at 0800 and had breakfast. Half the men are leaving on No. 7 to go back today. The rest of us are going to catch No. 6 when it comes in this afternoon. This morning we went on a photography expedition to the German graves over by the old hangar. We then went out on the desert track and took some pictures of the old road markers. We took pictures of the places where we used to live and also the wrecked aircraft. It is terrible to think of men killing each other for no reason at all. And when you look at the graves it makes you stop and think. This one small graveyard on the field is German and has over 200 graves. There are men that were killed just on the field along and they had time to give them a decent burial. There are, I would say, over 100 shallow graves scattered around the field where men were killed and they only had time to scrape back the dirt, put them in and throw a small amount of sand over the top of them. Some have sticks in the ground over the graves and you find others that don't have any marker at all.

"The first time we were here at El Adem rats had dug into many of these graves and the smell was terrible. Also, lots of tanks had been shot and burned up and still had skulls, big bones and such in them. The odor in these tanks was anything but good. This time I walked over to a tank where I knew a man had been left, or part of him. He was still there, that is, what would not burn up. His shoes had been entirely burned off and all you could see was his heel plates. He must also have been blown to bits because parts of his skull were all over the front section of the tank. In one place about a mile form the field, beside a truck, a man's leg bone is still sticking up out of his shoe. He must have been hauling ammunition because there is practically nothing left of the truck. There are just a few of the things that you see in the desert. You can see tanks and trucks and airplanes destroyed by the hundreds around each airfield because there is always a battle over the airfields, and each airfield was always fortified. It is impossible to tell you of the destruction out here in the desert. Just picture flying over a landing ground and seeing 40 to 50 destroyed planes.

"No. 6 came in this afternoon at about 1330 and everyone threw their baggage on. Then we got ready to take off and I was going to take her off and fly back to Fayid. But when I checked the engine it backfired and I about jumped out of the ship. We found that the carburetor accelerating arm had broken. By the time we had it repaired the weather in the Nile Delta was bad and they would not let us take off. So we had to spend another night here at dear old El Adem. The do have some places scraped and rolled to make better runways and they have also put a roof back on one of the hangars."

The next day (April 3) the weather cleared and, "We took off at 0930 and if I must say I made a good takeoff. I have not made one for a month now. We flew pretty low over the desert and it is surprising how they are salvaging the old tanks. In about another two months there won't be any old tanks left out there at all. All of the men were glad to get back to where they could take a bath."

Bob's next letter to Toots spoke of what he had seen but not the full horror of it as he had recorded in his diary: "I have been up in the desert again on business. The dear old desert is changing. Where we used to be and where part of our outfit still is all of the debris is cleared away. I take it back – I mean old trucks

and destroyed tanks. They are salvaging everything so when I bring you over here there won't be anything but junk strewn through the desert. Now don't get me wrong, there will always be enough there to make your eyes pop out. But you won't be able to see all of the trucks and tanks. All of the airplanes are still there because when they are shot up there is not very much left. I have seen lots of tanks that have been shot up and then caught on fire and burned and you could still see parts of the men still in the tanks. But they have moved back and forth so much that it is hard to tell how long they have been there. In battle it is impossible for the men to find all of the dead and sometimes they miss some. Our old home does not smell as bad as she used to when we were there. You see around each airfield and landing ground there was usually a great battle and many men were killed and about half the time if they are in a hurry they have to bury them where they find them. So most of the graves are shallow and then the rats dig in and you get the best odor that you ever inhaled."

## On Leave in Palestine

By early April, the North African campaign was winding down. The Allies had the Germans bottled up in one corner of Tunisia and were closing in for the destruction or capture of Rommel's forces. The Libyan ports were gradually being reopened for supplies to reach the front and the need for further airlift support to the British forces was diminished significantly. That gave the 316th time to recover from their deployments and conduct some much-needed maintenance on their C-47s.

Finally with some free time, Bob was able to describe his recent leave to Toots. "Now to tell you about part of my trip. Leach, Fennimore, Pace and I left here on the 6th and flew in a British bomber to Tel Aviv, Palestine. There we went to the Yarkon Hotel. We stayed at this hotel for two days and the moved to another hotel that had a room that opened out on to the sea. In fact it was the most beautiful sight from our door that I have ever seen. From Tel Aviv we went to visit Bethlehem and went through the Church of the Nativity which is built on the birthplace of Christ. The place where he was born has a gold star and the place where the manger was is about ten feet away. A stable in those days was a place for a man and his horse to stay and they were mostly in caves and this is where Christ was born. We had to use candles to go into the place where Christ was born. It gives you such a funny feeling when you see the place where he was born. Bethlehem is a very small town and not too terribly clean. It is surrounded by olive trees and pasture for the sheep and goats. It is terribly hilly and all of the hills are terraced and it makes them look like something out of a fairy book. The ground is very stony. I said hills but they are like the mountains in Pennsylvania. The terraces with olive trees and plenty of green grass make them beautiful. They still herd their sheep like they did when Christ was a child. They use long staffs and dress the same. All of their sheep and goats are beautiful and very clean. Bethlehem itself sits at the top of a large hill overlooking the large valley where the shepherd's field is and also other points of interest. They have quite a few small workshops where they make things out of mother-of-pearl and olive wood. Well, so much for Bethlehem."

Bob's next diary log was April 14: "Not much doing today. We have No. 1 out for engine change and No. 3 with first valve check. We also have No. 4 up at El Adem out with a bad engine. Capt Farris flew No. 9 back from El Adem tonight and she has two new engines. Most everyone is away tonight to the free show of 'Star Spangled Rhythm' so I am writing letters home."

Now the desert heat was beginning to become severe and Bob described the unusual mirages to Toots. "I can sit here in the tent and see about three lakes out in the desert. But you could walk for days and not find them. I can see them and I am only half crazy so I know what men see when they are crazy from thirst. They are mirages. In fact we have talked some of the fellows into walking out to go swimming but as yet they have found no water. It is hard to explain how they look. But they look like the real thing. You can even see reflections in them and they are in the same place day after day. We just finished overhauling the motorcycle and she runs like a new one. It has the most power of anything that I have ever seen. All the news from the front is good and looks like it will all be over here in a short time. What we do then is beyond me."

His next letter to Toots was April 16. "This morning I worked and McMenamin was down and we all shot the bull. We are on the same field together now so we see a lot of each other. He comes down to the engineering tent about every other day and we talk about being back at Patterson Field. Tonight I am working and it is plenty cold. At night it still gets cold and also damp. We do most of our maintenance at night. They fly the pants off of them all day long and we then have to work most of the night on them."

Bob also had time to describe to Toots his impressions of his new Allies. "You can tell Peg and this is also for you that there are no women for over a hundred miles. We are plenty far from even a town the size of New Carlisle and in these towns there are only the native women. The only time that you see white women is when you go on leave. And as yet they have not bothered me but if I have to stay over here many more years I don't know. But I think I can last through it. Believe it or not I am beginning to like my tea. The English quit whatever they are doing to have tea and that is the truth. When we were up in the desert the antiaircraft gun crews would make their tea beside their guns and some of our ships were close by and they would have us come over in the mornings and it does taste good when it is cold. Another thing is no matter where you are or when they will never give or take a cup of tea unless it is hot. They all have what are called tea pots. In their troop only about one out of six comes from England. So you have a mixture and some are pretty good Joes. But I am not convinced yet by a long shot." Little did he know what was in store for him later in the war.

At the end of April, the 36th was alerted for another possible re-deployment. "Tonight (April 21) Major McMenamin told us to get out ships ready to move and we are getting engines for the ships that still have their old original engines. One engine in No. 11 has 775 hours, No. 4 has 738 hours and No.7 has 749 hours. We are also going to get new tires for all ships that need them, so it looks like we are on our way somewhere."

Two days later there was still no news on the potential movement. "We got No. 1 ready to fly today and flew it half an hour but it was so rough at 2000 rpm that we had to land. Neither magneto was rough so we checked the valves and found some of them without any clearance. We are now changing engines on Nos. 4, 7, 11, and 12. We flew No. 1 again and put another hour on it. She ran OK and I got to land her when we came in and I made a good one for a change. It is now getting pretty hot here. We have one tent for the Engineering Office, two for Equipment, one for Sheet Metal, Welding and Blacksmith, One for Dope, Fabric and Paint, and one for Electrical Test Equipment. We are set up pretty good here now, but I guess it is time for us to move. We have built all our own tables and most of our special equipment. Lumber all has to be shipped in, so as it is next to impossible to get any. We have to hunt all over the place and use just pieces of scrap."

Sunday, April 25 was Easter, but still a work day in a combat zone. "This is Easter Sunday but I had to get an engine ready to send up in the desert to put in No. 13 which had a burned out engine. So by the time I had finished this work it was too late to go to church. In the afternoon, F/O Pace and I took the motorcycle and rode over to the foot of the Mount 'The Big Flea' – so named by Cleopatra. We then climbed to the top, but once we reached it I did not have any feelings left in my legs. It is quite a climb. On the top we sat down and shot at rocks down in the valley with a .45 pistol and a German Luger. We met one English soldier on top who has a son two and a half years old and he has never seen him yet. Their mail service is terrible. In fact it takes about three months for them to get a letter from home.

"After we came back to the barracks we decided to go swimming in the Great Bitter Lake which is about a mile from the field. On the way you have to pass through Fayid which is nothing but a small native village. All the houses are made of mud and have palm tree leaves for roofs. The natives and all their livestock live in the same building. Close to the lake they irrigate the land and raise all types of vegetables, also wheat and sweet clover for their livestock. The plots are about 25 yards square and where they have clover they tie their stock to stakes and let them feed on it. Most of their work is done by oxen and camels. You should see the loads they put on their donkeys! We went swimming and then came home. This motorcycle sure is a god-send to us, but now it belongs to the squadron. But Major Garland lets me use it just as if it were my own. You see it is contraband, and it would have had to be turned in so we got permission to use it in the squadron. In the evening Cull played his accordion and the room was as full as ever. We then made coffee and finally went to bed at about 0030."

## The Desert Environment Takes Its Toll

After months in the harsh desert environment, the 316th aircraft required extensive maintenance and servicing. This was caused by continuous flying operations; but also by the extreme desert temperatures, sand and dust and rough airfield operations. Bob and his staff of maintainers were constantly busy with all manner of repairs. "Lt Coursen, F/O Wheeler, TSgt Middendorf and a crew of men and I left this morning (April 28) for Agedabia where Sgt Crook's plane is forced down with a burned-out engine. I am taking a crew of three men up. We are also taking a lighting plant so they can work night and day to get the engine done and change the engine on No. 13 because we are supposed to move on Sunday (May 9) and Major Garland wants all the planes back as soon as possible. The new engine was sent up to Crook last Saturday, April 24.

"Arrived at Agedabia and found that Crook already had the engine installed. He had help from our crew at El Adem. All that was left to do was put the propeller on and run her up. So while they were doing that, Coursen, Wheeler and I went to the little town of Agedabia. It is, or was, an Italian town, and looks to be about the size of Donnelsville (300 people), but the buildings, those that are left, are made of cement and stone. There are a few natives still living there, and I do mean a few. The flying field is at the edge of town. When we returned to the plane we found that they were having trouble with the propeller dome. I found that the old dome seal was causing all the trouble so we decided to fly up to Berka which is just outside of Benghazi. There is a B-24 bomber outfit there and a couple of Service Groups. After running around for about an hour we found what we wanted. We flew back and found camels on the field. The mechanics were throwing rocks at them, and kicking them, but they would not get out of the way so we could land. So we gave them a good buzzing and they got off the field. We then finished repairing the propeller and Crook and I ran it up and everything checked out OK. So we loaded Crook's old engine on his ship, along with the engine tripod and hoist. I left two men with Crook to help finish up.

"All that is left at Agedabia now is one Wimpy [Wellington bomber], one trainer and our ship. To think that a short time ago we landed a little while after the Germans had left and our fighters had started coming in. There was so much action that we had to fight out way in to land. Now it looks like a ghost field with just the skeletons of old airplanes to tell you there was once a field there. To think that so many men died at each one of those fields and to find them deserted now makes me shudder. The pilots of Crook's ship are at El Adem, so late in the evening we left and flew to El Adem to tell them to send the pilots back to Agedabia after the ship. There is still plenty going on at El Adem which is just outside of Tobruk. They are even building runways and fixing up the hangars, so they must intend using it for some time as a re-fueling stop. We still have three of our ships hauling truck parts to Castel Benito and hauling repairable parts back to El Adem."

Lieutenants Williard Boston (l) and Pershing Harmon (r) of the 36th TCS in North Africa. Lt Harmon is wearing British battle dress wool trousers. Since they were based so far from American supply centers, the 316th had to make do with whatever uniforms they could scrounge from UK forces.

By the end of April, the men had even been able to locate liquor in the desert. "Last night I visited 'Pappy' Street and 'Pappy' Singer and the rest of the men at El Adem. We had some beer and whiskey to drink. They go to Tobruk and draw liquor rations for 300 men and there are only 30 men here, so they have very generous whiskey rations. Plenty to drink! In the afternoon we flew home so I am ready to go back to work again."

In the midst of all this, Bob still found time to write home. "The sand is blowing just a little today but not enough to bother us. In fact the sand has left us alone for quite some time now. Last night most of the fellows went to a movie that the army had. They show them outside and everyone just goes and takes their own chairs. I finished work too late to be able to go. Another fellow and I are fixing equipment and a place where we can develop our own pictures. It will be something to do and I always wanted to do it. They just don't do good work over here and we are afraid to keep the film undeveloped for fear it will spoil in the heat."

On May 3, Bob was able to get a brief hop to Cairo. "This morning F/O Jackson and I caught the 1000 mail plane to Cairo. We caught a ride from the field into town. Then we went to the camera shop to see if his pictures of the Pyramids had been developed. They told him to come back at 1600. I looked at some cameras but I think I am going to save my money until I can buy a Leica. They are plenty expensive over here but this is the place where you need a camera. In the afternoon we went out to the Pyramids and rode horseback all around them.

"We caught a 37th Squadron plane back to the field this evening but had to go to Deversoir first to unload some freight. Received two letters from Toots this evening. Leach got one from Chick that is making him click his heels. Flight Officer Mobeley got one from home and by some unknown means it had been opened and a picture of an unknown girl had gotten in the envelope. Right now he is taking the beating of his life from the kidding the fellows are giving him. He probably will never live it down as long as he is in this squadron."

The next week found the Squadron preparing for another move westward to the area held by U.S. forces. "Plenty of work to do today. We now have all of our planes grounded for the move that we are going to make. The rest of the ships have come in from El Adem and we are working on them to get them ready to go. We had to transfer No. 13 tonight to the 37th Squadron. They have lost two ships so we had to give them one and the 44th Squadron gave them the other. The 36th and 44th are the only Squadrons moving right now. The 45th Squadron is still in Tripoli. We loaded all of our supplies and part of our Engineering equipment tonight. The night crew is going to have to work very late. We are supposed to move to Oran (Algeria), the other side of Algiers and we are only taking the Air Echelon.

"The best news since we came over here just came in. The Allies have defeated the Germans at Bizerte and Tunis and as those are the last two towns in North Africa, the war in North Africa is over except for a little cleaning up around. It certainly makes you feel good to know it."

**Another Day; Another Country**

On May 9 the 36th Squadron was finally on the move, but it was a welcome change. "We were up this morning at 0400. We packed our bedrolls and hauled them to the planes then took off about 0700. We stopped at El Adem for gas and had to make some repairs to No. 11. We then took off and our next stop was Castel Benito where we spent the night. No. 6 got stuck in the sand and it took us about two hours to get it out. Only one other ship needed maintenance and it only needed a solenoid.

"During the evening we had music by F/O Cull, and singing by most everyone. There are also quite a few poker and blackjack games. We now have 104 men with us in the Air Echelon. We have two gasoline stoves, so every time we stop we set up our own Mess. Everyone is in good spirits and having a good time. Before I went to sleep I lay and thought a long time about Toots and how much I would like to be with her. It was a beautiful night and the moon shone in the freight door of the ship onto my bedroll. Most of the fellows slept under the wings of the ships."

It was a long flight to their destination, requiring several "hops" to get there. "We got up this morning (May 10) at 0400. Then after breakfast we just hung around the airplanes until takeoff. We left at 0600 and Leach and I went to sleep on my bunk at the back of the ship. Once in a while I would get up and take a look

at the country which was very beautiful with just a few mountains. We are now getting into country where things are green and there is no sand to worry about. It is also cooler.

"We flew over Algeria and landed at Oujda in French Morocco. The field had dirt runways but they are plenty long. Everything is green and we are surrounded by mountains that run up to about 4000 feet. On the field here there are all types of wild flowers and they are very beautiful. They are all colors of the rainbow here and best of all we can hear birds singing. It makes you feel like you were back in the States. On the far side of the field we can see roofs of hangars and other buildings. Major Garland rode the motorcycle over but no one seemed to know anything about our coming. We don't even know what we are supposed to do. There are only two officers assigned to this field and they live in town which is about eight miles away. The 313th TCG is also here. They just arrived from the States yesterday. As yet we don't now what type of people live here but we will if we stay awhile. Some of the fellows are over now talking to the men from the 313th to see what they have in the way of parts and also if what they have is worth stealing for Engineering.

"Just came back from chow and I guess we will have to move this afternoon. We are now beginning to get a little news. We are moving alright, in short order. Love is staying behind with a couple of men to change the flat tire in No. 2. We then flew up to Nouvion, Algeria. It is a beautiful valley with green fields all around. The barracks are on a small hill about a mile from the flying field. Part of the men have to sleep in the ships tonight because there is not enough room in the barracks."

After settling in at Nouvion, the men resumed exploring their new location. "We were up early this morning and we went to the motor pool and got a truck from the other Troop Carrier Group that is here. We then unloaded the Engineering equipment and supplies and also the men's baggage. This took up most of the day. The 63rd Group that was in England is here now. I saw MSgt Collier for the first time since we left the States. We shot the bull all afternoon and part of the evening. He says that he put in for a commission but was turned down.

"For our Engineering equipment we have two metal shacks. We also set up our three tents that we brought along. It only took us about two hours to set up our equipment and be ready to operate. We are becoming past masters at this moving business. We can pack and are ready to move in nothing flat."

In spite of the hectic series of moves, Bob was still writing to Toots on a regular basis and keeping her updated on his activities. "Well I guess you are wondering what has happened to me. We have moved again for a short time and I am now much closer to home. We are now where the country is beautiful. Everything is green and the valley we are in is surrounded by mountains. The ones in view are just a little higher than the ones is southern Ohio. There plenty of grapes, oats, wheat and all types of farm produce. Sitting here on the barracks porch you can look across the valley and if you did not know it, you would believe that you were in Ohio. Their horses here are the most beautiful animals that I have ever seen. It is so good to see something green again that everyone about has a fit. It is also much cooler here. I saw Collier yesterday and we are now together. He is just the same old Collier and is still a Master Sergeant. I am so busy now and will be for a few days that I won't be able to write very long letters."

The following day, May 12, 1943, started unremarkably for the Troop Carrier men, but was to be remembered as a momentous day for the Allied team. "We got up this morning at 0630 and had breakfast then we went up to the line and the aircraft started flying at 0800. After the ships took off we started to fix things up. We have our Supply, Engineering Office and Welding all in one of the metal shacks. We chiseled some lumber from the packing crates for the WACO gliders. We use it for flooring in the other metal shack. We are going to keep our extra baggage in it. We are all so cramped for sleeping space that it is not possible to keep much of our baggage with us. Between the two metal shacks we have our two storage tents and one tent that we are going to use for the Flight Chiefs and Line Chief. They are going to have to sleep on the line.

"Today it was announced over the radio that the Germans in North Africa had surrendered. So this puts the finishing touch on the whole business over here." Bob's last diary entry for that historic day was particularly insightful and marked a significant transition for the 316th TCG. "We are over here for training with parachute troops and gliders, so it looks like we are getting ready for the invasion."

At the end of the North African campaign on May 13, 1943, General Sir Harold Alexander, the British commander, sent a brief but stirring message to Prime Minister Winston Churchill, "Sir, it is my duty to report that the Tunisian campaign is over. All enemy resistance has ceased. We are masters of the North African shores."[17]

**The Reckoning**

During the Desert War, the planes and crews of the 316th played their part in the epic drive of the British Eighth Army from El Alamein to Tripoli. "Montgomery's immediate pursuits of Rommel after the Panzer Army's collapse at El Alamein may have been disappointing operationally, but his 1500 mile advance from Egypt to Tunisia in little over two months was a logistics triumph."[18] This logistics triumph was made possible by using airlift as an integral component of a ground advance – a brand new concept born in battle. In part it was made possible by the use of the rugged DC-3/C-47 airplane. The C-47s flew rations, fuel, ammunition and other supplies daily, directly in harm's way, to both U.S. and RAF fighter units in forward areas and there the air evacuation of wounded men was pioneered. At the same time, other crews were sent on missions to Eritrea, the Congo, and other far-flung outposts in the Middle East and Africa.

It was in Egypt, while supporting the Eighth Army, that the 316th developed the concept of air evacuation. The group is credited with establishing the first air evacuation system in any theater of war. The 316th had the hazardous job of flying vital supplies and equipment into the battle zone for British forces. The aircraft landed on fields still sown with German mines. As Bob Uhrig vividly described, the fields used by the advanced echelon were bombed and strafed at night. The group lost one aircraft and one was severely damaged by ground fire but was able to land safely.[19]

As the desert fighting progressed, battle casualties began to show up at the 316th field at Agedabia. For the return trip to the base at Tobruk, as many litters as could be laid on the floor of the C-47s were taken aboard and ferried back for treatment in rear areas. Major Reinus, the group surgeon supervised the medical evacuation operation. The operation was a success, but unfortunately two of the more severely wounded casualties died on route. Eventually, after results of this ad-hoc evacuation had been studied, it was adopted throughout the AAF resulting in the saving of many more lives. Tributes to the group came from Air Vice Marshall Sir Arthur Coningham, RAF, "This job [the Libyan Campaign] could not have been completed without the assistance of the American transport."[20]

Later Bob summarized the difficulties of operations in North Africa:

- No [US] Air Force Supply base in Eastern Africa.
- The Group had to operate with only the Air Echelon of 90 people for 4 months.
- Water was rationed because the Germans salted any well in the desert.
- British rations were bad. We had to send hunting parties into the hills to hunt gazelles to have some meat to supplement the rations.
- Any time the Squadron arrived at a new location in the desert, they had to remove all the land mines before they could operate.
- They were always under the threat of German air attack and were contantly strafed.
- The C-47s operated without any fighter cover. They had some desert camouflage paint that helped to hide the aircraft and they flew at extreme low levels so they were hidden from the Germans.

Certainly the 316th deserved their share of battle honors for the North African campaign. The role of military airlift had been fundamentally defined. American paratroops had been used in combat for the first time, and a large land force was kept supplied by airpower. But there was little time to rest on their laurels. As Bob had clearly recognized when the 316th arrived in Algeria, now it was time to prepare for the invasion of Sicily, and the first large-scale test of Allied Airborne forces.

# Chapter 7

## Training and Tragedy

*Of course, really, those that stayed were entitled to precisely as much honor as those that went.*
*Each man was doing his duty, and much the hardest and most disagreeable duty was to stay.*
*Credit should go with the performance of duty, and not with what is very often the accident of glory.*
Theodore Roosevelt, *The Rough Riders*

The North African Campaign was over. The Allies now set their sights on their next major objective in the Mediterranean Theater. Following the agreed-to "soft underbelly" strategy, the next blow was to fall not in France, but in Italy. Specifically, the Mediterranean island of Sicily. The plan as it developed was that the forces in-being in Tunisia would be augmented by additional troops. This large force would then embark directly from Tunisia and proceed with an amphibious landing on the southern coast of Sicily, less than 200 air miles from the coast of North Africa.

The invasion operation, code-named Operation HUSKY, was centered on the capture of the Sicilian port of Syracuse. Although it was further from the Italian mainland than the key city of Messina, it was easier to reach by sea from Tunisia and had long been a strategic crossroads. In fact, as far back as 415 B.C., the Greek general Alcibiades remarked during the Peloponnesian War that, "if Syracuse falls, all Sicily falls, and Italy immediately afterwards." British forces under Montgomery would land just south of Syracuse, while American forces under Lt Gen George Patton would land in the area around Gela, and capture key airfields from which air support could be provided to all the Allied forces in their drive north to Messina.

A major feature of HUSKY was the first large-scale employment of airborne forces – both paratroopers and glidermen. Their objectives would be airfields and key bridges in the lodgment area that would secure these crucial points and deny them to the Italian and German forces on the island. The original American paratroopers in the North African campaign were the 509th PIR. They were a separate parachute regiment. In April 1943, they were joined by the entire 82nd Airborne Division, the "All Americans," recently arrived from the U.S. This division of approximately 8500 men included airborne and gliderborne infantry, artillery and engineers. All these airborne forces had a role in HUSKY.

The USAAF C-47 Troop Carrier units now turned their attention from the mission of logistics re-supply to a more direct role in combat operations – dropping the airborne troops on these key objectives. Unarmed and unarmored, they would have to "fly to the sound of guns;" navigating directly into hostile skies at the leading edge of the invasion.

The 316th TCG was now part of the 52nd Troop Carrier Wing, and, along with other Troop Carrier Wings began training for the mass airborne drop. The 52nd Wing was assigned to carry the 82nd Airborne Division into Sicily. Within the 52nd, five groups would carry elements of the 505th Parachute Infantry Regiment. The 314th, 313th and 64th Groups would carry the three infantry battalions and Bob Uhrig's 316th would carry the 505th Regimental Headquarters. Some histories say that the Troop Carrier units received very little training in night operations, in spite of the efforts of the wing commander, and that the airfields around Oujda had not been completed until May 25, delaying the joint the Air Corps/82nd practice drops. This meant that the 52nd was not ready for training missions until June 1, only a month before the invasion was to start.[21]

In addition to the paratroopers on board and their individual equipment loads, many of the C-47s were outfitted with additional equipment for the paratroopers in "parapacks" slung underneath the aircraft on bomb racks. These equipment bundles, which carried heavy weapons (disassembled mortars, machine guns, bazookas and howitzers) and ammunition, medical supplies, demolition materials and communications gear would be dropped by separate parachutes along with the paratroopers when they jumped. The soldiers could then recover the parapacks and reassemble their equipment as soon as they landed. Installing these and other equipment on the C-47s was the job of Bob Uhrig's Engineering Department.

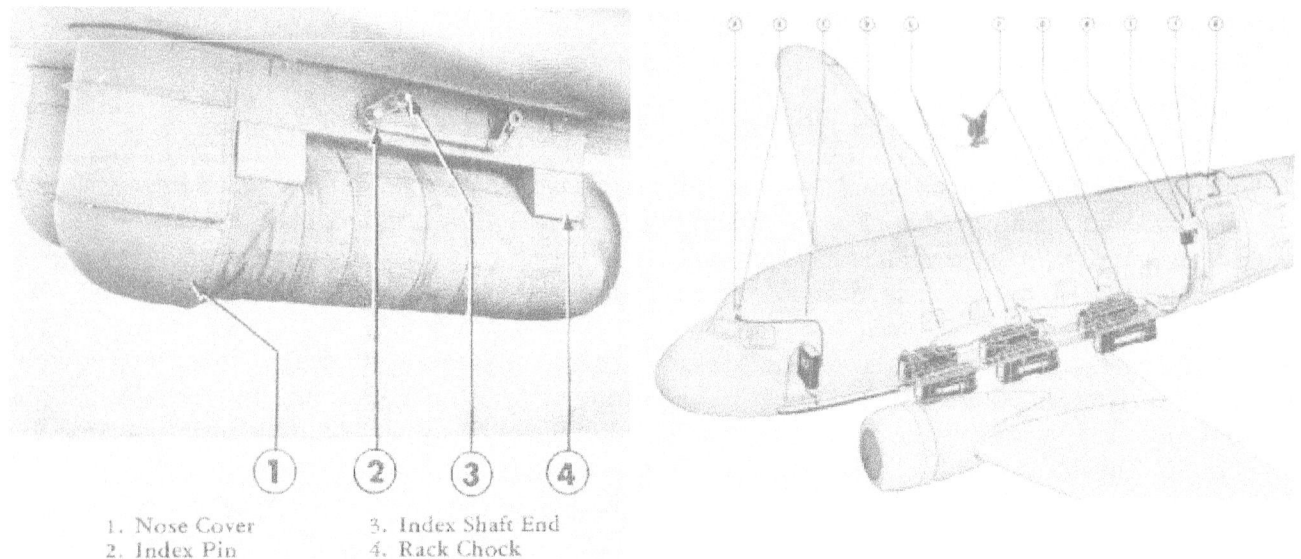

1. Nose Cover
2. Index Pin
3. Index Shaft End
4. Rack Chock

*Above:* Diagrams of the parapack installation from the Air Force Technical Manual on the C-47.

The shooting had hardly stopped in North Africa when preparations for HUSKY began in earnest. Bob recorded in his diary that on May 15: "Two days have passed and in that time we have done the following things: dragged two glider packing crates down here. The Flight Chiefs and Line Chief have fixed one so that they can live in it. They have electric lights that they run off of our portable lighting plants. One end they fixed so the whole end lifts up for ventilation during the day and then they close it at night. At the other end they built a door and then covered the rest of the opening with mosquito netting. It is just like a small house. Also, we now have a shower. We took a crew chief stand and put two 50 gallon drums on top of it, then mounted our valves on each drum. One shower at each end of the crew chief stand. There are also duck boards (planks) to stand on and a dressing platform. Alongside the platform we have a bench and a clothes rack. The sun is hot enough to keep the water warm. We have also constructed a latrine of our own. It is a hole six feet deep and six feet long with a 2x4 long ways across the top of the hole and this is what you sit on to do your business. There is even a toilet paper holder just in front of the rail. We now have 'Old Glory' flying over Engineering. She is becoming a little faded from that sun back in Egypt. Our ships are flying every day now on our training program. It is just like being back at Lawson Field in training again. I don't know what they want us to go all through this again for."

*Left:* A "desert lilly" improvised latrine in use.

*Right:* Bob using the homemade field shower his men had constructed in North Africa.

Bob soon found time to write to Toots to give her the "latest." "I now have some time that I can write and tell you all about us. As usual, engineering has everything well in hand. We have two metal huts about two miles from anything. We have a double storage tent and one personnel tent. We then went over to where they uncrate the WACO gliders and bummed their lumber. We were able to floor one tent and one hut. We then pulled over one entire box that the gliders come in and let the Line Chief and Flight Chiefs set up housekeeping in it. We have made a shower, in fact a double one, by using a crew chief (maintenance) stand. So now in the middle of the afternoon or when the men have time they run out and take a shower. We have done all of this in two days. It is wonderful to have these old boxes around; the men don't know how to act with all of this old lumber to work with. We have a Desert Lilly but I don't know what to call it on this side, also a latrine. Just give us a couple of days anyplace and we will make a home out of it. Some of the men are going to a town today which is about 40 miles from here. I don't know how they are going to get along because all of the people speak French. They will probably have to use the sign language to make the people understand."

Although the preparations may have seemed excessive to the men in the squadrons, they were essential to a successful airborne operation. The next day, Bob went along on one of the training missions. "Most of our ships went on cross country today for the training they are getting. Today they were supposed to fly by dead reckoning. So, I climbed into No. 12 to ride along. We flew to Biskra, Algeria and landed. From the airport the city looked very beautiful and is built in the Egyptian style. The other part of the town is all mud huts. The English have the flying field but there are very flew planes stationed there now. From the air the town looks like a large grove of trees. It is the last oasis before entering the desert. In the background and to the north are a large range of the Atlas Mountains. They have mountains to the north of them and desert all around, but they have plenty of palm trees for shade. An oasis is a very funny thing to see – all desert and in the middle you will see all this green vegetation and you will think you are going nuts. The one at Biskra is about three miles square. Tonight we have all the ships night flying with the exception of No. 1. I am writing this in the Flight Chief's hut and they really do have it fixed up like home."

An American crew loading an aircraft engine into a C-47 in Algeria, demonstrating the versatility of an aircraft built originally to carry passengers. (US Air Force)

On May 17, Bob recorded his first experiences seeing the Operation TORCH campaign area and the "American" side of the front. "One ship with about eight of us flew to Oran today. It is about 60 miles from Nouvian. The boys on this side have had it pretty good because they could get about anything they wanted in the way of cigarettes and American candy and such. But they sure forgot all about us 'desert rats' on the other side. But we are glad to know that we were the only Transport Group assigned to the Eighth Army

and we hauled all the gasoline that the fighters used on the push through the desert. We went to Oran to buy some cotton clothes because most of ours are worn out and we were using English summer uniforms. We bought what we needed and then proceeded to look the town over. The mountains run right up to the beach on the west side of Oran and on the last peak is an old castle and it helps set the town off. On the outskirts are some very modern apartment houses. I would say that the town has about 100,000 residents. It is a very clean town and the people are about 90% French. Their stores are almost like the ones in the States. Food for the civilian population is very scarce. When you eat in town you have to eat at the restaurants which the Army approves and operates.

"The women are just about the most beautiful things that I have ever seen. They dress so very modern and every one of them has her hair just so. It looks like they spend most of their time on their hair. And when they look up at you with their big dreamy eyes you just about blow your top. About 75% of them have perfect shapes. Their legs are beautiful. No wonder everyone raves about the French women. After we ate we all went to a bar and had three bottles of champagne. That fixed us up pretty good so that we could see the sights better and admire the women more. Now, coming from where we did with very few white women you might picture what we looked like standing there on the street. At 1500 we went to the Red Cross where they have a cold snack bar. You can make your own sandwiches and they have all types of pickles and cold meats and coffee. They also have ice cream. There is a donation box with a sign on it saying you can give two francs (four cents). Now what do you think of that. The Red Cross might not have done much in the last war but they are sure making up for it in this one.

"Some of our pilots met men that were in their class at flight school so there were some long bull sessions. They have only been over here a week and say they are ready to go back to the States. I just thought to myself I would like to take them out in the desert and make them live awhile. The place they have now would seem like heaven on earth to us. And to just see American soldiers around is something in itself. All we have seen are English soldiers and all their other countries soldiers. We flew back here at about 1730 and were ready to go back to Oran as soon as we hit the ground."

After their trip to Oran, Bob wrote to Toots about his visit to the Red Cross facilities. "Put in a good word for the Red Cross. If they did not do much in the last war they are sure making up for it in this one. They do everything in the world for you. This is where you need things done for you and not back in the States like the USO. So if anyone wants to give, give to the Red Cross. It beats me all that money spent in the States by the USO. In the States the men can always find things to do but over here it is different. I want to praise the Red Cross once more."

On May 18 the intensive training continued. One concern about the upcoming operation was that they Troop Carrier units would have to fly a long, out-of-the-way course over water to avoid flying over the invasion ships, which might expose them to friendly anti-aircraft fire. As a result, the squadrons needed practice on formation flying over water. "I am now staying with the flight chiefs and Line Chief in their house until I get mine fixed. We use the squadron radio every night after the radio men go home. We bring her right over in the shack with us and have music until we shut the generator off. Tonight nine of our ships are on a night formation flight over water and they are going within a few miles of Spain. They won't be back until midnight so I am catching up with my writing. Just read that in the North African Campaign there were 93,000 Germans killed."

When the ships returned, Bob and his men spent the next day modifying them for additional navigation equipment for the new missions. "We are now moving our navigation compartments up forward in the airplanes to where the baggage compartments are. It is quite a job. We have to tear out all the baggage compartments and install the navigator tables, and also cut through the side of the fuselage to install the drift indicators.

"Tonight I went up with the flight. There were twelve ships and we flew formation and made dummy parachute drops on an island close to Spain and then few back. When we reached the coast of Algeria, about ten miles on our left there were about 30 anti-aircraft guns firing. We did not know if they were firing at us or not. But it was just too close for comfort. We kept on flying straight on home but when we reached what

was supposed to be home, the field was covered with low clouds and haze. There are mountains within two miles at the east end of the runway so you can only land one way at night. So, a haze over the ground, the mountains sticking up in it and twelve airplanes flying around makes it a little bad. So Major Garland told the rest to stay above it and he would go on down through. Of course I had to be with him, so down we go, and I could see those mountains in front of me all the time. We broke out of it at 50 feet above the ground. Even at 50 feet there was a light fog. After we landed, we turned on our radio and let the other ships up above turn on their radio compass and then one flight at a time would home in on us. After we got them overhead where we could hear them and tell which way they were headed, we would talk them down by telling them to turn 90 degrees to the left or right and push the nose down and come on through. It took two hours to get them all down. Sometimes they would head over you in such a direction that you could not tell them to turn and let down, so they would have to fly out and home in on you from another direction."

The next day, after their harrowing adventure, the 36 TCS men learned it was even more dangerous than they knew. "We found out that the Germans had bombed Oran harbor and also the town and killed 100 people, so that is what all the shooting was about. They shot down three German planes in the raid. We were talking about what kept the Germans from seeing our formation last night and knocking us off, which is a mystery."

On May 20 Bob penned a long letter to Toots. "A ship flew in today from our old home with the first mail for us since we moved. It makes you feel like you started living all over again when you get a lot of mail. Thanks for the large flag. I think we have plenty of thanks for you. We are going to try and make the original one last through the war, and then maybe afterwards. She has a couple of little scars from high winds and moving but she still flies over Engineering wherever we are. Yes, I destroy all of my letters. I hate to do it, but it is just impossible for me to carry all of them with me. So your worries are now over. So you saw *Desert Victory*? Do you know we covered every inch of the territory where that picture was being made? At no time did we haul gasoline more than 30 miles from the front and we were stationed about 100 miles back all the time. We moved with the Eighth Army all the time until they reached Tripoli. What do you mean you were so glad that we were where we are? We only left them and moved back after they took Tripoli and did not need us anymore. How would you like to change tires on a ship while dog fights were going on over the edge of the Landing Grounds? Wait until I get home to tell you about it. But we had nothing like the poor infantry or tanks. They had nothing for cover or to hide behind but a prayer. When they came after us we could get in slit trenches but they were in the open with nothing. The men that we would haul back would have been shot about 45 minutes or an hour before. You don't know how fast things move during a big push like that one was. That is why I would like to see the show because we were in the push from the second week on. On the fields they would just clear enough space of mines so you could land and unload, then the fighters would come in and start operating immediately. After this they would start clearing the rest of the field of mines. They get most of the men. If you did not know that we were fighting in the Libyan Desert where did you think we were while the push was on?"

On May 21 Bob was still scrounging for parts when he met up with some old friends. "McMenamin, Kawalski and myself few to Oujda this morning to see about some parts from the 313th that just came over. As usual we were unable to get anything. I saw Red Ryder and Major Biggerstaff. They are both in the 313th TCG. Biggerstaff told us that Joe Corcia is a Warrant Officer and Lagana is a Master Sergeant. After visiting for a short while, we borrowed a jeep and drove to Oujda which is about eight miles from the field. We had lunch and wine and looked the town over for a short while and then returned to the field. We met up with the pilots and came on home."

The next day Bob wrote to Toots that he had decided to send her the completed pages from his diary for safekeeping. One of his squadron mates was headed back to the States and Bob had asked him to courier the book for him. "I am sending my diary home with Major Matthews so you can read it and correct the misspelled words. He is sure getting a break. By the time you get this he will be home. Over on this side (Algeria) I am beginning to see a bunch of the old gang. Of course they are all split up but I find quite a few of them at different fields. This is a lot better than being with the English all of the time. Everything we see

here is American and that is a sight for sore eyes. We even get four bars of American candy once a week. I hope my camera is on its way by now. It makes me sick to think of all the good action pictures that I missed being able to take. I will never get a chance to take them out in the desert again. It is almost deserted now and nothing but memories left."

The 36th Squadron went on another long cross country on May 23, this time to Casablanca. It was a town the men had been anxious to visit for some time, particularly with the popularity of the recently-released Humphrey Bogart movie. "We were up this morning at 0600 and all the ships but one which has to remain at the field took off for Casablanca. I took all but five of the Engineering personnel along. We flew to Casablanca and remained until 2000 and then returned home. At the Casablanca airport we got three trucks to take us to town. Farris, Leach, Pace and I stuck together all day. First we went to a restaurant on the beach. It was a very beautiful place but was for officers only. It is built in a half-moon shape and very modern. It sits very high on cement stilts so when the tide is in the building hangs over the water and makes a very beautiful sight. After lunch we took a walk along the beach and the men and women wear less than nothing for a bathing suit. So we had a good look at all the sights. A lot of men change their bathing suits right on the beach. They put their shirts on and then pull off their bathing suits. I would like to see them try that back in the States! But a lot of people rent the beach tents to change clothes in. Here, as anyplace else where the French are, you see beautiful women with great shapes.

"After taking in the sights on the beach we caught a gharry (horse and carriage) into town which is about one and a half miles away. There are no privately owned cars because of the gasoline shortage. They also do trucks the same way. On the gharries used for trucks they use four horses to pull them, and on these they can haul from 24 to 36 people. Some gharries are made like and old spring wagon but almost all of them have pneumatic tires. Downtown we walked around looking at the stores and the sights. On the main square downtown they have a sign up with the mileage to Berlin, London, Tokyo, New York and Washington. I bought a leather hassock for Toots and it is very pretty. There are plenty of leather goods here. We drank a little wine, but not too much, because we did not want to get one of those good headaches. There was nothing doing downtown and most people were out to the beach, so we went back out and walked around for the rest of the day. We returned to the field for dinner and then took off for home."

After his trip to Casablanca, Bob wrote to Toots with the latest on his expected promotion and the status of his diary. "Major Garland returned today and he was unable to give my diary to Major Matthews so now I have it back. Matthews has left for home. Major Garland also stopped at Cairo to see why my promotion has not come through for 1st Lt and they told him they would not promote anyone unless they had three months in grade so I will have to wait until June 3rd before they can send in my recommendation again. I am beginning to get a pretty fair tan because I work in nothing but shorts and shoes and socks. It is not too terribly hot here. It is just about like Ohio. We have plenty of work to keep us busy and this helps the time to pass more quickly. Captain Nagle just made captain so he borrowed the jeep and came down the line to bring us a box of cigars to pass out to the engineering men. He is a navigator and one swell fellow. We had him in our shack for tea and shot the bull. He has typed on a piece of paper inside the box of cigars:

*To the Engineering Department, Compliments of Captain Nagle*
*(Master Sergeants Do Not Out Rank Second Lieutenants)*

"The crack about Master Sergeants was for Sgt White. Do you remember how Stracke and I used to say that we out-ranked a Second Lieutenant? White has always said the same thing. Well in Casablanca when we were there a 2nd Lt stopped him for not saluting and gave him a good going over. So now everybody in the Squadron is giving White a good going over. He is leading a miserable life just now. But everyone was stopped in town that day. Even one of our Captains was stopped because he had the wrong type of hat on. You see the other branches of service hate the Air Corps so that is the only way they can get back at us. I think most of them have just come over anyway."

The incidents in Casablanca may be the result of the 316th running afoul of General Patton's uniform regulations. To tighten up discipline among the American Army men in North Africa, Patton enforced some strict rules on dress, appearance and military customs and courtesies. The Air Corps men fresh from 'the desert' had been living with the British for most of their time in theater and the British had more casual approach to uniform items than the Americans did. Bob's Group was in for a bit of a jolt as they gradually transitioned back to U.S. Army control.

With little over a month to go before the start of the HUSKY operation, invasion preparations intensified for the Airborne and Troop Carrier forces. For the 316th, this included the addition of more new equipment and crews. "All our ships flew until 0230 last night (May 26) and as usual the fog rolled in just before the ships could land and they were flying around just off the ground trying to find the runway. I don't see how they miss the mountains all the time. After they landed we towed them up on the hill. We received three new aircraft yesterday. They were sent over to replace the crews and ships that we have lost. Before they left the States they had only 15 hours on them. They have lots more and better equipment on them than we have on our old ones. Two of the crew chiefs used to be in my outfit at Patterson Field and before they came across they were in Marcum's squadron, the 9th, that was formed at Patterson and then moved to Mobile Group. The three crew chiefs are Robinson, Jones and Poppleton. Group called us at 2300 and told us we had to have the new ships ready to go to Fayid at 0500, so Love, White, Tucker and I had to unload them. We finished at 0315. Then at 0330 we took the crew chiefs to the mess hall. We all had breakfast and then we picked up the pilots and crew's baggage and took them to the ships. They all took off at 0515. They are going to Fayid to bring over some tentage, parapacks, parachutes and cots. So we did not get to bed until 0545."

**Algeria to Morocco**

Amidst all the pre-invasion preparations, the 316th was now due for another move. "We got up again at 1220 but none of us felt so very good. After lunch they told the Advanced Echelon to get their bed rolls and put them on the ships. You see they have loaded three times to go to our new home and each time it has been called off. We have all the Engineering equipment on the ships and they have been flying their missions with it on board. We were supposed to move three days ago, and today as usual they are ready to tear their hair out. Major McMenamin feels the same way. They promise him tents for the Group and then when he arrives they won't give them to him.

"We still belong to the 9th Air Force in Egypt so we can't get anything over here. But this is nothing new for us since we have had to fight for everything we have since we came from the States. Tonight we packed all the equipment and baggage for Engineering. We are always ready to go before anyone else. We have nine ships flying tonight and they won't be back for another two hours. So I am going to sign off and write to Toots."

Finally on May 28 the 316th moved again, this time to Morocco. "This morning we were up and putting our bedrolls on the ships. We all ate early chow at noon and left at 1300. Our new home is Guercif, Morocco, which is 100 miles west of Oujda and is in the desert. It is not sand, just cactus and small patches of grass. After we landed, all of the men's bags were hauled to the tent area. We still do not have any transportation but that is as usual. We are using trucks belonging to the Ordnance unit, but they are moving out so then we won't have any. The Advanced Echelon did not have enough tents so some of the mechanics and me are going to sleep under the airplanes. The rest of the tents are supposed to be in from Fayid tomorrow. There are only 36 men on the field, so one thing is no one but Group will bother us.

"The field where we are is in a large valley surrounded by mountains to the south of us. One about 50 miles away is snow capped. Other large mountains are within 20 miles of the field but they do not have any snow on them. The country out towards the mountains looks flat but it is a series of small valleys. They are very wide and only about 30 to 40 feet deep. The country here is almost the same type as around Tucson, only the cactus is not nearly so tall. The ones that grow here are very small and very different from those in Arizona. They are very pretty and there are plenty of them, along with small patches of oats and other grains.

"The fellows that have been on the field for some time told us that there are gazelle in the valleys and in the foothills there are wild boars. So this evening Love, Keating, Babbitt and I went for a walk. We took our rifles along but never did expect to see anything. But after we had walked about 15 minutes, I saw what looked like two gazelles on a small hill. When we walked towards them, they saw us and began to run. We tracked them for some distance over the hill. They were about 200 yards away and running when we say them the second time. We shot at them but did not hit any of them. The look just like deer and are just as fast but a lot smaller. The natives tell us if we drive about ten miles towards the mountains that we can find them in herds. The men that have been stationed here hunt them in jeeps, drive up close and shoot at them. Where we are now it is just as hot as it was in Egypt. "

The following day Bob and his men got to work on what now must have seemed like a routine task – setting up camp in a new location. "I had a good night's sleep under the ship last night. This morning we started setting up Engineering. We did not have any transportation so we taxied the ships up close to where we were going to unload and then after we had unloaded we taxied them away again. By noon we had set up three Engineering tents, Supply, Engineering Office, shower latrine, lister (water) bag, and had all the equipment put away.

"This afternoon the three new ships came in from Fayid with the tents so I had to let all the mechanics go to the tent area to help set them up. About the middle of the afternoon the men came back and told me that the Major sent them back because there was no transportation to haul the tents. Boy, if this Group don't go off half-cocked! We don't have any way to haul water except for borrowing the trailer from the Ordnance unit. There is a well here on the field but no way to haul the water. There is also no gasoline, so we have to fly to Oujda which is about 100 miles away. Now how about that for efficiency?

"Love says next time we move, the Group will probably move and leave the cooks and rations behind. So far they have messed up every time we have moved. They never do make arrangements beforehand. The flies here are pretty bad, but they are like our flies at home, you can at least scare them away.

"Tonight we start our Mess with just us and the 45th Squadron. I am glad of that because with the entire Group eating together the lines are terrific. Our ships that came back from Fayid also brought our mail back. I received 22 letters and have spent most of the evening reading them. At about 2130 three of our crew chiefs came in with a gazelle. They were sure proud of it and I would have been too. Just think of it, the second night here and meat on the table already. They look like deer but are about half the size and have horns like a mountain goat. I have now moved into a tent that we put up close to the Engineering area. We have our lighting plant hooked up and our tents wired already. Some speed, eh? Also transportation came in this afternoon and we now have three trucks for each squadron. On the flight line we have our own truck and trailer. Boy is it plenty hot here now and they tell us it even gets worse. So if I melt and run away don't be surprised. The letters from home help a lot but not like being there in person."

Once settled in Morocco, Bob was finally able to write back to Toots on May 30 and answer many of her letters. "Well, here we are at a new home again. We don't even stay in one place long enough to take our shoes off. So that is why I am three days late with this letter. After we arrived we set up engineering in record time as usual and what I mean the shower is something to talk about. It was 120 degrees here this afternoon. Most of the country here is like Tucson Arizona but cactus does not get even half as large. I am sleeping on the line here and I have one of those good old English tents. They are about the best tents that I have ever seen. One of our ships came in from Egypt yesterday and they finally brought out mail with them."

As June dawned in North Africa the American forces destined for HUSKY were putting the finishing touches on their operational plans and equipment. Exercises were still underway, but the men sensed that the date for the invasion was drawing near. "For the last three days the wind has been blowing like it did at El Adem. It is not good. Dust is in just about everything and we have one hell of a time keeping out tents up, especially our large Engineering tent. Today we are putting the pararacks on our ships. Our mission for tonight starts at 0205 in the morning. The group flies out to sea and then hits an unfamiliar point on the coast. So you see, we are getting ready for the real thing. Everyone you talk to is on edge as to when it is going to

happen. Everywhere you fly you see preparations being made. Landing craft are practicing landings both night and day. Also there were 9,300 parachute troopers landed in North Africa in one day."

In his next letter home Bob shared some of the difficulties of settling in at their new field. "Well, this place is like being back in the desert at El Adem. Boy, we have had plenty of wind and dust for the past three days. We have had one hell of a time trying to keep the tents from blowing away. We have had to modify our shower today. The wind has been blowing so hard that every time we turned the shower on the wind would blow the water away before it hit you. So we dug holes and put two-by-fours in them then we nailed small boards from one to the other. After we had done this we put an engine cover around the whole thing, now we have a little room for our shower. One good thing here, we can get water about five miles from here. Another is our new latrine. The American Army wants you to dig a hole and do it in the ground, and the British put seats on large buckets. So we are now with the American forces and must do it their way. So we took two British seats and put them on a stand and then closed it in with boards. It is one fancy job. So now we are going to take it down over this small slope and put it over the hole that we dug. When it is almost full we just dig another hole and move our shed house over her. Only one drawback - our new one is so comfortable that we are going to have to put a time limit on it.

"TSgt Tucker and SSgt Babbitt took the motorcycle and went hunting this evening. We have just finished fixing the cycle. The clutch went bad on it but she is in fine shape now. Do you remember Robinson? Well he is in our squadron. He came over with a new ship and we got the crew and all. There is just about every type of bug in this country that has ever been listed in any book. One other thing that I forgot to tell you was about the snails, the ones that are in the shells. Well I thought they were just dried up shells like we see at home until I see them on small weeds and bushes. About ever other bush is full of them. Well, the boys returned and did not have any game. They shot at two of them but missed. Also they parked the cycle and then could not find it and they had to walk back to camp. We then took the truck and went out and looked for it and found it about midnight."

The long practice missions continued without letup for the next several days. These included flights from Oujda to other desert locations. Along with this work there was some additional excitement of escaped prisoners in the local area. "Sgt White and Sgt Crook have gone out in the desert to pick up two escaped German prisoners. There are 50 of them on the loose around here so everyone is carrying a gun and they have brought in a bunch of airborne infantry to guard the field."

When his men returned Bob got the rest of the story. "The Germans escaped from a train just a short distance from the field and they headed for Spanish Morocco which is only 30 miles from here. There is a signal burning in the mountains there that we can see from here. So you can see it was all planned. White and Crook did not catch any but the other party did. So far there have been four captured and two killed. The Germans have given them the slip and some of them have held up an Arab camp and forced the natives to feed them. Some Arabs were with White and Crook to show them the way to get through the ravines. I don't know if they will do anything in the morning or not because if the Germans walk all night they will be in Spanish Morocco by morning." [Although Spain was officially a non-belligerent in the war, the Spanish sympathized with the Axis cause and the German prisoners likely knew they could be returned to their own forces if they could make it to Spanish territory.]

Bob was always looking out for his men and on June 3 he set out to find a suitable swimming hole in the local area so the troops could get some relief from the heat. He also was becoming enamored of one of the greatest inventions of WWII – the ubiquitous Willys Jeep. "This afternoon I traded one of Engineering's trucks for Operations' jeep and took Doc Hamilton to inspect a river that runs from the mountains through this part of the desert. One of the men has reported a place big enough to swim. I took Doc Hamilton, Flight Officer Jackson and Pvt Keating. We picked up our guns and cameras and water then we headed out across the desert. After driving for about four miles we caught up with a couple of gazelle and chased them for a short distance but did not shoot at them because we did not want to haul them on the rest of the trip. After we found the river and looked it over we decided that it was not deep enough to wade in, let alone swim. It comes from the mountains that are about 30 miles away so we decided to head in that direction to see if it got

any bigger. After driving around dried up ravines and up and down hills we still could not find a place to swim. We found quite a few Arab camps and we would stop and take pictures and they would try to give us milk and food, but we would have to refuse. They live in low roof and long tents just like the ones in Libya do. They also dress and look just the same. These have a few cows though. They also have many sheep and goats. They would say 'gazelle' when they saw we had guns and pointed in the direction where we could find them. One old boy pointed to a long deep valley so we decided to hunt a while then go home. We only had two rifles for the four of us. Finally we got one of them in a long flat valley. We finally ran him down and he hid behind some brush. So we stopped the jeep and jumped out and he began to run again and I missed him twice and so did Keating. By the time we got back in the jeep and drove over the hill he was gone and we never did find him. So we looked around and chased some more this went on for about two hours and we never did find one. But I had the time of my life. After this war is over and they make jeeps we are sure going to have one. They can do everything but cook. And you can take them just about anyplace. It is so much fun to just take out cross country like that. It started out with us looking for a place to swim. The major told us that if we found it and Doc said that it was OK I could take one of our trucks each afternoon and take the men swimming. It is so hot here that we don't work in the afternoons."

The men of the AAF were pretty resourceful when it came to operating and maintaining aircraft in austere conditions as demonstrated by the remarkable work done by the 316th in the desert campaign and at a dozen different remote airfields. They were also pretty resourceful when it came to making creature comforts for their quarters. Bob describes one of their inventions in a letter home to Toots. "When we come home we are going to run Don out of business. Today our welder built a washing machine. None of us have had any clothes washed for at least a month and he said he was going to fix that. He took a 50 gallon gasoline drum and welded a small axle on each end. They he cut out a square hole in the middle of the drum and made a lid and lined it with sponge rubber so that it is water tight. Then he bolted small paddles on the inside. He took the strap off his rifle and used it for a belt. It goes around the barrel and then to the flywheel of our little compressor motor. The compressor motor turns the barrel and you get your clothes washed in nothing flat. We did about six washings this afternoon. He is now putting a crank on it so we can turn it by hand when the motor is being used. We build or fix anything in the book. I sure wish that you could see all the stuff that we have made. If you don't have things you either make them or do without. We don't even have to worry about hot water. We just sit a can of it out in the sun and in about a half an hour you can hardly stand to put your hand in it."

The desert living was catching up with a number of men in the 316th, as Bob noted on June 5: "We have been having about twelve men a day sick from dysentery. They have to go so often and are so weak they can hardly walk. A couple of them have been unable to make it in time and had to clean their pants. Most of them have been terribly sick. I have it also but have not been sick at all. The ships took off tonight and are going to fly without any navigator or formation lights. It is a very dark night and cloudy, so we don't know how they will come out. It is going to be plenty tough."

In addition to the almost daily formation flight work, the Troop Carriers now had a chance to exercise with the 82nd Airborne units at Oujda on a regular schedule. Bob recorded the tragic results of one exercise on June 7, now just one month until D-Day for the Sicily invasion. "We were up this morning at 0245 and took off at 0440 for Oudja. We picked up the parachute troops at 0700 and dropped them at 0830. The paratroopers now have green parachutes and plenty of equipment. These are just practice jumps but it is not far off when it will be the real thing. One of them was killed in the jump today. His static line got caught between his legs and his parachute did not open. They were only jumping from 600 feet so he never had time to open his reserve parachute. There were 32 planes from our Group in the flight. We returned home at 1030.

"This afternoon we went to work on the ships. We have to get the formation lights on as soon as possible. Our maintenance on these ships is getting to be something terrible. They are pretty much worn out from the desert campaign and they will never be the same again. The bugs and varmints are something terrible here. There are some of the most horrible looking creatures that you have ever seen or could ever imagine seeing.

There are so many different types that it gives you the chills to think about them. The Mess Hall is getting ice now and it is sure good to be able to have a cold drink. We have had very few drinks in the desert that were not at least warm and some were even hot."

Another bit of improvisation experienced in the field was *al fresco* barbering as Bob described to Toots along with some of the things he missed about the States. "You should be here and see the crazy things that go on. My hair was so long that it hung about two inches below my ears when it was not combed back. It was about to drive me mad so I got one of the mechanics to cut it. We took the motorcycle out under the wing of the airplane and used it for a barber's chair and then he went to work. He did a pretty fair job. There is always something like that going on that gives you a good laugh. Some of the officers gripe about living like we do but I like it. It gives you a chance to see what you can do with practically nothing. I think what I miss most of all just now is cold drinks and ice cream. Our mess hall truck drives 40 miles to get ice so we get one cold drink a day and that is something that we have never had before. But all other times of the day the water you drink is even more than warm, it is almost hot at times. To think how we used to holler at home if we did not have the pop, beer and water cold enough. If anyone would have ever told me I would sometime drink warm beer I probably would have told him he was crazy. Very few cold beers have we ever had since we left home. I will try to do better with my letters. I try to write every other day but with the heat and work it keeps me busy. I have had no time to write anyone else. Every day I think I will catch up and then they start calling me about this and that and before I know it a whole day is shot. And then we work and fly at night too.

"I had a good trip yesterday even if it was only 65 miles. A Spitfire was out of gasoline down in the desert so we flew some down to him and I asked to go along. He had the field marked by a white cross made out of white cloth. He was caught in a storm and got lost, and then he found this small town and landed. He had sent a message over two days ago and we had just received it. He had never been through this country before and did not have a map. On top of all that he got caught in a thunderstorm. When he finally got out of the storm he found this small town in the desert so he landed. In fact he had to because he was almost out of gas.

"There were a couple of officers with us that could speak French. When we landed I think everyone in the town came out to see us. The town is about the size of Donnelsville and is the seat of government for that section of the country. It is controlled by one French captain, one adjutant, one sergeant and their staff of French civilians. The civilians on the staff are the only ones allowed in the section, all the rest are Arabs. There has always been too much trouble when civilians move in so the French decided to keep them all out. This is a small group and their native troops control 25,000 Arabs and run the government. The Arabs are what they call Riffs or desert tribes, and they want to rule themselves and run the land by themselves. The French captain has been in charge for the past 14 years. The name of the place is Outat Oulad El Hadj (French Morocco). Now try and find that on the map. You will have to look south west of Guercif.

"There is a fair size river that runs through that part of the desert and it comes from melting snow in the mountains. Back from the banks a short distance there are very high cliffs. On the cliffs and in them you see small villages about every ten miles. Their houses are all built out of mud brick. Then at the base of the cliffs you see about 50 to 100 acres being cultivated and this part is irrigated by hand. They have some of the most beautiful horses that you have ever laid your eyes on. And it seems they and run forever. They are very long winded.

"You get so damn mad at the wind around here that you could go crazy. If you don't put a weight on everything it blows away. Yesterday I had a knife on my notebook and it still blew away and all my stamps were in it. It never stops all night and that is for about three hours. Even my bed - I cover the top part with a raincoat so when I pull the covers up over me at night I won't choke to death from the sand. Aircraft No. 10 came in and had two overhauled engines on board for No. 5 airplane. McMenamin had got them at Casablanca and sent them back on one of our ships. So we stayed to pull the engines."

By June 13 the men could finally get off post for a while. "Today is our time to go to town, so after we had roll call, everyone went to get dressed. We left the orderly room at 1000 and we even had some pilots and radio men with us. The town of Taza, Morocco, is about 43 miles from here. After driving for about 45

minutes, you reach the foothills and things begin to look greener. This is because they have some water and it also rains once in a while. Here you see what the natives grow. One hill after another is covered with wheat. They were cutting it today. They cut it with a hand sickle and tie it. At one place we saw a binder machine and it was being pulled by two horses, two donkeys and two oxen. It was sure some sight. The machine does not tie it, it just cuts it and they still tie it by hand, with the wheat straw. That is only the second binder machine I have ever seen over here. Almost all of it is done by hand. After they cut it and tie it they haul it to their camps on the backs of small donkeys. At times when they unload them you can only see the donkey's head and feet. How they carry the loads they stick on them is still a mystery to me.

"When we reached the town we drove around for a short time to look the place over and then picked out a restaurant and had lunch. We had roast beef and spaghetti with sliced tomatoes. You never have a meal that they don't serve wine. Their wine is very good and also powerful. In the afternoon we were able to go swimming. The town of Taza has only about 5000 people but they have a beautiful swimming pool. It is as nice as or even nicer than any in the States. It is like a dream to drive 43 miles through the desert and find a swimming pool like the one in Taza. It has palm trees and weeping willows all around. The bath house has a series of locker rooms. You receive a key to one of the rooms and leave your clothes in it while you swim. Each person gets their own dressing room. The pool is located at the base of three large hills and is fed by large springs. Taza itself is located at the base of a large snow-capped mountain so they have plenty of water. We had dinner afterwards and then sat and drank cognac until it was time to start for home at 1945."

After returning from their short leave Bob wrote to Toots. "The heat is wonderful, just about as hot as I have ever seen it. But as yet I have not passed out once but I don't know if it will continue that way or not. We have two trucks that we use in engineering so I asked the Major to let us send one into town the other day because none of the men have even been off a day, let alone go to town. They are off in the afternoon anyway. So I split engineering in half and sent Sgt White in with half of them day before yesterday and then I took the other half in yesterday. The town is about 43 miles from here but it is the closest one and it is about the size of New Carlisle. But there are about three restaurants and a few places to drink wine. The attraction is the swimming pool. It is something to write home about. [Apparently, as Bob is now writing home about it, ed.] It is a cement pool and is fed by springs so the water is plenty cold but with this heat it is just the thing. It sits at the foot of three hills, in other words it is just a small hole in between hills. Instead of having sand all around it has grass. The grass is very thick and soft. There were 23 of us in our group and the other three squadrons had men there also. A couple of times I was so tired I could hardly pull myself out of the water. The closer you drive to the town the greener things begin to get. This is because you are getting into the foothills and they have some water and they also get rain once in a while. Everything in the town is very green. We ate lunch and dinner (supper to you farmers) both in the town. Everyone had a good time and now we can work for another couple of weeks."

The next day saw the resumption of paratroop training and invasion preparations. "I am so stiff and tired this morning that I can hardly move. The parachute troopers landed here at 1100 this morning and started loading their equipment packs on the ships. There have been too many men killed and injured in training so they have stopped the big group jumps. They now just put one man in each plane and enough equipment for a ship load. It gives the pilots practice on hitting the fields [drop zones] and the chance of only injuring one man from each ship. The rest of the troopers are already at the field where the jumpers are going to land. After the jump they rush out and get their equipment and start their mission from there.

"Tonight is going to be a big jump and the whole wing will participate. We had 12 ships take off tonight. Twelve out of thirteen is pretty good we think, at least after all the flying we have been doing and with one ship out for an engine change. The ships all landed at 2330 and we went to the Mess Hall for coffee and doughnuts after."

During the time that the 82nd Airborne troops were stationed at Oujda there were only two practice night jumps due to the delay in finishing the airfield. These practice jumps were primarily intended to give the Troop Carrier crews additional practice in close formation flying and night jumps. The first practice jump at Oujda involved troops from the 505th Parachute Infantry. As Bob mentioned in his diary, each aircraft hand

only one paratrooper on board and he would jump following a three-hour flight which simulated the long overwater flight to Sicily. Once the drop zone was reached, each man would exit his assigned aircraft and when they landed they would plot their location on a map to measure the accuracy of the aircrew's navigation. The first of these practice missions was unsuccessful. "The jumpers had been scattered all over the desert and landed everywhere but on the DZ. This was pointed out to the aircrews very firmly, but politely, by Colonel James M. Gavin, the Regimental Commander, at the critique of the problem that afternoon."[22]

The second practice mission involved the 504th PIR with two men jumping from each aircraft. That mission was more successful, but still not as accurate as the airborne troops had hoped for. On June 16 the training program for the airborne troops at Oujda was completed and on June 19 a parade was held for General Eisenhower and his staff. The 82nd troops moved to Kairouan Tunisia on June 24.[23]

By June 18, final preparations were being made in the Troop Carrier Groups for the Sicily landings. The last of the equipment and supplies needed for the invasion were issued to the men and Bob snatched a small souvenir which he sent to Toots. "I am sending you a gold seal dollar. It is invasion money. The object is to keep the Germans from getting United States money and using it to pay their spies in America. It is no good in the States unless you take it to a post office and swear as to where it came from. The wind here is about to drive me crazy. The sand is not as bad as it was in Libya but the wind is just as bad. Everything you lay down you have to put weights on. And your beds and clothes get full of sand. It just keeps it up day after day until you could scream. The small weeds, cactus and grass are so dry and hard that when you touch it, it breaks into small pieces. Every once in a while we get together and talk about what we have done without all of these months. Also about things we used to gripe about at home and we have to put up with here. As yet we have never tasted fresh milk since we left the States. Also, we used to gripe when things were not cold enough; for us now if it doesn't burn our lips, it is plenty cold enough. I have ridden these cussed trucks until my back is about broken. How good it will feel to get home and take a ride on a smooth highway and in a good easy riding car."

June 20 was a big day for the Group and also for Bob Uhrig. The 316th was alerted to prepare for their final move to the airfield in Tunisia where they would stage for the paratroop assault on Sicily. So preparations had to be made to relocate one last time. In the midst of this, "Major Garland had returned from Egypt and had mail for us. He also stopped at Cairo to see about my promotion and brought it back with him so I am now a 1st Lieutenant and a very proud one also." Bob immediately sent the news in a letter to Toots, but he was so busy it was almost an afterthought. "Major Garland came in last night with the mail, so today I received ten letters. It has been twelve days since we received any mail. It was on the 16th that we put everything behind us. I wish that we were headed the other way. Major McMenamin and I flew over to see Coates the other day. He is a 2nd Lt and is doing much moaning and groaning about everything. On our way home McMenamin said that Coates had a lot to learn over here just yet. No, you don't have to send more checkers and now about the things you have been sending. No one has received anything for about three months. Oh, yes - I am now a 1st Lt."

The next day he was able to give Toots the rest of the story on his promotion. "Well the wind is in high gear and I am full of sand and just about to go crazy. You would think that it was impossible for the wind to blow like it does. Now, about the stores and things you want. Remember we are out in the desert and I have only been to a town three times in two months. You just can't buy things over here like you can in Egypt. Also you have to have coupons for almost everything. If a fellow bought anything he must have been stationed near a large town and was able to get some coupons. Major Garland just returned from Egypt with our mail and he stopped at Cairo to see about my promotion. Over here you have to wait six months now to be promoted but my recommendation had gone in before they started the six month deal, so Major Garland had them look it up and after about 1 ½ hours they found it and they sent it right through so he brought it back with him. You just have to know the right people. Boy am I ever glad and surprised to make it. It doesn't give me any more money because I am already drawing 1st Lt pay anyway. But as you say it gives you prestige."

## Final Preparations - Morocco to Tunisia

On June 21, the aircraft movements started to ferry the paratroopers to their pre-invasion sites. In the case of the 316th, it would be Enfidaville, Tunisia. "Eleven of our ships took off at 0400 this morning. They are going to fly to Oujda, Morocco and then move paratroopers to our new field. We are moving in a few days which is nothing new for us. It is about five hours flying time from here (Guercif) and is close to Kairouan, Tunisia. We will be at our last field before the invasion. Each Group will have their own paratroops on the same field with the airplanes so that when the time comes there will be no waste of time. And it looks like it won't be very long now. There will be at least 500 transport planes up there in a radius of 50 miles, so you can see the concentration of airplanes. This is without counting all the other types of aircraft. She is going to be something when she breaks. The Germans dropped 21 paratroopers on the airport at Oran the other night and one of them had 100 pounds of TNT on him to destroy airplanes. But they caught 20 of them and killed the other one before they could do any damage."

With only a few days delay, the 316th was finally on the move on June 24. "Last night we finished packing and everyone slept under the wings of their ships. We ate breakfast out on the line and then had to pack the kitchen stoves and equipment. We took off and flew to Tunisia which is about five hours flying time. We have a field close to Enfidaville which is only about seven miles from the coast. This is the last move, unless we go to Italy, which we are very close to now. We are living in tents and each tent is 200 yards from each other. Our airplanes are about 300 yards from each other and if you don't think that is plenty far, just park 13 airplanes 300 yards apart and see where your last airplane is."

Bob sent Toots a letter with some impressions of their "new," new home. "Back in some green country now with a cool breeze all day long and none of that cussed high wind. We are also able to go swimming in the afternoons. This afternoon we are going to start driving our truck to take the men swimming. It is like living again. I was watching them thresh wheat yesterday. They take three horses and run them around in a circle over the wheat that has been cut and laid on the ground. After they have done this for about half a day, they stop driving the horses over the wheat. Then they take wooden forks and throw the wheat and straw up in the air, the straw blows away and the wheat falls to the ground. There is plenty of destruction and old equipment here; almost as much as there was in the Libyan Desert. We have about five large tents which we use for a mess tent and day room both. They are all fastened together so it is like a large room. It makes a very large place. I am so glad to be away from that wind and sand that I could jump up and down with joy."

But the move to the Mediterranean coast meant that the Group was now much closer to the war than they had been in the comparative safety of Morocco. "Last night she was just like dear old Tobruk because the German planes came over and bombed two fields, one on each side of us, and also the harbor at Sousse. They dropped 24 flares on the harbor and gave it a good bombing. The sky was lit up like a Christmas tree from anti-aircraft fire. It lasted for about two hours and kept us on our toes.

"This morning, all our ships left to go and pull gliders back to a field about 15 miles from here. We are hoping that they don't stick us with towing gliders because they are terrible to pull and burn your engines out very quickly."

Bob's next letter to Toots was June 28. "Well it is like old times again. My hands are full of blisters from digging slit trenches and Desert Lilies. This dirt here just comes up in large chunks and is plenty tough. Plenty going on here now and it makes you feel like you are back in the swing of things again. The natives here live in stone huts with grass roofs, or should I say straw roofs. After they thresh they pile the straw in long slender piles. It is chopped up so fine that they used it for feed. I made a mistake the other day; they don't use a wooden fork to throw the straw up in the air, it is a wooden shovel. Also they may use one horse or a number of horses to walk over the wheat to thresh it. I also saw little donkeys and horses working together and it is a very funny sight. We are going swimming every afternoon and it makes you feel good to have enough water to swim in close to you and we are making good use of it. The small towns close by must have been beautiful before the war, but now there is not much more than a memory left. Our ships (U.S.) are the ones that have done the damage. They never even left one small bridge standing. And where they caught

convoys on the highway and bombed and strafed them you see the destroyed equipment and then the graves of the men alongside it. The 37th brings our mail up to where we are now so we received mail every other day and that makes everyone feel much better."

That night there was an alert on the base for additional German attacks. "Last night we had to double the guard because there was information that the Germans were sending parachute troops over. But all our airplanes are OK this morning and also all the men are OK, so it looks like the Germans did not make it over.

"Around here the natives live in small mud brick huts with roofs made out of straw. You never see one hut by itself, but there are always small camps of about five huts or more. They have to irrigate all the land where they raise corn and garden vegetables. This is done by camels or donkeys or horses. They have large stone wells and there are large round leather buckets with a leather tube about six inches in diameter and about three feet long. They use a beast of burden to pull the bucket up out of the well and a small string holds the tube end up as high as the bucket. So after the bucket is pulled to the top, the tube is then pulled over the side of the well and the water runs out into the irrigation ditches. This is done by all the farmers.

"All the ships came in last night from pulling gliders and they had one hell of a time. One glider had its tail come off and the pilots lost control. It also broke the tow rope holding it to the C-47. The glider was loaded with freight, so the pilots had to kick out the side panel and jump. Then our ship landed on the first level place they could find, but the crew chief and navigator had to walk for about three hours before they found the glider pilots. One of the pilots hurt his back but not too bad. It was then so late that they had to bring the men on into our field. Another ship lost his glider and had to land and pick him up again. There are so many landing fields up here that some of the ships cut their gliders loose at the wrong field."

A few nights later there was another German paratroop scare in Tunisia. Ironically, the men sent to Tunisia to prepare for an invasion were now guarding against an invasion by the Germans. "Because of the parachute scare and doubling the guard the night before last, they brought our own paratroopers to guard the field last night. Now it is not safe to step out of your tent after dark. So far they have caught over 100 German paratroopers in this vicinity, but none of them have destroyed any ships yet. So you can see that we are all sort of jittery. Tomorrow the bombing starts on a wholesale style. So far they have only used heavy bombers on Italy but tomorrow all the light ones start also. Even pursuits start tomorrow to bomb and strafe."

An invasion date had finally been set and on July 3. "They grounded all the ships this morning so we can work on them. There are about 50 things that have to be done to each of them and all these things must be finished for the ships to be ready to go by Tuesday, 6th July. So it begins to look like the invasion is getting pretty close and everyone is talking about it, and are a little jittery. We went swimming this afternoon as usual and I am getting pretty suntanned. I have lost my bathing suit and my bottom got burned in the sun."

Independence Day 1943 brought more work, but Bob found time to describe the area around their new airbase that was the scene of some grim fighting near the end of the desert campaign. "Well this is the 4th of July but here it is just another day with no changes. We still have plenty of work to do before the 6th but I think we will be finished.

"There are mountains to the west and south of us and then in the foothills there is one large hill that the Germans and Italians built into a fort. It is almost solid rock and they cut the top of it and made emplacements for the 88mm guns and then all around the rest of the hill they planted mines. The hill commands the entire view of this valley and all the airfields and the two main highways, so our troops had to clear this hill of all guns before they could advance. They tried to take it with 180 men the first time and all but 20 were killed. On their way up the 20 remaining men could see two wounded British soldiers lying on the side of the hill, one with his legs blown off and the other with his stomach blown open. Neither could move. Then they saw an Italian run out of a dugout and put a hand grenade between the two of them and then run back. In a short time it blew them both up. The men could see them trying to get away but they could not move. Just think, laying there with a hand grenade beside you and not being able to move. The Italian that did it was protected by shell fire. After the hill was taken there were only 16 Italians left alive. In revenge for the 'trick' with the hand grenade, the British soldiers put 16 numbers in a hat and let each one draw a number. The first ten men

they threw off the cliff at the top of the hill. They threw one off every five minutes so the Italians could have time to think it over and sweat it out. The six that were left were 'accidentally killed' bringing them down the hill, so they say, but I don't think it was much of an accident. There is only one patch cleared of mines to the top at present so they have been unable to bury half the men yet and there is a terrible odor near the top."

D-Day for Operation HUSKY was now pushed back to July 10. The plan to begin with the airborne assault the night before, followed by the amphibious assault of American and British forces on the Sicilian beaches early that morning. That meant that combat operations for the 316th and other Troop Carrier Groups would commence on the afternoon of July 9. They would load up with the paratroopers and their equipment and fly a long dog-leg route to Sicily to avoid flying over the Allied invasion fleet. This was planned to lessen the chance of confusion and exposure to friendly anti-aircraft fire.

Bob and his men had worked feverishly to get the C-47s repaired and configured for the airborne assault, in spite of the distractions of German air raids. All those transport aircraft clustered on the fields in Tunisia made a tempting target for German bombers. "We have had plenty of air raid warnings but they have not dropped any bombs on us yet. The Germans have flown over the field time after time but as yet no bombs. The anti-aircraft gunners have orders not to fire until they drop bombs. This is to keep from giving away the position of the fields. Every time you take off from here you can just look down and see at least two airfields. The paratroopers have been assigned to the airplanes that they are going to jump from and they even know where they are going to jump. We are all just waiting for the takeoff now.

"The following is a list of installations and modifications we have had to do in the last three days: British static line, British parapack, glider towing equipment, exhaust flame dampeners, carburetor air scoops, VHF radios installed, litter brackets removed, 25 life vests, 5 life rafts, 25 parachutes, navigator's compass changes, jump master light changes, formation light installation, drift meter aligned, navigator's cabin light dimmed, jump and dome lights wired and taped, vesicant paint, navigator's table light, servicing pump, rations, guns and ammunition, flare pistols, flyer's subsistence kits, first aid kits, yellow circle insignia, British jump mats, decontamination units, Coleman stoves, oxygen bottles, alcohol tanks filled with water, flares, call letters, extra equipment removed, parapacks installed and chalk numbers.*

*Left:* Sudanese guard by a 316th TCG C-47. The yellow circle has been applied to the U.S. insignia on the fuselage.
*Above:* Engine maintenance in the desert was never-ending job.
(All photos: Mike Ingrisano collection)

---

* Some explanation on the technical terminology: exhaust flame dampeners reduced the glare from the engine exhaust to make the planes less visible from the ground and aid formation flying; vesicant paint was applied to the aircraft to detect the presence of chemical warfare agents in the air; "yellow circles" were painted around the U.S. star insignia on the wings and fuselage to aid in recognition. Chalk numbers were just that – numbers chalked on the side of the aircraft so the paratroopers could find the ship assigned to their "stick" for this jump.

"Yesterday I did not get a chance to send a money order home so today (July 6) I rode into Sousse in the jeep and bought mine and also left about 25 others there for the rest of the men. We pick them up tomorrow. Sousse has a very good harbor so she has been bombed plenty. Many of the homes have been destroyed and much damage done. All along the roads there are destroyed vehicles and equipment."

The work on the aircraft and other invasion preparations took place in the worst period of summer desert heat in Africa. Bob's diary records that the temperatures were over 130°F for several days in a row. Coupled with that, a fierce wind blew in from the desert and brought the usual sand and dust along with it. On July 8, Bob was able to get a letter off to Toots, but he was not able to mention the invasion related activities due to security. "You remember how I always said I could take all of the heat they could put out? Well yesterday it made me take that back. It gets plenty hot here, but most of the time there is a little breeze from the direction of the sea. Now yesterday she went up to 138°F, now that alone is bad enough but now listen to this, a wind of about 35 miles an hour started blowing off of the Sahara Desert. Now if that don't make a blast furnace I can't tell you what would. It was impossible to expose any bare part of your body to the wind for it would burn just like a flame. Men with metal wrist watch bands and rings had to take them off because the metal burned them. The only possible relief we could get was from wet towels wrapped around our heads and necks. The evaporation was so great that it made the towels cool and then we could breathe cool air through them. All the water was almost to the boiling point. It was just miserable and unbearable. They tell us that your body could not stand many days of it. Everybody got sick and all of us had terrific headaches. But this morning everything is OK again, but we feel like someone has beaten on us."

*Left:* 82nd AB paratroopers learn how to load parapacks onto the belly of the C-47. (US Army via NARA)
*Right:* Two "sticks" of paratroopers ready for a training mission on a C-47. (US Army via NARA)

## Operation HUSKY

It was now July 9 and the invasion of Sicily was ON. The Troop Carrier aircraft, crews and paratroopers were as ready as they could be. All were tense and keyed up. The German air raids of the last few weeks must have indicated to the men that the Germans knew they were massing for a big operation and that either Sicily or Italy was the intended objective. The operation would not have the same benefit of surprise as had Operation TORCH.

For the Sicily jump, the Troop Carrier Groups would be employing a newly-developed combat formation, eventually known as the "V-of-Vs." Nine C-47s, (three Vs of three), a lead plane and two wingmen for each V, would be carrying one company of paratroopers. The lead planes in each V would carry navigators and the others would guide off their wing. Each V of V would fly one behind the next about a minute and a half apart. Four or five nine-plane formations would make up one so-called "serial." A serial of aircraft would be carrying a full parachute battalion. To maximize the number of men dropped in the smallest amount of time, the serials would follow each other at ten-minute intervals. This formation would eventually be standardized for all future airborne operations, but the interval between serials would be adjusted for each operation. For the Sicily operation, the 36th TCS would supply 12 aircraft, or 4 "V's" and one spare in case of mechanical failure. In total, 250 C-47s would be assigned to drop the 82nd Airborne troops. Another 109 C-47s and 30 British aircraft would be assigned to drop the 1st British Airborne division in their assigned assault area.[24]

Bob recorded the events in great detail in his diary, knowing how significant this operation would be. "I started a new page for today and tomorrow because they will be two days that go down in history. The invasion starts at 0006 tomorrow morning. All of our ships take off from here at 2035 tonight with paratroopers. There have been waves of bombers going over one after the other for the past six days. There has never been less than 40 in a flight and we see all of them from this one spot. So you can imagine what it is like with them flying all along the African coast like that.

"But tonight Sicily will catch more hell than they ever knew could break loose. Where we make the drop tonight will be between a town which is three miles from the coast and an airport which is three miles from the town. The 12th Bomb Group has been flying over the town each night so they won't think anything strange to hear so many ships coming over. Pursuit aircraft are supposed to knock out the searchlights and anti-aircraft guns and the bombers overhead will drop dummy parachutes about ten miles to the north. There will be more than 4000 airplanes on this invasion and how many boats and barges I can't tell you but they are going to hit from every place in North Africa as well as Malta and the small islands that have been captured. This is going to be one of the best planned and biggest things in history.

"The paratroopers have three hours to knock out one fort, but the rest of them will go after the pill boxes and bunkers on the beach as they have only one spot safe for a landing. If they are successful, they will flash a message out to sea and then the barges will come in for a landing. For the past few days everyone has been jittery and high strung and not talking too much."

Bob was correct about the objectives for the night. The All Americans were supposed to be dropped just north of the town of Gela, near the airfield at Ponte Olivo, as well as near the city of Niscemi. Their job was to secure these areas in advance of the landings of the U.S. 1st Infantry Division on the beaches at Gela. What the American paratroopers did not know was that units of the powerful German Hermann Goering Division were stationed immediately to the north of their intended drop zones. This division was equipped with a significant amount of tanks and armored vehicles which the paratroopers were not fully equipped to deal with.

All this would later come to light, but for now; "It is 2312 and all the ground crew are sitting here in the Engineering tent sweating it out. The Colonel led the Flight and the 36th Squadron followed him with the exception of five ships. Three of these took off last because their paratroopers have a mission of their own which is blowing up railroads and bridges. The other two are being kept for spares.

"We were talking to most of the paratroopers here and they are in the best of spirits. One sergeant was matching coins with a lieutenant to see which one of them was going to blow the bridge and which one got to knife the sentry. While our ships were taking off, two other Transport Groups flew over and also about 200 bombers. Tonight the sky was filled with airplanes and it was a welcome sight. But the paratroopers were the most interesting part. Each one of them knows the job he has to do. There is a war correspondent jumping with them tonight and he has a small typewriter strapped on him and the rest of his equipment is with the parachute packs. Also a photographer is jumping. Both are civilians and the photographer has never jumped before.† It was sure a thrill to watch them take off. I asked the Major again today to let me go along and he again said no, so I am just sitting here with my fingers crossed. I cannot explain what a feeling it is to watch them take off and then have to sit here and wait for them to come back home. We are using one of our tents for a First Aid station, but I hope we don't have to use it."

---

† The war correspondent referred to here was likely John "Beaver" Thompson of the *Chicago Tribune* who did jump with the 505th Headquarters Company. He had previously with jumped with the 509th PIR in North Africa.

Early the next morning the planes did return and the stories of the operation were grim. The Troop Carrier aircraft met heavy anti-aircraft fire over Sicily. "Nine of our ships came in on time this morning and the other two made our hair turn grey. But one of them came in at about 0400 this morning. He had got into some heavy anti-aircraft fire and became lost. The other landed at Sfax (Tunisia) and then came on in this morning at 0730. All of our ships are now here on the field and are OK. The 44th Squadron lost two planes last night. One blew up in the air and the other was forced down on the beach. The one that blew up was carrying bags of TNT in the parapack racks and received a direct hit. Once this happened the ship just went to pieces in the air. The men reported much anti-aircraft fire and searchlights but they said the pursuit ships made short work of the searchlights. The Germans and Italians were shooting plenty close to all of our ships but were unable to hit them. Our Group became lost and flew between Italy and Sicily to the north side of the island, then cut south across the island to where they were to drop the paratroopers. They reported that almost every town on the island was ablaze from the past six days of bombing. Only one paratrooper in the Group refused to jump and they had to bring him back. What will happen to him now I am unable to tell you."

Colonel James Gavin, Commander of the 505th PIR in Sicily with war correspondent John "Beaver" Thompson of the *Chicago Tribune*. (US Army Signal Corps)

One of the 505th officers, 2Lt James Coyle, remembered the details about the flight and the jump. "The flight to Sicily was a rough one. The planes flew low over the water and seemed to bump around more than usual. Most of the men in my plane became airsick…I spent the flight in a seat by the open door, throwing out paper bags which a smart crew chief had provided to prevent the men from throwing up all over his plane. I was not aware of any navigational problems. The formation seemed to be holding together which made me feel that everything was going well. The red light came on. I was barely able to get the men's static lines hooked up to the cable in the plane and get an equipment check before we crossed the coast of Sicily. The green light came on and we jumped out into the dark."[25]

The crew chiefs for each C-47 were maintenance men assigned to Bob Uhrig, so he was particularly worried about them as he knew each one intimately. "When my men returned I had a quart of whiskey (my only one) waiting for them and they were sure glad to have it. I have each crew chief's name and also the name and number of his ship on the bottle and I hope to be able to take it along home. At the top, I have the date of the invasion (July 9, 1943.)"

The aircraft did indeed return and the crews were filled with stories of their first airborne operation. Bob recorded some of these in his diary.

From TSgt Cecil Cooke (crew chief): "We were flying along at about 850 feet above the ground just over the target when we were caught in the beam of a searchlight which they kept on us for about one and half minutes, but it felt like hours. The next thing I knew they shot the ship down that was flying in back of us and I saw him blow up and crash. When they crashed it lit up the whole sky."

From TSgt Walter Kawohl: "We were just at the edge of the target and I was standing in the doorway with one of the paratroopers and we had an equipment pack ready to throw out when there was a terrific explosion and debris few into the ship and into our faces. We found out later it was the ship in front of us that had blown up. When I looked out I then saw an equipment bundle go by with the parachute unopened. I watched this until she hit the ground and then there was a terrific explosion which lit up the whole sky."

2nd Lt James Leach saw the loss of both aircraft from the 44th Squadron: "After seeing the ship next to us blow up in the sky and crash I saw another shot down but he maintained control and used his landing lights to land on the beach. About this time the pill boxes started giving us hell."

Finally, TSgt Roscoe Best shared his experience with Bob: "We got separated from the rest of the flight over the target but we dropped our troops and started back. I then went to the rear to pull in the static lines when the shooting started. The ack-ack shells went between the wing and the tail and shook the ship plenty. Then I ran to the front of the ship just as the pilot dove the ship to within about 100 feet of the ground to get out of the fire."

Bob's final diary entry for July 10 mentioned another mission for that day. "Another group of paratroopers came over today and loaded the ships. Tonight they are hauling heavy guns. About 1600 they told us that the flight would be called off."

## Tragedy Strikes the 36th TCS

On July 11th, the 316th and the other Troop Carrier groups were preparing for HUSKY II which was an airborne landing intended to reinforce the U.S. troops already on the ground in Sicily. On paper, this mission was supposed to be easier than the drop on July 9, since the aircraft would only be flying in airspace controlled by Allied forces, and the paratroopers would be dropped into an area already held by the landing forces.

Patton had originally suggested the idea to Major General Ridgway, the commander of the All Americans. Although he supported the concept, Ridgway was concerned that the Troop Carrier aircraft might be subjected to friendly fire from the Allied ships or ground troops anti-aircraft units. He therefore tried repeatedly to get assurances from the naval force commanders that they would enforce a friendly air corridor near the invasion fleet and near the coast of Sicily. Reluctantly, the Navy said they would comply and cooperate, but only if they were assured that the Troop Carrier aircraft would follow a strict flight pattern, and that the last leg of the mission before they reached the drop zone was over land.

The airborne plan as it developed, was that the aircraft would fly through a corridor only two miles wide at 1000 foot altitude. They would cross the Sicilian coast about 30 miles east of Gela (the extreme eastern end of U.S. landing beaches.) The planes would then fly inland briefly before they turned northwest for their run over the drop zones near the Gela-Farello airfield which was held by U.S. troops. In theory, the airborne force would be over 35 miles of Allied-held territory all the way and would therefore have lesser exposure to enemy anti-aircraft fire. In practice, however, almost everything went horribly wrong. Not all anti-aircraft units received the word that about the U.S. Troop Carriers and even those that did had already been exposed to continual enemy air attacks that same day.

Bob Uhrig's diary for July 11-12 documents the tragic events of that night and the effect on the Troop Carrier units. "Last night (July 10) about 2230 they got all the paratroopers out of the ships and arranged a search party to look for German gliders that were supposedly sighted going over, but they did not catch any of them. I slept in Engineering all night because of the alert, which continued all night. I slept there in case orders were received to haul the paratroopers to Sicily.

"Tonight (July 11) the ships took off for Sicily at 2015 and as usual the sky was filled with paratroopers. The invasion is supposed to be a success so far. All of our ships are flying tonight with the exception of No. 6. At 2315 and SOS came in from one of the ships in the 44th Squadron so we all knew something had gone

wrong. The first ship landed at 0045 (July 12). It was No. 8 and when he taxied up on the line they called for the ambulance. From then on for the next three hours everything was like a bad dream and it was horrible.

"Before they landed, the Wing called up and said that a couple of the Groups had been shot up bad so we were all hoping that it was not us, but I guess our luck had just run out and the cards were stacked against us. After they all came in, we unloaded the wounded and dead and we counted the ships. We had Nos. 1, 2, 5, 7, 8 and one borrowed new ship. The rest: Nos. 3, 4, 10, 11, 12 and 13 are missing. I am telling you we were all dazed and could hardly believe it. What makes it hurt even more was that the mission was uncalled for. They were going to drop them over friendly territory at night just to give them practice. Because they did not need them for front-line action they were going to drop them in the same place as we dropped the paratroopers the first night.

"But now comes the part that has just about made us crazy. The Germans had been bombing the ships and ground troops just before our ships flew in. They must have thought it was the Germans coming back so they opened fire on our Group! It must have been terrible. The ships were flying so low and all the guns on the ground shooting at them. There were two ships that blew up, again from direct hits on the demolition equipment in the parapacks. The ships could not stay in formation under the terrific fire. Ship No. 00 was flying in the same element as Nos. 12 and 4 and they reported seeing 12 and 4 collide in mid-air and then crash in flames.

"No matter which way they turned, they were being shot at by anti-aircraft fire, and also small arms fire. When they flew over the drop zone (which part of the flight was never able to reach) some of the paratroopers had been wounded, but most of them jumped anyway. One was hit so hard in the leg that he had to crawl to the door, but still jumped. Another one was hit in the leg and his boot was so full of blood it was running over when he jumped. The only ones that did not jump were the ones wounded so badly that they could not get to the door. Some ships brought their paratroopers back because the ships had been hit so hard that they could only just keep them in the air. After this they flew out towards the sea and all the ships fired on them and this was even worse than the ground firing. All the ships were flying just above the water and some of them even hit the water trying to get out of the fire. They flew through fire for 15 minutes and we got any back is a miracle. They say the sky was lit up like daylight from so many guns shooting."

One of the 505th paratroopers who had landed the day before, 2Lt James Coyle, saw the horrifying spectacle from the ground. "That night, 10 July (sic), as we were hiking up the road, we saw in the distance a tremendous sustained volley of anti-aircraft fire fill the sky. We took this to be an attack by German bombers on Gela. At about the same time several planes flew in from the sea directly over our heads. No one fired on them; in fact, the men were calling out up and down the line 'C-47s!', 'C-47s!', even before they were able to spot a plane or two against the moon. I never knew a paratrooper who couldn't identify a C-47 just by the sound of its engines."[26]

The final aircraft losses and casualty figures were staggering. But it would be some time before the whole picture of what happened that night was finally understood. "Of all the ships that came back from the flight, *only three out of the whole Group were undamaged* [emphasis added]. The only ship that came back in our squadron untouched was a new one that had been loaned to us for the mission. So we still have No. 6 here OK. Just think, our Squadron is 93% out of commission from one night's mission. We will have to salvage (scrap) No. 8 and possibly No. 5. I just received information that the borrowed ship was hit three times with machine gun fire. So none of our ships came back untouched. If the holes were not in the range three inches to three feet, I've never would have noticed them."

The accounts of the mission in Bob Uhrig's diary are surprisingly accurate, given that they were made in the very few hours after the aircraft had returned. Later an investigation of the incident pieced together the facts of the tragedy and they were close to Bob's on-the-spot documentation. Although the word to "hold fire" on the C-47s was supposedly passed to all anti-aircraft units, that particular day the Germans and Italians together flew almost 500 sorties against the Allied beachhead. There were several sustained air

attacks all throughout the daylight hours and one blew up an ammunition ship in Gela harbor. The gunners on the ground and on board the ships were understandably on alert and nervous about more attacks. Into this hotly contested airspace flew the 144 Troop Carrier C-47s that evening, including the entire 316th TCG. They were carrying the bulk of the 504th PIR of the 82nd Airborne Division. The aircrews had been following their strict flight plan ever since leaving their North African airfields. The first checkpoint and formation turn was over Malta, and then they pressed on for the Sicilian coast in their V of Vs formations. Unfortunately, that is when the tragic events unfolded.

About 2150 as the American aircraft were approaching the Sicilian coast, another raid of German aircraft appeared over the beachhead at Gela. Almost as soon as this attack was beaten off and the Germans retired, the C-47s crossed into Sicily and began their turn towards the drop zones. The first few airplanes were able to drop their paratroops in the correct spot, but then the U.S. anti-aircraft units on the ground began to open up on them, assuming they were yet another approaching German formation. The airmen vainly tried to dodge the ground fire and as a result the formations broke apart, reformed again and scattered. Some aircraft few out to sea and returned to North Africa with their paratroops still on board. The entire formation was savaged by friendly anti-aircraft fire and the unarmored C-47s were completely helpless. When it was all over, the airborne forces counted their horrific losses. Of the 144 aircraft dispatched for Sicily, 23 never returned of which 6 had been shot down before the paratroopers could jump. Of those that did return, 37 were badly damaged. The brunt of the losses seemed to fall on the 316th TCG and the 36th Squadron in particular. The airborne troops on board were also caught in the aircraft before they could jump. "The disastrous friendly fire incident had cost the 504th RCT 81 killed, 16 missing in action and 132 wounded. An estimated 60 aircrew of the 52nd Troop Carrier Wing were killed and another 30 wounded."[27]

Of all the 36th Squadron aircraft that did survive the mission, probably the most harrowing story came from Bob Uhrig's roommate, Captain Jim Farris. Farris, the Squadron Operations Officer and his co-pilot, 1Lt Joe Baxter, were flying in aircraft 41-38704 (aircraft No. 8), nicknamed *Geronimo*. In formation with the rest of the 36th Squadron, *Geronimo* was hit repeatedly by fire from the American ships. Flak tore off the rear fuselage cargo doors and damaged the left stabilizer. Farris and Baxter dove the airplane towards the ground, hoping to get out of the sights of the gunners on the ground and on the ships. As they flew on, violently maneuvering to get out of the searchlight pattern, they saw directly in front of them an equipment bundle or parapack that had come loose from one of the other C-47s in front of them. The parapack missed hitting the cockpit by inches but smashed into the upper fuselage, tearing a large hole in the side and flying through the tightly packed paratroopers on board. Fortunately the equipment bundle exited the aircraft through the area that had held the now-missing cargo doors. The collision caused the aircraft to gyrate wildly and Farris and Baxter were barely able to recover just as they reached the drop zone.[28] They were able to release their paratroops but then had to fly the crippled aircraft all the way back to Tunisia. After landing Bob and his men determined that the aircraft was too badly damaged to be repaired.

The last entry in Bob's diary for July 12 was the total accounting of lost ships on the HUSKY II mission of July 11-12. Over 20 men were now Missing in Action from the 36th TCS alone‡:

| Aircraft Number | Serial Number | Crew Chief | Aircraft Nickname/Nose Art |
|---|---|---|---|
| 3 | 41-18511 | Onstine, Howard C. | *Texas Tornado* |
| 4 | 41-18606 | Cooke, Cecil R. | *Hoosier Hot Spot* |
| 10 | 41-38620 | Cull, Dalton W. | *Down & Go* |
| 11 | 41-18525 | Best, Roscoe M. | *Sandy* |
| 12 | 41-18522 | Singer, George A. | *Briar Hopper* |
| 13 | 41-25513 | Jones, Carl W. | *Lovah Boy* |

---

‡ Also in the diary for July 12 was a listing of the KIA/MIA members of the 36th TCS and their temporary grave numbers in Italy. The remains have all since been removed from these cemeteries and some are still buried overseas. Appendix B gives a listing of the current burial locations, if known.

## 316th TCG C-47s Damaged in Operation HUSKY

Damaged Aircraft (*Clockwise from top*):

Flak tore through the cargo door of aircraft #12.

Captain Jim Farris' *Geronimo* showing severe damage to the fuselage.

Aircraft #25 with a damaged rudder. Insignia below the serial number is the RAF-style "fin flash."

Aircraft with flak holes through the rear fuselage.

(All photos AFHRA)

## 316th TCG C-47s Damaged in Operation HUSKY

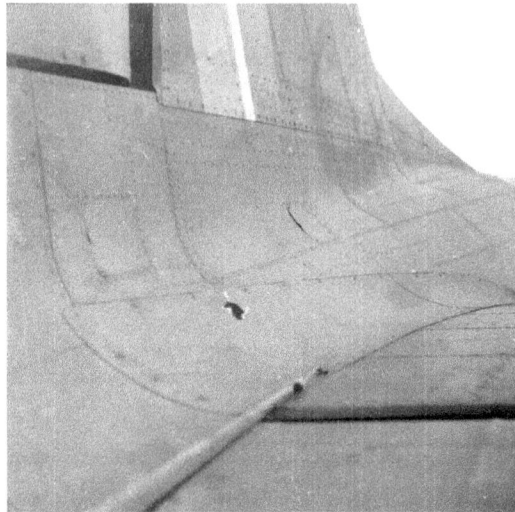

*Left:* Flak damage to the tail of the C-47s. (AFHRA)

*Left:* Damage from a gear-up landing (AFHRA)

*Right:* Damage to the control surface. (AFHRA)

*Right:* Flak damage to the wing, fuselage and cabin. Caption to the photo states: "Paratrooper got it." (AFHRA)

# The 316th Recovers from Their Losses

The next day and in fact for the next few days, the men of the 316th TCG went about their duties but were still stunned in disbelief at the tremendous losses they had sustained in that one mission. Bob had lost so many friends and comrades, men he had known and served with for many years. "It is impossible to work yet today because all the men in the Squadron are feeling so low. Major Garland had a meeting this morning and told us that they were sending him to Group as Operations Officer, Captain Farris is going up as Assistant Operations Officer and Captain Milstead is going up as Intelligence Officer. It is bad enough to have your squadron shot up with the loss of 29 men, but then to have Group pull out three more is just too much. He could not finish his speech because he felt so badly.

"They told us all to go swimming this afternoon so we went to the beach, but could not enjoy ourselves. This evening we went to a show over at the 340th Bomb Group. It was a USO show and it was just as good as most stage shows that you see in the States. Every bit was good and it was a surprise to all of us. The women in the cast were very good looking. Then they showed a movie. This was almost more than we could stand after not seeing any type of entertainment for such a long time. After seeing that we came back to the field and found that the crew of aircraft No. 13 had been saved; all but the navigator. They were shot down but picked up by a destroyer. The pilot and co-pilot are in the hospital but we don't know how bad they are injured. The crew chief (Jones) and radio operator are OK and are on a field about 50 miles from here. "

Bob's first letter to Toots after the HUSKY missions gave only the smallest hint of what the men had gone through. "I lost out on two days writing but in the past 48 hours I have had 3 ½ hours sleep and I am just about dead. They are keeping us ever more than busy just now and by the looks of things it will keep up for a while yet. I only wish I could tell you the news. We have not had any more of those heat waves like I wrote to you about. I made one mistake it was about 143° about three miles from here."

In the days that followed, more news and more missing men trickled back into the 316th TCG, which helped to lift spirits. "More good news this morning (July 14). Major Bowman is in a hospital with a broken leg so he is still alive. Also the Group Operations Officer and Assistant Operations Officer turned up last night, so Major Garland, Captain Farris and Captain Milstead are coming back to the Squadron. We have been given three new C-53s and three new C-47s but two of the C-47s were on the last mission and were shot up so they will have to be repaired. The rest are OK to fly. We are hoping that some more of the crews will show up.

"About 1100 Captain McCullough (pilot) and Sgt Ouellette (radio operator) from aircraft No. 3 walked into Engineering and surprised the hell out of us. Both of them had their clothes all torn and they looked like someone had beaten them up. They told us that the ship had been hit plenty and the last burst of anti aircraft fire had caught the left engine. They flew on and in a few minutes it blew up and they landed in the water about 17 miles from land. The ship broke up and started to sink."

Captain McCullough finished the story of their ditching and rescue in his own words,

> "We all got out as soon as possible with just our life vests on. A couple of them would not inflate so they had to blow them up themselves. One paratrooper was dead and the other one jumped out into the water with all his equipment on and no life vest. We tried to hold him up until we could get his equipment off but he could not swim so there was no hope for him and he drowned in a few minutes. We then decided to swim towards land. We kept calling to each other during the night. McGregor, the co-pilot, was wounded and during the night he became lost from us. We kept in contact with Onstine the crew chief until just before daylight then we lost him too. Ouellette, Morgan and I were able to stay together. During the night we could hear someone hollering in the distance but we were unable to make him or them hear us because they were upwind from us. It was terrible there in the water and not knowing if you were ever going to be picked up or not. Just when it was beginning to get daylight a destroyer came by and Ouellette waved his jacket. We thought that in the poor light they

had missed us and that it when I was sore because I had shot all my ammunition in the air to attract the attention of the men's voices we had heard in the night. We thought maybe they had a boat and would pick us up. Anyway, after passing us, the destroyer slowed up and made a circle and came back and picked us up. Ouellette had used all his strength waving his jacket and passed out. Morgan had passed out a short time before and I was holding on to him. After they took us on board ship, I passed out. We had all been in the water for seven hours. After regaining consciousness they told me Morgan had died from exposure. We then buried him at sea. They were never able to find Onstine or McGregor. The destroyer brought us into Sousse and then a truck brought us on up to the field."

On July 15, Bob noted in a letter to Toots that, "Eight months ago tonight we were ready to leave the good old United States and that was something hard to do."

The next day, the men of the Engineering Division continued the difficult task of repairing the large number of damaged aircraft and cannibalizing the ones that would not fly again. The Troop Carrier planes were now urgently needed to fly supplies to the American forces fighting in Sicily. "We now have three of the new ships in flying condition. We have taken the wings off of No. 8 [Farris' damaged ship *Geronimo*] and also the stabilizer and radio equipment. This afternoon Sgt Love and TSgt Middendorf were taking off for Sicily when Middendorf's ship (No. 6) blew a tire and had to pull off the runway. Love's ship, No. 7, went on with the mission. There were 8 transports in all and they picked up an escort of P-51 fighters. Love says the island is beautiful but there has been so much terrible damage done. He says that you could walk for miles along the beach just on top of the landing barges. As quick as the barges unload they go right back for another load. Also the harbor is filled with larger boats. The flying fields are covered with destroyed Axis planes. He says he could see a lot of our destroyed C-47s from the air."

Bob came down with a case of food poisoning on July 17 and was out of action for the day. He was assigned a new tent with his old roommates, Doc Hamilton, Jim Farris and William Coursen as the Squadron tried to recover from the large losses in personnel from the HUSKY operation.

On July 18, Bob went along on a flight to Bizerte to recover one of their men who was injured on the HUSKY mission and in the hospital. "This morning we took a ship and flew to Bizerte to pick up Lt Bourne. The hospital said that if we would leave Bourne there, they would put in some new teeth for him. So Doc decided to leave him there. We then decided to look the field over. All of the hangars are blown apart and most of the roofs are gone, or have large holes in them. You can see all types of destroyed German planes. I saw my first Focke-Wulf 190 fighter. There were plenty of them lying around. They are a beautiful ship and are powered by twin-row radial engines. The cowling comes to a sharp point as it does on a liquid cooled engine. They have a cooling fan that is geared to the propeller and this fan is what they use for cooling the engine.

"Bizerte has a natural harbor and the inland bay is about five miles across. The docks are just everywhere you look and each one is full of boats and landing barges. Fenimore talked a captain into letting us fly an Aeronca light plane in the afternoon. The captain invited us over to the navy area for lunch so we were all set for a nice day. But then Doc came back and said he was not going to take Bourne back which put a crimp in our plans so we just flew on back home instead."

That night, Bob mentioned in a letter to Toots that, "I just can't find much to write about unless I tell you things that I am not supposed to tell you and that would be bad." He was clearly still depressed about losing so many friends in recent days. But security regulations would not allow him to give any details of the mission or the fate of the men in his squadron.

The next day he organized his men for a short field trip to get their mind off of recent events. "There were 21 or 22 of us and we loaded into a 2 1/2 ton truck and then we started for fishing country. The road winds through beautiful mountain country with plenty of timber but after a short time you have climbed out of the timber and are above the timber line where the mountains are very rugged and also plenty rocky. After driving about 20 miles we turned off onto what looked like a cow trail but it was not too terribly rough. Now we are getting into some real wild country. After taking this path for about five miles we came to a

barrier across the road and there were two French soldiers there guarding it. There was a sign at one side which was written in French and read 'This is the boundary of security.' From here on you are on your own because there are Berber tribes from here on. We only drove about 20 miles into this country and the natives are friendly this far.

"By the time we reached the stream we had climbed to about 10,000 ft. This stream is supposed to be the best known in that part of the country and where it is located it should be because very few people can get to it. It is a beautiful mountain stream that really moves and there is enough trout in it to feed all of Springfield [Ohio]. The fishing tackle that they gave us at the hotel is not very good for trout so we used rifles. We caught and cleaned over 200 that one day. I caught four on hook and line and then I started shooting them myself. It was sure funny to see everyone with rifles and pistols fishing. All up and down the stream you could see fellows up on rocks in deep holes and shooting just the big fish as they came out. The concussion would stun them and we would pull them out before they came to. It rained when we started home and everyone got good and wet but before we reached home the sun had come out and we had dried out. We all had fish for breakfast at the hotel the next morning."

More news began to filter back about some of the missing 316th men on July 20 and also about the paratroopers they had dropped on the fateful night of July 11.

<u>Aircraft No. 4</u>: Three paratroopers jumped but all the rest on board were lost
<u>Aircraft No. 10</u>: The paratroopers report the aircraft was OK when they jumped. After that no one has reported seeing them again
<u>Aircraft No 11</u>: Crashed. The graves of the five crewmembers were found a short distance from the wreckage, northwest of Gela.
<u>Aircraft No. 12</u>: One aircrewman jumped but they have not received any information on who it was. All the rest are presumed lost.

As the men recovered from the Sicily disaster, reminders of the earlier desert campaign still surrounded the field at Enfidaville. "MSgt Santomissino was up on the hill the other day. He can speak Italian, so he found a native that could speak Italian and had him take him up on the hill and show him around. He said there were a great number of shallow graves of Italian soldiers and some not even buried. There is plenty of equipment and bodies on top. He found some letters beside one body. The letters had not even been opened so Santomissino brought them back and read them. One was from the dead man's mother telling him how terrible things were at home. She had sent him a religious medal to protect him through the war. One of the others was from his girlfriend with her picture in it.

"We went swimming this afternoon and on the way home we bought some watermelons from the natives. There are very few melons raised here and what are have to be irrigated so this makes them very expensive. They are about the size of our real large cantaloupe melons but they cost about $1.50 each and a very large one will cost you $4!"

Bob's next letter to Toots mentioned one of the men in the Squadron who was sent home and was supposed to be bringing things back for the family members and also his brother in law, Omer Fultz, who was now also in the AAF. "So all of you are plenty pissed off at Matthews, well I can't say that I blame you very much. You see we moved and we were over a thousand miles from him and when he was sent home none of us were able to give him a message. So Omer came home and he does not like the Army. Well maybe when he gets in a squadron and learns to know some of the fellows and goes around with them he will like it much better. I sure hope that he does because I know how terrible you can feel if things don't go right.

"You should have seen these caravans that we have to go by. They consist of camels, donkeys and horses, some of them sure have a mixture. They take things from the farms to market and vice versa. They also haul things from one town to another. The load that they put on the donkeys is something terrible. Sometimes you can hardly see the little asses. The people in the native villages have a grass that they make baskets out of. Basket making and pottery is about all the industry these natives have. We all have water jugs that they have

made and they keep the water pretty cool. The water seeps through the pottery and the hot air evaporates it and thus the water is kept cool from the rapid evaporation. If we stay over here much longer I think that we shall all go native. They raise quite a few olives here. They have them in large groves like our orchards at home. They keep them very clean and the trees well trimmed."

As the Sicily campaign dragged on, it became apparent that the 316th would be staying in Tunisia for a while longer, so Bob and his men began to settle in and wait it out. "We have a name plate up in front of Engineering which says, 'Ali Baba and His Forty Thieves.' Now why they call the 36th Engineering a name like that I don't know because I have just 40 men with me. We won't steal anything that is tied down or too heavy to lift. We now have an Italian truck and a German bus that we got out of a field and fixed up so they will run. The only drawback is that we have no batteries so we have to push them to get them started. In this outfit you might see anything drive up in front of the tent. We are all looking for a German jeep (*Kubelwagen*) and if we are every any place where they have some and no one is looking we will have one. We had fresh meat for the first time in two months with the exception of when we went hunting and got it ourselves. We are all so dammed tired of canned rations. And if you ever utter the word Spam or Vienna sausages I will run you out of the house. Even in the Libyan Desert we received better rations than they are giving us here. But we all manage to get things from the natives for ourselves. We are not supposed to eat the tomatoes but we do and we trade for eggs and then hard boil them. One package of cigarettes will get you three eggs, or three lumps of sugar will get you one egg."

In the higher reaches of the Allied command, there was now serious soul-searching about the effectiveness of airborne operations. The friendly fire losses, missed drop zones and other calamities of both the American and British airborne troops raised serious doubts in the minds of the American generals in particular. The crisis was serious enough for General Eisenhower to write to General Marshall, the Army Chief of Staff and report that, "I do not believe in the airborne division."[29]

One report, prepared by Major General Joseph Swing, an airborne officer from the U.S., provided a useful critique on what went wrong over Sicily:

1. Insufficient time spent in coordinating the air routes with all forces
2. Complexity of the flight route for the low degree of training of the navigators.
3. The rigid naval policy of firing on any aircraft.
4. The unfortunate timing of airdrops directly after extensive enemy air attacks.
5. The failure of some army ground commanders to warn all antiaircraft gun units of the impending airborne mission.[30]

The final Air Force report on air operations in Sicily pointed to lack of training for Troop Carrier crews in advance of the invasion. It also found fault with the still-developing doctrine on how to employ airborne forces in combat:

> "The troop carrier operations which spearheaded the assault were perhaps the least satisfactorily executed phase of the air participation in HUSKY. The salient shortcoming here was in night navigation. Although a satisfactory standard had been achieved in practices and rehearsals, navigational errors were made on the night preceding D-day which resulted in the dropping of paratroopers in the wrong place and in a number of gliders falling into the sea. The dropping of a parachute brigade within Allied lines on the second night of operation purely as reinforcement was unsound in principle. In this instance a failure to arrange a safe corridor for passage resulted in high casualties inflicted by friendly troops and ships. Airborne troops should be employed only on missions which cannot be accomplished more expeditiously or more economically by some other means, and their use as reinforcements should be confined to serious emergencies."[31]

If airborne forces were to be used in large numbers in future operations, they needed better training, navigational aids and tactics to ensure the troops were delivered safely to the drop zones. The Sicily disaster led to many changes in the procedures for airborne employment. These were accelerated as the date for the invasion of France loomed.

But back in North Africa, Bob Uhrig still had a large group of men to lead and they would be out of the action for a while as they rebuilt their Squadron with new aircraft and personnel. On July 26 he gave them a short break in their work. "Yesterday I sent TSgts Love and Tucker to Tunis on a one-day visit. They took about three-quarters of the engineering men with them. So this morning, TSgt Street and I took the rest of them. After arriving we had lunch and then took the truck and drove to the old city of Carthage which the Romans had destroyed. The city itself overlooks Tunis and the harbor. All the ruins are being excavated. You can see the columns of the palace and their sewage system, and where they had running water. At the south side of town there is an amphitheater. Most of the columns are still visible but none are left standing. It seated 50,000 people and all the columns and seats are made out of marble. It must have been very beautiful. We then visited the museum and saw all the valuable pieces that have been removed from the old city, including statues, money, lamps and writings on marble. They even have cannonballs made out of stone that they used with a catapult.

"On our way back to Tunis we stopped at an airport that the Germans used and had a look at the largest transport aircraft in the world. There are three of them that have been destroyed, sitting on the field. It just does not look possible that they could fly. They were made as gliders but then they put six engines in them and made them into transports.§ Even with six engines they have to pull them with another ship to get them off the ground. They have six wheels, three on each side. The nose opens on hinges and is divided in the middle. It has regular steel I-beams across the fuselage and the rest of the airplane structure is made out of large steel pipes. This alone makes the ship heavier than any other airplane. It can haul 120 fully equipped men. It is just so monstrous that you would never believe that it could ever fly.

*Left and Right:*
The wreckage of German Me-323 *Gigant* aircraft in Tunisia.
(Mike Ingrisano collection)

"We spent the balance of the afternoon drinking wine and in the evening we had dinner and returned to the truck at 1930. We found that one of our men had taken on too much wine and the Military Police had just put him in the guardhouse. It took about two hours for me to find him and get him released. Then when we did start for home, the truck began to cut out. It would only work in second gear and by the time we drove around looking for two spark plugs it was 2330 and we were all tired and sore. It was 0130 before we reached home.

"After arriving home we were told that aircraft No. 9 had crashed in Sicily and was a total wash out. MSgt White was crewing it. They were bringing it in for a landing at 90 mph when it stalled out and crashed. None of the crew was hurt and one of our other ships brought them all back. Captain Roberts and 2Lt Leach were flying it. So now we have only 11 airplanes and only 7 of them flying.

"The Chaplain returned from Sicily today and reported finding the graves of Captain Churchill, F/O Mobley and TSgt Singer. Aircraft No. 4 has been found in a swamp half covered with water and as yet they have been unable to get the bodies out. There is still No. 10 missing and no reports of it ever being seen after

---

§ These aircraft were the Messerschmitt Me-323 *Gigant*, a six engine glider-transport used by the Germans to ferry supplies from Sicily to North Africa. On April 22, 1943 off the coast of Tunisia, several squadrons of Spitfires and P-40s attacked a squadron of Me 323s and their fighter escorts. From 14-16 Me-323s were shot down with the loss of all their crews and 700 drums of fuel.

the mission. Two days ago F/O Quisenberry and TSgt Johnson were awarded Purple Hearts for the wounds they received on the night of July 11. Lt Bourne and Lt Harmon have returned to the squadron but they are still in pretty bad shape and also look plenty bad."

Bob wrote to Toots on July 27, still despondent over his unit's losses, but still unable to give any details. "I have been in such a spin lately that I don't know what I am doing half the time. Have you seen anything in the papers about our doings over here? You should find something in them about us because things have really happened to us lately. We think that we are going to have to retire the original 'Old Glory' because she is worn almost down to the stripes. She has been through too many wind storms. We are going to have to put up one of the new ones that you sent me. We wanted to make the original one last the war out but the old girl just can't take it any longer. I will send it home to you and you can exhibit it and let her tell you about all of her experiences. She has seen a lot of things happen and also a lot of country. I see McMenamin about every day now. He comes over to check on us. We compare notes from the letters we receive from you gals. He had me over for a few drinks the other night. Where he got the whiskey I don't know because it is next to impossible to get. Beer costs (and there is none, only in _____[deleted by the censor]) $3.80 a can and that is only a ten cent can back in the States. They are out of everything to drink in Africa. We get native beer once a month and that is not too bad."

By the end of July the Allies had gained enough ground in Sicily that more units, particularly air units, could be sent to the island to reinforce the progress made earlier in the month. Patton had taken Palermo on July 22 and the entire western side of Sicily was now in American hands. On July 26, Italian Fascist leader Benito Mussolini was overthrown and arrested. This threw the Italian forces into further turmoil. There were still weeks of hard fighting ahead to capture the rest of Sicily, but both Patton and Montgomery now had more maneuvering room to force the Axis troops back to the northeast corner of the island near Messina.

The 316th TCG was not ready to make the move by that time since they were still well under strength, but they would help deploy other American air units to the island. "We sent five ships to help move a fighter squadron to Sicily today (July 28) and they have just returned. Their reports of the flying fields in Sicily are all alike. They all say that you land one way on them and then turn around and take off the other way. This is caused by the hilly condition of the island. Every field runs almost into a hill, so they have to land towards it and take off away from it. While our men were there they saw General Montgomery land in a B-17. They say he looks just his pictures. I am Officer of the Day today and it is the first one that I have ever pulled in my life."

**Engineering Retires the Colors**

A formal retirement ceremony was held by the 36th Engineering on July 29 for their "Stars and Stripes" flag that Toots had sent to Bob many months ago. The flag had seen a lot of action and was well worn by the desert weather. "She is so torn and beaten up that we decided to take her down and I am going to send it to Toots. We are putting up a new one that Toots sent me. Engineering had the first and only flag in the Group for a long time. Toots sent it to us when we were at El Adem and we have had her flying over Engineering every place that we have been."

Bob finally got a chance to see Sicily first hand the next day. This would be his first trip "out of Africa" since his arrival the previous November. "Three of our ships took off this morning for Sicily. After so many days I talked Major Garland into letting me take one mission over to Sicily. First we went to Tunis and landed at the wrong field so we had to take off and fly to the correct field. We picked up the freight and headed for Sicily.

"We crossed over Pantelleria Island which we took about a week before we invaded Sicily. We flew into a new landing strip just north of Gela. Gela is where the drop zones were for the invasion and also the place where we lost all of our ships. After we landed about five Sicilian boys came over to our ships. They dress a lot like our American boys and they love to have their pictures taken. Sicily has its share of mountains and the country is very beautiful. They raise a lot of grapes, cotton, sheep and goats. All the pens for their livestock are made out of bamboo. The towns are built with most of the buildings connected. They build the

house and the walls around them out of a grayish stone. When we took off for home we saw all of the sunken ships that were hit in the invasion. All the harbors are still full of ships bringing in supplies."

The week that followed was a quiet one for Bob and his men. Other than the work to repair the C-47s, there was the usual gambling and games of chance that came right after payday. The highlight of the week was being able to acquire some excess German equipment to make their work go a little faster. "We knew some fellows in a Service Group that moved to Sicily so before they left they gave us some of their German equipment that they could not take with them. The best finds were the two German rifles that MSgt White and I picked up. Also spare parts for our motorcycle so we are now overhauling it. They also gave us another BMW motorcycle with a sidecar so when we get ours overhauled Engineering will have some transportation.

One of the German BMW motorcycles used by Bob and the mechanics of 36th TCS.

"We now have ships Nos. 1, 3, 4, 5, 7, 8, 11 and 00 back in commission. They look like someone had thrown paint spots all over them. This is due to all the shell hole patches that they now have on them. We are changing the engines on No. 6 and were able to run it up for the first time last evening. One engine just about jumped out of the mounts so we decided it was the propeller that was out of balance. We had put two new propellers on with the two new engines. The difficulty is that we have to assemble our propellers out here in the field, so if they are not balanced right at the factory it makes for a lot of extra work for us. We had one spare propeller this time, so this morning we assembled it and put it on the engine and everything was OK. Those engines that we put on No. 6 are brand new and having new engines is something special in this country."

Bob's August 4 letter to Toots described his first visit to Sicily. "We now have another motorcycle and also a side car. Two can ride on the motorcycle and two in the side car so it is just about like having an automobile. I don't know what I am going to do when I get home and get into a good car on a good highway. Riding over here just about beats your brains out. Even if we only ride from one ship to another by the time you have been to all the ships in the Squadron you will have driven two miles and it is the worst two miles that is possible. You should see the boys in Sicily. They remind you of our farm boys back home. To see white people in that country is something new. The country is very beautiful with many mountains. In fact, I would like to move there for a while. The boys just love to have their pictures taken and pose every way possible from holding up goats to riding horses. I have a pretty good cold just now. It is the first one that I remember having since I have been over here. Do you know we have only seen four rain storms in the nine months we have spent in this part of the world?"

By early August, the Sicily campaign was drawing to a close. The remaining Axis troops were bottled up around Messina and were desperately trying to evacuate the island before the Allies could capture them and the strategic port only three miles from the Italian mainland. Allied planning was progressing towards the next objective – Italy itself. Another seaborne invasion was in the works, supported by additional airborne operations. This required the 82nd Airborne troops to be withdrawn from combat in Sicily and be refit and prepared for this next operation. It also required the Troop Carrier units to be brought back up to strength, re-trained and re-organized for combat operations in Italy.

On August 7, Bob and his men continued preparations for the next airborne operation. "Five of our ships left yesterday to go to Oran to bring back gliders. Also we just got in 26 glider pilots and 9 glider mechanics.

We now have all the replacement crews for the ones that we lost. We have all our ships in service with the exception of No. 2. But as usual there is extra work besides the maintenance. Now we have to put all the British pararacks on the airplanes. We have to make and mount hoods on all the American-type formation lights. They also have us changing the Air Corps insignia again. This is the second time in the same number of months. It now looks something like this [Stars and Bars figure in original letter]*

"They keep us busy with junk like that and then for us to try to keep up with the maintenance is just a little more than we can do. In fact, today at times I thought I would scream. They just called and told us to send our glider mechanics to Oran to help get the gliders in commission so our ships can bring them home. I have now traded things for some pistols – one Luger, two Berettas and one Mauser. Hedy Lamarr is playing at our theater tonight in *White Cargo* so I guess most of us will blow our tops when we see her."

Bobs' next letter to Toots described some of their minor comforts and the anticipation of movie night. "We went swimming yesterday afternoon and found a new place which has a nicer beach and the water is as clear as crystal. It is in a little cove with a high cliff in the back of the beach and is a very pretty place. Every time we go swimming we just about drown each other and act as crazy as a bunch of kids. The only bad part is the ride over. We go in a 2 ½ ton truck and the road is so rough that you have to hang on every minute. It is 18 miles of which 8 miles are terrific. Once a month we get our ration of cigarettes, candy, toothpaste and like items. We put it all in the mess hall on the mess tables and everyone files by and gets his rations for the month. You can have three cartons of cigarettes, six cigars, two chocolate bars, three tootsie rolls and three packages of chewing gum. They also have mints. The other items are razor blades, towels, toothpaste, soap, pipes, shaving cream and matches. Say, I might not come home because we are going to have Hedy Lamarr here Saturday night in *White Cargo*. Did I tell you that after eight months our group got a picture machine and they are trying to have a show each night? But each picture is on for two nights. The screen is at the foot of a small hill so you can always see it very well and the speakers are very good. Just finished another engine change so I let Love take the men into town. It is 73 miles from here. All there is for them to do is drink wine, but that is good enough. [Hedy Lamarr played a character in *White Cargo* named "Tondelayo." That name later appeared on the nose art of a 36th TCS C-47 shortly after the film was shown.]

A group of 36th TCS men, Bob among them, was granted leave to visit a rest camp in Morocco. They departed from Enfidaville on August 11 and dropped eleven enlisted men off in Algiers and then six of the officers flew on to Meknes, Morocco where the rest camp was located. They were met by a French interpreter who hosted them at the French Officer's Mess and then made arrangements for their transportation to the camp up in the mountains in the town of Ifrane.

Bob described their vacation to Toots in a letter on August 17. "Just returned from the rest camp and I don't believe I have ever been in a more beautiful place. It is run by the Red Cross and the Air Corps. They can only take care of 150 men at a time, so you see there is not too large a crowd. They tell us that the resort is known all over Europe. It is both a summer and winter resort. Just think of all the pictures that you have seen of resorts in the Alps and then you can tell just how the place looks where we were. All the buildings have steep, sloped roofs to keep the snow off in winter and most of all they are built out of stone. The top parts of the houses are made out of wood and each house has the fancy shape of the houses in the Alps. We landed at the closest airport which is about 40 miles from the resort and the French interpreter took us to chow and while we were eating he called the rest camp and they came after us in a jeep.

"On the road out were passed through the richest land that I have seen in Africa. They had wheat fields as large as ones out West. They also raise acres and acres of sunflowers but don't ask me what they use them for because I do not know. But soon we left the fertile valley and began climbing, after a short time we hit the timber line and here the mountains are beautiful. The timber is very large and you would think that you were out West. All the time you are driving you are passing one mountain stream after another. They rush

---

* This was a change to the U.S. national insignia on the aircraft from the existing blue circle and white star or the yellow-bordered blue circle added for North African and Sicilian operations. On June 29, 1943, the second major design change for USAAF aircraft was adopted. A white rectangle or bar was added on each side of the blue circle with a red border surrounding the entire insignia. While the new design was estimated to be more recognizable, the red border was only used for a few months and was changed to blue for the rest of the war.

down out of the mountains at full speed and the water is as cold as ice. All the time you are climbing and by the time you reach the hotel you are above 5000 feet and at that altitude the heat does not bother you. The temperature is just right.

"Where we stayed was the largest hotel in town and was built about two hundred yards up the side of a hill and we could see over the entire resort. At the base of the hill that the hotel sets on is the resorts park and a mountain brook runs through it. The park has tennis courts, ball diamond and refreshment stand and the largest swimming pool that I have ever seen. It is fed by the mountain stream so you know cold the water is. The town consists of nothing but hotels, private lodges and restaurants. Our days consisted of breakfast at 0830, and then we would play tennis, baseball or ride bicycles. At 1100 about every one in town goes swimming so we did too. After lunch we would start drinking champagne which cost us the sum of $1.20 a quart. Lt Conroy and I met two other officers from another group (Crawford and Bellco) and the four of us were together from the time we got up in the morning until we went to bed a night. Crawford is from Oklahoma. Bellco can speak French just like a native. After a couple of quarts we would go swimming. Dinner was served at 1900. The food was prepared by Frenchmen and was wonderful. Steaks and chicken and pork chops and ham. It was something like a dream book.

"Since I am back home again it seems like a dream because it was so wonderful. Each day we would have fresh meat and all of that we could eat. Twice we had ice cream and the other times it was fruit and watermelon. In the evenings there was always a moonlight picnic, a picture show or dance at the hotel. The Sultan of Morocco has a shack about the size of Donnelsville which was just to the right of our hotel. Two Red Cross women and three Air Corps officers run the hotel and everything is free. All we had to pay for was the drinks that we bought in the hotel. It is sure one ritzy place and no one can afford to vacation there unless they have plenty of dough. We were there four days and five nights and we drank plenty of champagne and beer and still the whole trip only cost me $18.00. If anyone had told me there was a place like that in Morocco, I would have called them a liar. There are also trips to different large towns each day. Only one bad feature, it takes a full day to fly there and a day to fly back but it is sure worth it. I enjoyed myself more than any place I have ever been since I left the States. I feel so much better that I did before I left."

**All Sicily Now in Allied Hands**

When Bob and his mates returned to Enfidaville on August 17, they found that there was further reason to celebrate. The Allies had conquered Sicily. The official announcement reached them the next day. "Sicily fell this morning at 0800, so the entire island is in Allied hands. We all think that we will start on Italy in the near future. We are still on alert to move at a moment's notice. We are still towing gliders and also making plenty of runs to Sicily. The rains have not started as yet but it won't be very long now until they do and then this place will be one hell of a mess. We have received another new airplane. It is a C-47A and on this one they have the rear door fixed so we don't have to remove it to load a jeep. This brings our total of aircraft to 12 ships. It still gets plenty hot here every day but turns cool in the evenings."

As soon as the fighting ended in Sicily, the 316th began preparations for yet another deployment. This time the men were sure the next destination was either Sicily or the Italian mainland. "We were working on movement plans last night (August 19) and at about 0200 two British fliers walked in and asked for transportation to look for their buddies. They had jumped out of a bomber and had landed very close to our tent area. We gave them a jeep and by morning they had all been found. One of them landed in a cactus patch and he was not feeling too terribly good. The bomber crashed and burned but no one was in it when it hit."

Unknown to the men of the 316th, General Ridgway was informed in late July that a plan had been developed for the 82nd Airborne to conduct an airborne operation in conjunction with an amphibious landing in Italy. The initial concept was for the operation to commence on September 9 in the Bay of Salerno. In early August, the details of the plan were crafted -- two parachute regiments were to be dropped on the north side of the Sorrento Mountains to prevent the Germans from reinforcing their troops in the Salerno area. Other airborne troops from the division were to be landed on the beaches. To prevent the recurrence of the

problems that plagued the airborne operations in Sicily, there was a critical need to improve the accuracy of night parachute drops. The solution proposed to address this problem was the "pathfinder" concept.

The pathfinders would be a small group of paratroopers specially trained to operate electronic navigation equipment that could communicate with the Troop Carrier aircraft. They would be dropped some 30 minutes before the main airborne force arrived over the drop zones. When they landed they would set up their navigation gear to help guide the C-47s to the exact spot where the remaining paratroopers needed to drop. Their initial training was conducted in Sicily in advance of the invasion of Italy. It was hoped that the employment of the pathfinders would lead to much more accurate navigation by the Troop Carrier crews, even in bad weather or in the presence of anti aircraft fire.

The equipment to be used by the pathfinders and Troop Carriers was called the Rebecca-Eureka system. The transmitter device (Rebecca) would be mounted on special pathfinder C-47s and the corresponding homing beacon (Eureka) would be set up by the pathfinders on the ground close to the intended drop zones. Several hand-picked men from both the 82nd and Troop Carrier units were trained in the use of this equipment during the first two weeks of August in preparation for the drop in the Salerno area.[32]

Bob's August 21 letter to Toots spoke of his unit's preparations for the next deployment and how much they had learned in their year overseas. "I think there is a curse sent on us because no matter how many times we move or where we go sooner or later we get high winds and sand or silt. This is nothing like the last place but it gets a little worse each day. We now have the motorcycle fixed up again and now we have a side car for it. One that some kind German could not run fast enough with. So we can haul three people. Also there is a small rumble seat in the back of the side car and we can haul another passenger or freight. She has saved us a lot of walking and she has also covered plenty of country. We are now beginning to get a lot of grapes from the natives. They are plenty good. In fact, better than what we call California grapes. Our office consists of: wooden horses with boards on top; gravel on the floor to keep the dust down; wooden boxes which contain all our forms and papers; one portable typewriter, a field phone and a small status board. We have to have everything collapsible because some times we must move in a hurry and we can't be bothered with large heavy items. You should see us move when the word is given. Each man knows his job in engineering and we can be gone in nothing flat. I would sure like to tell new outfits what to leave at home and what to bring. It sure would help them a lot because they just don't know what to expect or what they must be able to do."

Until they could move for good, the 316th was now flying regular cargo runs from North Africa to Sicily. Allied air superiority over the island made it safer for the C-47s to keep up a regular schedule. "We sent seven planes to Sicily again today (August 22). In fact, hardly a day passes that we don't send some ships to Sicily. In the afternoon we went swimming and came back about 1800.

"Every evening we have hard boiled eggs and some canned goods. Doc Boyer keeps us in eggs. All the natives that step on land mines and also the ones that were hurt during the fighting close by our camp area come to him for treatment. They all give him eggs in exchange for treatment services. He now goes by egg-days. For example, Saturday was a nine-egg day and Sunday he was unhappy because he only had a three-egg day. We are going to start advertising for him."

On August 24, Bob took another opportunity to ride along on one of the cargo runs to the island. "Today we had seven ships to go to Sicily. So I went in one of them because they went to the north shore of Sicily and I have never been there before. We took off from here at 1210 and flew to Bizerte to pick up rations. While at Bizerte I saw a fellow who used to be in the 3rd Transport Squadron at Duncan Field. We had a long talk about the fellows that we both knew and how good the Transport Squadron used to be. His name is Priceson and he is now a Warrant Officer. It was 1510 before we were loaded and ready to take off. Because each ship had taken off as soon as they were loaded, both of our ships really headed for Sicily in a hurry.

"I was flying in ship No. 6 with Lt. Coursen. After we reached Sicily the weather began to get a little hazy. We flew the full length of the island and it is plenty mountainous, but most of the mountains have vegetation on them. The country as a whole is very rough. The Sicilians have a funny idea about towns because every one of them is built right on the ridge of a hill. Every bit of land that can be used is under cultivation but about 90% of the fertile ground is used for grapes.

"We flew past Mt Etna and we were in and out of the clouds most of the way. Our estimated time of arrival was off so one time we looked down and we were over the tip of Italy! You should have seen us wheel her over and head for Sicily. In our minds we could all see about 100 German fighters after us. The Messina Straits are only about two miles wide.

"After much flying up and down the coast we found the field where we were to unload the rations. The field was a new one and was just a strip scraped in between two hills and through a grape field. You come in to land from the sea. When we landed it looked like 500 kids were there to meet us. Each one of them had a bunch of grapes for us. They let us go into the fields and help ourselves. We brought back plenty for the whole squadron. The grapes there were tall stalks and they tied them up to bamboo poles like we tie up tomatoes back in the States. The name of the place we landed was Falcone and it is one pretty place with green mountains all around. Each house out in the country, around the vineyards, are just like farm houses back in the States, with the exception of their different shapes and most all of them are built out of stone or cement. Also they use a lot of brightly colored tile. Most of the hills are terraced and everything is green so you know it is pretty. Talked to one fellow who spent 15 years in Pittsburgh. He came back to Sicily on a visit, got married and just stayed. Every time a ship lands in Sicily the crew finds someone who has lived in the States.

"Coursen let me take off and fly home. We were pushing the old crate because we had to reach the coast of Africa before dark as they will shoot you down unless Intelligence is notified that you are coming. We passed over the field at Pantelleria. The town itself was blown to pieces from the heavy bombings. Between Pantelleria and the coast of Tunisia we had to fly around a convoy of 25 ships which had destroyers running around and in between them all the time. We stay plenty clear of the Navy since that night when they knocked half of our ships down. Those boys just don't miss. It was just dark when we reached the coast but we had a hard time finding the home field in the dark."

**"Spam for Victory"**

Bob's next letter to Toots that day mentioned some of the local produce they were able to find and one of the WWII soldiers' least favorite delicacies: Spam. "Tell mom I think of her chow every time they shoot the Spam and Vienna Sausages to us, so that makes it every day. You know that you are going to get one of them during the day but it always keeps you guessing as to which meal it is going to be. Do you remember how I used to like Spam? Well now it turns my stomach to see it on the table. We have a picture up here that shows the 'Final Victory' and it is B-17s bombing a Spam factory. After the war is over they might just as well close up all of the Spam factories.

"I make $318 a month over here but they take out for my insurance, allotment and chow. We spend the rest for food in the tent. Anytime a ship flies to a place where they can buy canned food it looks like a grocery store when it comes in. You see we can't run out and buy a sandwich when we get hungry. If we don't have anything of our own, we just go hungry until chow time. We don't get as much as we used to because they have cut our rations by 12%. We are able to get watermelons now and then from the natives. But they ask a fortune for them. Most of them of any size are $2 to $4. Canned goods are just like gold nuggets. Doc gets us eggs now. He doctors the natives that have stepped on land mines and also some got burned during the fighting. Ever times one rides in on an ass it means a couple of fresh eggs. I am so sorry when I read where you said that Singer's wife wrote Mildred that the latest rumor over here was that we were coming home, because Singer is never coming home and neither is Cull. They were both killed in action. You might know some more of them but I am unable to tell you any more." [Bob is referring to TSgt George Singer, USAAF, 36 TCS, 316 TCG entered service from Ohio. He is listed as Killed in Action on 11 July 1943 and is buried at the American Cemetery in Nettuno, Italy. TSgt Dalton W. Cull; same unit, KIA same date, is buried at the Normandy American Cemetery[33].]

On August 25, the paratroopers and Air Corps men still in North Africa did get one special treat: a USO show direct from the States with some famous entertainers. Bob was able to attend since his men had finally

completed all the repairs on the aircraft damaged in the HUSKY operation. "The last of our shot-up airplanes will be ready for service today some time and so once again we will have 100% in-commission. It is No. 2 and Lt Quisenberry and I test-hopped it this morning. He let me take her off and land. We have a practice mission tonight at 2000 so we have all the ships on the ground for a change to get ready for the mission.

"They had Bob Hope and Frances Langford over at the 505th Paratroopers camp. They had a home-made stage and put it up in a field and used different colored parachutes for the ceiling, sides and backdrop. It made a very pretty stage. Bob and Frances and two men with them have been over here for three weeks putting on shows for the troops. Before they came here they were in Sicily. The show was very good and everyone enjoyed it very much. Frances Langford is one gorgeous creature. She was dressed in shorts and a very brief jacket. Imagine a creature like that out here. I tell you it just about drives you mad."

Bob Hope, Francis Langford and the USO troupe entertain American soliders in North Africa. (US Army Signal Corps)

Some of the 505th paratroopers in attendance decided to go a little mad themselves with a recently arrived box of prophylactics. "E Company were there in force with the contents of their box. They proceeded to blow up the condoms like balloons. Throughout the show large balloon type objects floated across the audience much to the amusement of the paratroopers and to the bewilderment of singing star Frances Langford."[34]

The 505th also planned a demonstration jump right in the middle of the USO show. Paratrooper Bill Tucker was one of the men selected to make the jump. "That particular night Bob Hope and Frances Langford were entertaining the troops below and perhaps our jump was something of a show for their benefit. We overflew the show and (I) could see the lighted platform eight hundred feet below. At any rate, I wheeled into line and jumped lustily out the door, not paying much attention to anything. I gathered in my chute when I hit and tramped on back to sit down and catch the end of the show."[35]

The following day featured another stage show, this time from a French entertainment troupe. Bob was also able to attend this one. "We have another move coming up in a few days. This time we are moving to Sicily but just when we don't know. This afternoon we had another stage show. It looks like they are trying to out-do themselves giving shows for us. I would have liked for some of the people in the States to have seen this show. We brought their equipment in on one of our trucks. Also we pulled the stage in. It was a German flat-top trailer. We pulled it up beside a Nissen hut so the actors could use it as a dressing room. All the area is nothing but silt about three inches deep. They had a beat-up piano and a set of drums and a saxophone. They put this equipment up in front of the stage. The show consisted of a dog act, three girl dancers, a strip tease dancer and two very good girl singers. You have to give them credit, giving the show under such conditions. The wind was blowing about 25 miles an hour and at times you could not see the players on the stage. It was terrible for them. We were sitting on boxes, gasoline cans and trucks. You should have seen us after the show; I can tell you we were filthy. We all talked about seeing a show back home under those conditions. In fact, no

one would even think of putting up with that back home.

"We have a practice mission again tonight. We all think that they are getting ready for Italy this time. We have to transfer one of our ships now so that will only leave us 11. But it looks good to look up at the old status board and see all of the ships marked 'in-commission.' On the 12th of last month I never thought I would ever see it like that again."

Bob then touched on one of the serious issues he was facing as a supervisor – the impact of the stress of operations on his men and himself after so many months at war. This was made worse after their terrible losses in the HUSKY operations. "I bought 12 watermelons last night for Engineering and they cost me $12. After we ate them we had a good bull session just like the old times. I thanked them for their good work and told them we would have to look out for each other a little more. You see, we are all so high strung from being over here so long and wanting to go home, that we fight with each other at the drop of a hat. But we still get along better than any other outfit in the Group, and we are very proud of that. I think I have the best Engineering men that I have ever worked with. They are one fine bunch of men with no 'heels' amongst them. I hope we can all stay together through the rest of this war. I hope we can all go home together. Only one bad feature and it is a very ugly one, which is the men we lost on the invasion. It will always be on my mind. From two to three times a day it hits me for about five minutes, or until I can shake it off. I just can't get them off my mind. Also when there is music and everything is quiet, my thoughts always go to them and Toots. I hope to see her before long."

Like most people, Bob knew the formula for keeping the bad feelings away was to stay busy. And at the end of August, the 316th squadrons were still quite busy, preparing for movement to Sicily and perhaps another airborne operation in Italy. "Yesterday we had orders to transfer No. 9. It was to be replaced by a ship with radar equipment on it. But when No. 9 reached the squadron that it was being transferred to, they refused it because it did not have a door on it. Just about a month ago the same ship was transferred to us from the same squadron and it was short of a door then! So I guess you can only transfer them one way. Today the Wing said they would pick the ships that were to be transferred from both Groups. We lost No. 7 and No. 12. The crew chiefs and men are sure burned up because we had our ships 100% in-commission then they make us transfer two out and give us two in place of them. One is going to take at least three days to put into flying condition. The other is not in too bad a condition. But about as fast as a crew chief gets his ship fixed up like he wants, they make us transfer it.

"TSgt Jones (the crew chief on No. 13) is going home. He broke his arm when they were shot down and just got back to the Squadron. Col McCauley [Group Commander] has gone home† and some of our pilots that were shot down have gone too: Lt Harmon and Lt Bourne, and one radio operator, SSgt Mann.

"Sgt Hershberg and I took a jeep and went to Tunis this afternoon after some parts for the airplanes. After we finished we went into town and I got a haircut, if you want to call it that. If you don't speak French you are just stuck. Then we tried to get something to eat but they only serve food at certain hours, so we could not get anything. We bought a few quarts of wine and started for home. We picked up three RAF boys and gave them a lift. Last night a C-47 was shot down over Palermo, Sicily, but it was not one from our Group. We had a practice mission tonight (August 28). The whole Group flew to Malta, then home. It was to give them practice hitting their drop zones when we go on another mission."

The next letter to Toots spoke of the continuing difficulties with the desert weather and of the new Squadron mascot. "The wind blew terribly today and the sand and silt has covered everything. We had to work all morning because we had two ships out from last night's flying. In the afternoon we took off and went swimming. As usual, we just about killed each other. You should see the kids in Sicily run out to give you grapes when you land. The grapes they raise grow straight up and they tie them to bamboo poles just like we tie up tomatoes back home. They sure do have some swell ones for eating. They use most of them for wine. We just clipped PP Prune, our dog. He is a German wire haired terrier on of the fellows bought in Palestine. He is the only dog in the Squadron so he receives all the attention of everyone."

In their last days in Tunisia, the men took one last opportunity to view the carnage left over from several

---

† Col McCauley, who had been with the Group since it was formed in the U.S., was replaced by Lt Col Burton Fleet.

months of desert war prior to the German surrender. "We had a command car from Group operations this afternoon so Sgts Sanamireno, Pernia, White, Robinson, Vigianno, Lt Quisenberry and I drove over to the hill north of the field where the big battle was. We could only drive halfway up and it was terrible all the way. We started walking after we parked the car and the rocks are enormous and under every one you can see where the Italians and Germans were dug in and there are gun emplacements everywhere. There is a native village about three-quarters of the way up. It is filthy like the rest of the native villages and all of the shacks are made out of stone. We found a dead Italian in the cactus at the foot of a cliff and he had never been buried. His head was about two feet from his body and his scalp and hair was still hanging on the side of his helmet which is close to his head. One of his feet is about three feet from the body with a sock on and the rest of his body is in one piece and smells terrible. There are a lot of graves all the way up the hill and near all of them you can see part of the body sticking out.

"On the very top of the hill there is another native village with one of their temples there also. The top of the hill is about a city block square and it is a straight drop of 300 feet all around with the exception of the path to walk up. There were lots of men on both sides killed when they were trying to take it. After looking over the hill we returned to the car and drove on north on a dirt road. There are still minefields on both sides of the road and barbed wire entanglements. Every little hole or hill has gun emplacements on them. On one hill we went up to where they had five 25-pounder artillery pieces. These guns are plenty large and hard to move around so they had to leave them behind, but they had destroyed or damaged every one of them. Each barrel was split open at the end and peeled back like a banana skin. Every place there is dead donkeys, cattle and horses that were killed during the battle and then not buried. When they build a machine gun emplacement, they put two barbed wire entanglements around them and in between the two entanglements they plant landmines. So about the only way to put them out of action is by bombing or artillery. We all had pistols and plenty of ammunition along so we had target practice using Italian helmets as targets."

*Left*: Desert living, 1943.   *Right:* Jeep of the 37 TCS in Libya.

*Below:* The US Headquarters at Castel Benito airfield, Libya. (All photos: Mike Ingrisano collection)

On August 26, 1943, the 316th TCG was formally transferred from the Northwest Africa Troop Carrier Command to the Twelfth Air Force. Then on the 1st of September 1943, the 316th TCG started their pull-out from North Africa. The destination was Sicily and the Group would eventually be housed on two different airfields. Up until the moment of departure the Air Force command in North Africa was trying to juggle airplanes and groups in advance of what they thought was an impending airborne operation in Italy. The 316th Group Headquarters and the 36th, 44th and 45th Squadrons were to begin the Group's initial movement to Sicily in anticipation of these combat missions. Bob and his men prepared their equipment for the move. "Yesterday we started our move to Sicily. It was the advanced party. They go ahead and fix the latrines and so forth. When the ships came back from taking them over we got all the news from our new home and it was not good. They said it was still 'in the making.' The runway is new, only a little wider than the wingtips, though they are still widening it. Also our parking area is still being made. It is all in grape vineyards. They just take bulldozers and scrapers and go down through the grapes making taxi strips and roads."

The next day was to be their final full day in Tunisia and would ten months of operations for the 316th in North Africa. "Today we are loading all of our equipment and it is a mess. We only have four trucks and everyone wants one at the same time. You see I have the transportation stuck on me as well. To top it all off, Wing called and said we had to transfer two airplanes by 1800 and get two more. Boy, if that don't take the rag off the bush I don't know what would. I am about to go nuts at any minute.

"It is 2000 and we are about all finished but the Mess Tent, and I just sent my men up to take it down. We taxied No. 12 up through the bush to the Mess Tent so that we would not have to haul all the equipment around by truck. The two ships that we get on the trade are probably just able to fly. This makes five different ships in our squadron in less than a week. We just get them fixed up and then we have to transfer them. It looks like we are a Service Squadron and the men are getting plenty tired of it. All of our personal equipment had been loaded so we will have to sleep under the wings of the airplanes."

On the night of August 28-29, a test was conducted using the new Rebecca-Eureka equipped C-47s and the pathfinders near the field where Bob and the 36th TCS were stationed. Sixteen specially equipped aircraft outfitted with the navigation gear were able to pick up the beacon about 20 miles from the practice drop zone and were successful in navigating directly to the spot where the equipment was located. Special lights on the ground, visible from the cockpits of the aircraft were also used as backup navigation aids. On August 30, a second test was conducted, this time using a small team of paratroopers who did jump on the drop zones. The tests were both considered highly successful and there was much relief among the airborne troops that the next operation in Italy, code-named GIANT, would have a much greater chance of success. With that, the Allies firmed up their plans for the next invasion.[36]

Bob's last letter to Toots from North Africa on September 1 contains a fitting closure to their desert adventures. "This will have to hold you for a few days because we will be plenty busy. I told you the other day what McMenamin was looking for well he found it and we are going. I will write as soon as we get set up and describe the country to you. I am sending you some Channel #5 perfume. I had a fellow bring it from Egypt. There were five of us drinking farewell to this desolate hole and hoping the next is better but we hear it is just as bad. It is the first we have had anything to drink in a long long time. It came from Egypt. So tomorrow we take our houses down and put them on our back and look for new lands to conquer."

The actual flag flown by Bob Uhrig and his men over their headquarters tent in various locations throughout North Africa and preserved by his daughter Jan.
The note Bob attached to the flag lists the locations:
"El Adem, Libya; Marble Arch, Libya; Ismalia, Egypt; Fayid, Egypt; Nouvion, Algiers; Guercif, French Morocco; Enfidaville, Tunisia. Retired July 29, 1943."

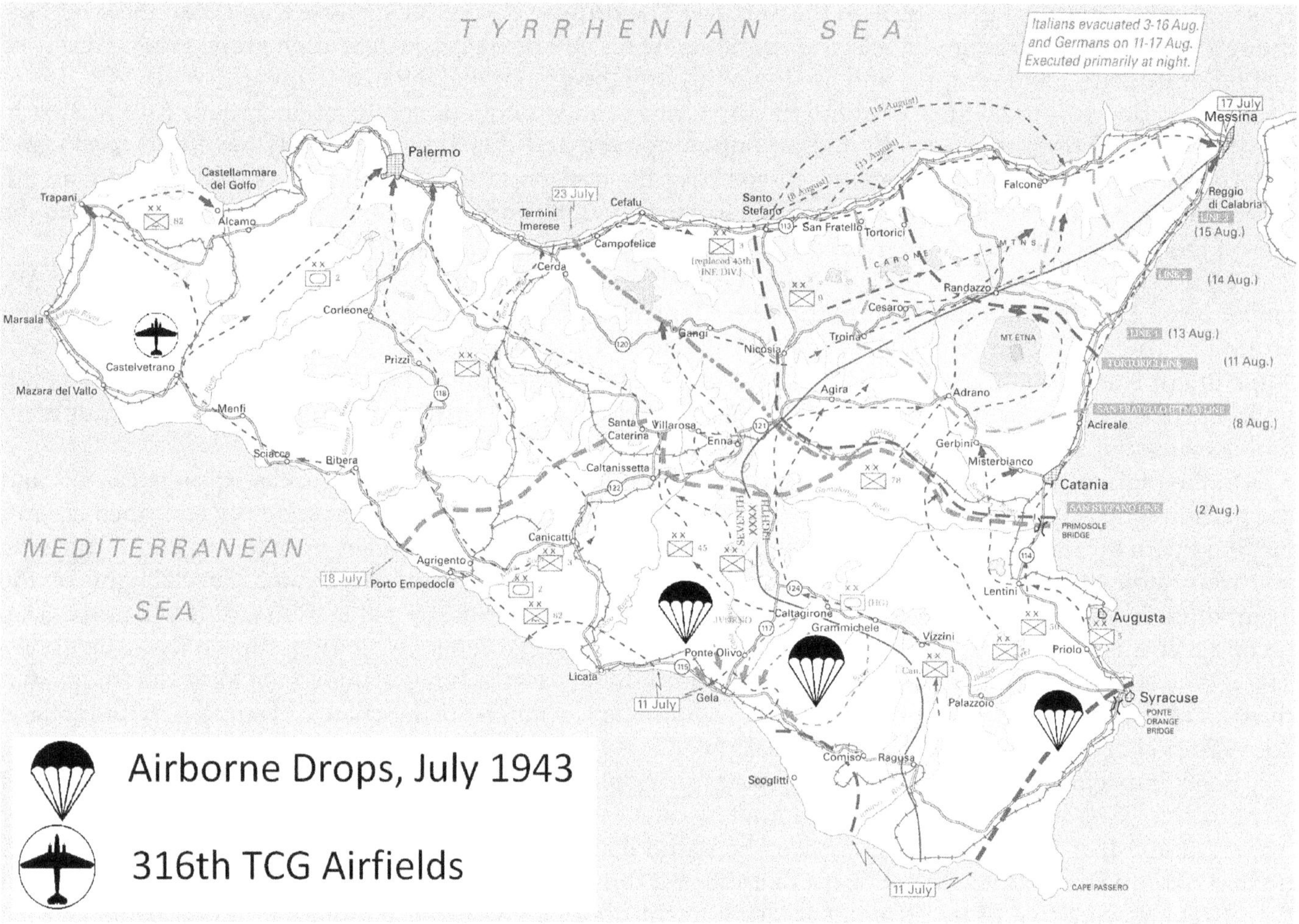

Map of Sicilian Campaign showing locations of the airborne drops for Operation HUSKY and the general location of the 316th TCG airfields after the capture of the island.
(Base map from US Miltary Academy Department of History)

# Chapter 8

## On to Sicily

*The people of this country are the most destitute and God-forgotten people I have ever seen.*
General George S. Patton, Jr., War As I Knew It

The triumphant departure of the 36th TCS for Sicily was announced by the Squadron poultry in Tunisia on September 3. "This morning TSgt Winnie's rooster woke us up this morning by his crowing. He has his three chickens in a box. We put Prune, our dog, in the ship with the chickens and he has a big time barking and trying to bite the chickens."

With their livestock and circus tents secured safely on board their C-47s, "We took off at 0830 and about every ship was overloaded. Some ships had close to 8000 pounds of freight on them. If the people in the States could only see how we load these ships they would be afraid to get in them, let alone fly them. We landed at this new field which is about six miles from Castelvetrano in the west part of Sicily. The field is an old lake bed which now has been farmed. There are wheat fields and plenty of grapes and figs around. But they tell us that where we are the water gets about three feet deep during the rainy season which starts next month.

"We had a road grader scrape out roads for us when we arrived. The runway has just been finished. It is just a clear strip that was made by a scraper through the grapes and other fields. After we landed and unloaded all personnel and their baggage we then taxied the ships to the dispersal area. After all the men's baggage was hauled to the tent area they all helped put up the Mess Tent and then they put up each of their own."

The following day the beddown continued, but by now it was an old routine for the 316th TGC after so many moves in North Africa. "This morning we had to give ten men to the Mess Hall to finish up. The rest of Engineering put up the two Supply Tents, Engineering Tent and Sheet Metal and Welding Tent. In fact, we were all finished and ready for business by 1530. On one side of the field we have grapes and on the other is an old wheat field where we park the airplanes. They all have to be 300 yards from each other so that makes 13 airplanes cover a lot of territory. The ground here is very dry and hard. It also has cracks two to three feet deep and it is so rough it is impossible to drive over.

"Everywhere you look you can see very small stone houses. In these houses is where the people stay when they come out from town to live during the wheat and grape harvests. In one of them down by Engineering, a former bootlegger from Detroit has started selling wine. He is sure a lot of fun to talk to. He told us all about his past and after he had been arrested 1000 times, they deported him. There are lots and lots of his same type on the island. Most of the gangsters come from Sicily and not Italy.

"We are surrounded by hills that are about 1000 feet high. To look at this country from a distance, it looks like farm country in Ohio. The people here are very clean, that is, compared to people in Africa. They look just like any Italian that you would see in the States. They are all friendly and will give you anything they have, which is very little."

On Sunday, September 5, the unit held its first chapel service in their new location. "They had church at the Mess Hall this morning at 1100, so Street, Love and I went. There were only six of us altogether. After the service, the Chaplain gave us Testaments, Prayer Books and War Crosses. It was very different from going to church in the States – dirt floor, Mess Hall benches and a table for a pulpit. With my voice and only six others including the Chaplain to cover it up, all made the singing pretty, yes, pretty raunchy.

"This evening we went through the grape fields and got what we call California grapes. Here you can walk only about three city blocks and find any type of grape you want. There are some as sweet as sugar and some as sour as vinegar. There are also figs that are now ripe at the top of the hill just behind our tent area. Also, when we are working and want some grapes, we just have to walk about ten feet behind the Engineering Tent to get them.

115

"There are a couple of kids that visit us every day and they can now say 'Piss poor' and 'Tucker is queer' in English. They will give you most anything for a couple of cans of rations. They tell us that 60% of the population here is friendly and the rest vice versa, so we have to stay on our toes. We have to keep the windshields up on the jeeps at all times because they string wires across the road. A couple of soldiers have had their heads cut off in that way close by here.‡ You see a lot of Italian soldiers put on civilian clothes when they saw they were going to be captured and most of them are still running loose. Most of the people here hate Mussolini. A lot of them want America to govern Sicily after the war."

One of Bob's first letters to Toots from Sicily mentioned the local characters. "We are still in turmoil but I have to take time out and write to you. We still have lots of excitement and everyone is in high spirits. I don't know and I don't believe anyone else as to how soon we will get to come home. So you will have to hope just like I do. First night here we got some local wine from a bootlegger that spent 11 years in the States and then was deported. Boy he is sure a case and he brings up anything we want and he can get. You should see the beating that equipment gets over here. It seems impossible for it to take what it does and still hold together. One month over here is equal to a year in the States on equipment and also me and the rest of the men."

Unbeknownst to Allied soldiers, there had been ongoing secret negotiations between the Allied governments and Italy prior to the capture of Sicily. The Italian government wanted to get out of the war and there was much back and forth between them and the British and Americans for most of the month of August. The main question was what would be the reaction of the German forces in Italy to an Italian capitulation. One operation that was proposed, but later rejected, was for American paratroopers to drop directly on Rome and capture the capital before the Germans could move in. Uncertain reactions on the Italian part and strong German reserves in the area led to abandonment of that plan. But negotiations to take Italy out of the war continued into early September and several of the other alternative plans involved parachute operations of one type or another. This planning drove the on-going exercises between the Troop Carrier forces and the 82nd Airborne units. These exercises increased even more after the aircraft arrived in Sicily for good.

**Settled in Sicily**

Once the 36th TCS was firmly settled in Sicily, Bob sent Toots a series of letters describing their new home. "I will now try to tell you some things about this country. It is very pretty country. They raise all types of grapes. We have a vineyard just in back of Engineering and everyone is so full of grapes that they can hardly walk. Every night we take a long walk looking for new types of grapes. The best ones are like what we call California grapes only these are twice the size. We sure get choosey about them from having too many to pick from.

Reconnaissance photo of Castelvetrano airfield taken while still in Axis hands. The runway is in the center, surrounded by an oval-shaped perimeter track. Dispersed aircraft parking is visible in the right of the photo. (NARA)

---

‡ This type of "booby trap" was commonly found by Allied forces throughout the European campaign. Eventually, many jeeps had a steel or iron pole attached to the bumper as a wire cutter so the vehicle could be driven with the windshield down.

"The Sicilians are picking the grapes now for the winery. They use one horse and a two wheeled cart with a bed on it that looks like half a large bathtub. They drive down the rows and pick them just like the farmers back home husk standing corn. The grapes over here do not grow on arbors like they do at home. They are a bush type and around the stock close to the ground are the grapes. They are so thick that you cannot even see the stock. This is the way wine making grapes are raised but the eating grapes are tied up to bamboo stakes. About 98% of the grapes are used for wine. Some of them are so sweet that they are like dissolved sugar. In fact, Sicily makes their own sugar from their grapes. Some vineyards have rows of olive trees planted in them others fig trees but most of them are just plain vineyards just as far as you can see. They are in the sides of hills, on the tops and the valleys. It all presents a very beautiful sight. They also raise vegetables and a lot of wheat. The olives now are about the size of the largest ones we ever get at home. I can just picture them in a bottle back home.

"Now I will try to tell you about the people here. They are very short and look just like the people you see at home. One thing though they have not been eating so hot since the war started. They all look half starved. I sure pity the kids they go around with clothes that look like they are nothing but patches and their shoes are terrible. Shoes cost $10 and they are patched up things that you see running around, or with string tied around them or any other way they can make them stay on their feet. There are a couple of kids that herd sheep and they bring their sheep down through our parking area. We have them come into the tent and there are a couple of fellows who can speak Italian very well. The kids live in those very small villages and they walk out to the farms to herd the sheep. They get $4 a month and they have to furnish their own food. We give them food if you call it that and they don't know how or when to stop thanking you. Most of them are 10 to 12 years old and they look like kids in the States that are nine years old. Everyone seems very friendly and act as if they are very glad the Americans are here. How about you getting me a paper and keep it for me about Italy surrendering. Also any other big news that they put out."

**Italy Capitulates**

As Bob pointed out in his letter, after protracted negotiations Italy formally surrendered to the Allies on September 8. The men of the 316th and most other Americans in Sicily had no idea of the complexity of the negotiations, but an airborne drop on Rome was timed to coincide with the surrender announcement. The mission (GIANT II) was cancelled at the last minute as C-47s with paratroopers were already in the air on their way to Italy. When it was clear that Italy was finally out of the war, the focus became the German troops on the Italian mainland and the upcoming seaborne invasion. Airborne planning switched from keeping German forces from Rome to supporting the landing forces on the west coast of Italy.

A few days earlier, on September 6 the 316th had already begun to ferry 82nd troopers from North Africa to Sicily in preparation for the upcoming operations. "Today is Labor Day but it doesn't mean a thing. Yesterday we started hauling paratroopers over here from Africa. Made one trip yesterday with the whole Group and another one this morning. There is sure something big coming off in a very short time. The Eighth Army has already invaded Italy but we all think that it is just to draw Axis troops down to the toe of Italy and then we pull off a big job close to Rome. The British have a pontoon bridge laid across the Straits of Messina and are just pouring equipment across.

"The anti-aircraft gunners moved in today and began setting up their guns. Ever since they came to Sicily, the Germans have blasted them out of their positions. We hope it does not happen to them here. As yet we have not been bothered by any enemy planes. The enemy air force is just about whipped we think because they are on the defensive. For the last week the war news has been very good the world over. We (Street, Crook, Quisenberry, McWhirter and I) went grape hunting tonight. At about 2100 there was an air raid alarm but no enemy planes showed up. They must have gone to another field. The alarm was called off at 0930 and everything was quiet the rest of the night."

Bob's September 7 letter inquired about the American flag he had sent home for safe-keeping. "Never did hear where for sure where Bowen and the rest went but it was to the Pacific and that is one place I would

just as well stay away from. We don't have to live in that damp jungle anyway. Omer is in England so you say. That is very good because all the men I have talked to and were stationed there liked it very much. It is like being in the States. Since Bones is not flying anymore he feels a lot better and shows a few signs of life. We have gone grape picking together for the last three nights. I am so full of grapes that I can hardly look at them. You can find any type or size that you want. Also the figs are ripe. They are very good also. So you received the flag. She is plenty tattered and torn is she not? The old girl could tell some high old talk. When we used to run for the slit trenches she would just stay up there and defy anyone to take a crack at her."

The paratrooper ferrying missions continued the next few days as well. "Today we worked on Nos. 6 and 12. The rest of them left last night for an overnight mission and no one knew what it was all about. It turned out that the mission was just to haul paratroopers over here to Sicily from Tunisia.

"September 8 – Italy Surrenders 1600. The whole Squadron took off this morning to haul paratroopers back here. This island has about all the troops that it can hold without sinking! At about 1430 this afternoon Major McMenamin came over and told us to get ready for a parachute mission in the morning. He gave us a list of things we had to have done to the ships. All the work had to be accomplished by 1000 tomorrow morning.

"We all came back to work at 1800 this evening so we could get the work done on the aircraft. But at about 1850 they called to tell us Italy had surrendered unconditionally and if you don't think everyone went sky high with joy you better think again. About half an hour later Group called and said the mission was off."

The week that followed was filled with confusion for Allied forces. Italy was officially out of the war, but there was uncertainty about the actions of individual Italian military units. And the Germans were still fully in the fight, but it was not known if they would move to fully occupy Rome or keep their troops in place in the south. "All day today (September 9) Italian planes have been flying over. They are landing on all the airports in Sicily. They are then sending them to Africa. It is good to know that Italy is now finished. They say the Italians will assist us in driving the Germans out of Italy. Also we had news today that the Italian navy (what is left of it) is sailing for Allied ports."

**The Next Operation**

The buildup for the invasion of Italy continued. Some type of airborne operation was a certainty, but no one knew just yet what the objectives might be. "We are still hauling paratroopers from Africa over here. They are going to have enough troops on this island before long to sink it. Last night Nos. 8 and 10 were unable to come home. No. 8 had magneto trouble and No. 10 had a flat tail wheel tire. So this morning I flew along to see what was wrong with No. 8. When we landed at Tunis there sat No. 5 with a knot as big as a pumpkin on its landing gear tire. It kept getting bigger all the time and the cords began to break and in a very short time she blew out. So tomorrow we will have to send a main landing gear tire and a pair of jacks over. If things keep up like this we will have airplanes all over the world.

"After fueling our ship and getting the pilots from No. 5, we flew on to Mateur, Tunisia where we picked up the paratroopers. Everywhere you look you can see them in small groups with their equipment. After we arrived back home we found that a stove had blown up and one of the cooks had been hurt badly. Part of the stove blew off and hit him in the leg. As yet they don't know if it is broken or not but it put an awful hole in his leg. Captain Roberts was to fly him over to Palermo so I asked to go along because I wanted to see what was left of our No. 1 since they had salvaged it. The minute we landed I went to see the ship and found that the salvage crew had left a lot of things that we could use. When we came back to our field I asked the Colonel to let me take a few men and go over and get the parts that we want."

Bob and his crew returned to the crash site the next day, September 11. "I took five men and went to Palermo to finish picking over No. 1 which had crashed there. After it had been salvaged what was left had been put into a junk pile and they said we could have anything that was left. There were plenty of things that we needed. About 1530 we caught a ride into town which is about a mile from the field. We only stayed about one and a half hours and then had to start for home. Palermo is the largest town in Sicily. It also has one of

the largest and best harbors. The town is surrounded by mountains running up to 10-12,000 feet. They start at the edge of town and go almost straight up. From the air the town looks like it's in a cove in the mountains as well as the harbor is in a cove."

**The GIANT III Mission to Salerno**

On September 12 the remainder of the 316th Group began the final movement from North Africa to Sicily. "Our ships all took off this morning for Enfidaville to fly the Ground Echelon over here. They started arriving just before noon and from then on everything was a mad house. About 1530 Maj McMenamin called and told us we had to get our ships ready today for a mission, so that caused us to work most of the night."

The reason for the alert was that at that moment, unknown to most of the 316th, German troops had broken through the American lines near the Italian coastal city of Salerno. By the morning of the 13th, the Germans had driven a two-mile wedge in the beachhead line and had advanced to within three miles of the water's edge.[37] The situation was so desperate that Lt Gen Mark Clark, commander of the U.S. 5th Army contacted Maj Gen Ridgway and put his 82nd Airborne on full alert for a jump into this hotly-contested battle front. Clark expected the 82nd and the Troop Carrier forces to be ready for an immediate operation to reinforce the American troops at Salerno. "I realize the time needed to prepare for a drop, but this is an exception. I want you to make a drop within our lines on the beachhead, and I want you to make it tonight. This is a must."[38]

That day Bob and his men knew something was up, but they went about their normal routine. "We are still on alert for the mission but that is all we do know about it. This afternoon they had a USO show for us. They had three girls and one Master of Ceremonies. It was a pretty good show. Italian planes are still coming in. This afternoon about 60 Italian pursuits (fighters) flew over and 24 transports. It gives you a funny feeling to look up and see enemy aircraft flying overhead in squadrons and nobody bothering them."

At the same time the 316th was enjoying the USO show; airborne planners on Sicily were huddling with the Troop Carrier units to devise a plan to reinforce the Salerno salient. The operation (code-named GIANT III) that resulted had units of the 504th Parachute Infantry dropping on the night of September 13th from aircraft of the 61st, 313th and 314th Groups. The 316th would be called to join the action on the following night to drop the remainder of the paratroops from the 505th PIR and airborne engineers.

Bad memories from the Sicily jump were still fresh in their minds as the 36th men prepared for the mission on September 14. "I am Officer of the Day today and I am a tough one too. This afternoon at 1600 they called us and told us to get our ships ready for a mission. We were pulling two 100 hour inspections so we had to throw the airplanes back together. At 1830 the pilots were briefed and then at 2000 they reported to the flight line. We had coffee for the crew to fill their thermos flasks and drink before they took off. This was done in the lights of a jeep and I took a couple of pictures of it. It gives you a terrible feeling to know they are going on a mission and maybe some of them are not coming back. Everyone was kidding each other about not coming back and what each other could have if he did not come back. But down deep there was a different feeling. I have just about all their money here in the desk and other small things they wanted me to keep for them.

"They left for their ships at 2045 and they took off at 2100. They are going to another field to pick up the paratroopers and then take off at 2300. They are going to drop them just south of Naples where the Germans are giving the Allies a lot of trouble. We received news tonight that the Germans have pushed the Allies back five miles and that is not good. The fellows on the mission tonight from Engineering are: No 1. Robinson; No.2 Johnson; No. 3 Kestner; No. 5 Boggs; No. 6 Middendorf; No. 7 Ridgeway; No. 8 Kawohl; No. 10 Fenimore; No. 11 McDonald; No. 12; Winnie and No. 00 Richey. Young and McPherson had to go as crew chiefs on a couple of ships that Group borrowed. They did not like the idea because their own ships, Nos. 3 and 4, are sitting here ready to go. Now we must sit here until 0430 to see what the outcome will be."

But the GIANT III missions would not be a repeat of the Sicily disaster. On the contrary, the weeks between the two operations allowed for increased training of the Troop Carrier crews and the equipping of the C-47s with much-needed precision navigation gear. Bob was relieved that the operation was successful. "All our ships

returned this morning at 0530. We were all very pleased that there was very little fire and none of our ships were hit. The 5th Army is really getting hell and our mission last night was to reinforce them. There were eight Transport Groups that took part. The 5th Army has a beach head just south of Naples but it is surrounded by mountains. The Germans must have known about the invasion because the mountains are alive with large guns which are playing hell with the 5th Army which is down in the valley. Some paratroopers were dropped behind the mountains so they can come in from the rear and wipe out some of the gun positions."

**The Pathfinders**

The Troop Carrier Groups and airborne forces participating in the GIANT III operation debuted the concept of the "pathfinders" to avoid the confusion experienced on the HUSKY operations. Three planes carrying pathfinder teams equipped with radios, Krypton lamps and Rebecca/Eureka radar direction finders preceded the main body of aircraft to mark the drop zones. In addition, ground troops on the beachhead put out burning cans of gasoline in the shape of a "T" to further light the way to the drop zones. The pathfinders dropped squarely on the designated drop zone and within minutes had radio communications with the approaching C-47s. These preparations, along with a jump altitude of 600 feet resulted in accuracy in the jumps never before achieved. No aircraft were lost and by the morning following each night's jump, the paratroopers had assembled and were in the front lines. The arrival of these airborne troops, literally in the nick of time, saved the American beachhead at Salerno from being overrun.

In spite of the hastily-planned operation, Troop Carrier Command had done outstanding work and had demonstrated what properly equipped and trained airlift and airborne forces could do, even on short notice. The lessons from Sicily had been learned and learned hard. The success of the GIANT III operation had shown what was possible, now that the concept of airborne warfare had matured.

**A Crisis is Averted**

With the front line in the American sector of Italy now stabilized, the 316th turned its attention to helping their old British comrades, the 8th Army, link the American and British forces together. By September 16, the two Allied fronts were less than 60 miles apart and if the British could close the gap, it would help ensure a continuous Allied advance up the Italian peninsula. On that day, Bob recorded: "Nine of the ships took off this morning to haul equipment to the 8th Army. They are advancing 50 miles a day and now have the entire 'foot' of Italy. Everyone is hoping they reach the 5th Army in time to save them."

Once the frantic pace of resupply operations was over, Bob had time to write to Toots. "Well no mail as yet but it will come in a gob before long. Went to town yesterday afternoon and came back at 1900. Well I don't see what the Italians over here were fighting over because there is nothing here but grapes and wine. The people have less than nothing. It looks funny to see Italian soldiers walking on the streets with their uniforms and guns on. They are sure happy because the war is over as far as they are concerned. All the buildings are built out of stone and are all connected together, so for drainage the roofs must slope in all different angles. All the entrances are right on the street and the doors are just like barn doors back home only a little worse. No fooling, we have much better doors on our barns. They all have large padlocks on them. Most of the people are clean and their clothing and table covers and such are plenty clean. But the buildings and how they are laid out just defeats you. No matter where you go some kid is holding a piece of paper in front of you with 'spaghetti, eggs and chicken' written on it. There are no restaurants to speak of so the Italians have meals at their homes. And the food is very good and everything is so very clean. But there are hundreds of street urchins and beggars and they do not form a very good picture. I want to buy you something but there is just nothing here. Some of the fellows have found table covers but when they do find them, the people only have one. McMenamin said the order had been cancelled about coming home. I keep telling you about getting all worked up about such stuff. I don't care who they are, no one knows about coming home too much in advance."

Street scenes in Sicily taken by Bob in late 1943.

*Left:* Bob with one of the local residents.

*Right:* Postcard Bob sent home to Toots. The censor had even clipped the caption "Sicily" from the bottom of the card - as if that would fool either the Italians who lived there or the Germans who had just vacated the island.

Bob and the men also found time for some exploring in the Sicilian countryside. "We visited the catacombs today. They are long underground passages where they used to put the dead. They stopped this type of burial in 1885. Some of the bodies date back to 1600. After a person died, they would have them mummified. This would take a year to do, then they would bring them to the catacombs and here they were stood up in little alcoves along the wall or laid on small ledges. Some sitting up, others in wooden caskets. All have names and dates on them. It is a horrible looking place because of the bodies everywhere looking at you. Some are preserved very well and still have most of their features. Most of them have hair still on them because skin is still on most of the bodies. In some places bones show through the skin. Bombs have blown two large holes in the catacombs and have strewn bodies all over the place. There are over 18,000 people buried in the one set of tunnels we were in.

"In the evening, Cull was at our tent and so we had accordion music until about 0030. The operator left the telephone hooked up to Engineering so that Love, Tucker, and Miliman could hear the music. We also cooked bacon and eggs. Eggs are getting very hard to come by because the Germans ate most of the chickens when they went through here."

On Saturday, September 18th, the exploring continued. "Well today Toots is as old as I am (29). I hope she has a happy birthday. This afternoon I took Tisdale, Johnson, Wallace, Robinson, Tomcheck and Friend to Palermo to get some more parts off of crashed aircraft No. 1. The pilots, Lt Quisenberry and Riggs, also the crew chief Fenimore and I went to town. We went to the docks to see if any of the American boats had cokes or anything else to drink, but none of the ones docked had any. I bought Toots a piece of lace that is pretty nice. I tell you, these people have less than nothing.

"The capital building is beautiful but bomb fragments have broken all the windows and part of the building has been destroyed. On the waterfront there is nothing left standing. Most of the buildings are about five stories tall and when one of the big bombs hits, it goes clear to the basement. You can walk along and look right through a house on the top floors and on the rest you can see things still hanging on the walls and furniture that was left sitting on a section of floor that was not destroyed. Just picture a large ragged edged knife cutting a house in two or a corner off of it and you have a bomb-hit house. But the houses on the waterfront are leveled to the ground."

Bob now had a little more time for writing Toots about their living situation. "We had a picture show last night and it is only the second one that we have had since we have been over here. It was *The Philadelphia Story*. It has been so long since I saw it that I had forgotten most of it so it made it even better. We have been making coffee every evening and some nights we were able to have eggs, but they are very scarce because the Germans took all the chickens. The grapes are just about gone now and we don't know what to do about it. It is becoming a problem since we depend on grapes for a snack every morning and afternoon. I guess we will have to drink wine because that is where all the grapes are."

"This morning we had creamery butter for breakfast. This is the second time since we have been overseas that we have had it. I ate my share of it and was it ever good. We don't have much dust here so we are all able to breathe again which is something new for us."

On September 25 the 36th became involved in a rescue mission at sea. "One of our planes on their way to Africa sighted a raft with five men on it. They could see where the ship cracked up. They circled for two and a half hours and then a British bomber came out and took up the circling and our plane went on to Africa."

The next day was Sunday, and at services that morning, many of the men reflected on the comrades they had lost in the Sicily airborne operations. "This morning at church I asked the Chaplain if he knew where the men were buried and he told me he did, so I asked the CO for a ship to take us to Gela, and got it.

"So this afternoon 25 officers and men climbed into No. 7 and we took off for the 45 minute flight to Gela. The cemetery is about a half a mile from the Ponte Olivo airfield. We chiseled a truck to take us to the cemetery. There are over 1700 men buried there and they are bringing in more every day. We have never had any trace of No. 10 which Jackson, Loffredo and Cull§ were flying, but yesterday we found Jackson's grave. He was found floating about two miles out to sea about a week ago. So they must have cracked up at sea and sank before anyone could get out. We found Singer's grave and most of the other men. The cemetery is located on the side of a hill with an American flag at the entrance. It is surrounded by a barbed wire fence. The graves are laid out in plots with crushed stone all around the graves. The gravesites themselves are dirt and each one has a cross at the head. They are fixing up the cemetery about the best way they can and when it is finished it will be beautiful. It overlooks a beautiful valley and is located close to where most of them were killed."

The activity in Sicily began to slacken as Allied troops moved up the Italian peninsula. The 316th had now more or less settled into a routine at their new location. "Today (Sept 30) was payday and they flew to Licata in southern Sicily to pick up the money. We got paid at 1700. Then Quisenberry and I flew to Sciacca in southwest Sicily to check on a C-47 that cracked up. They told us that the 329th Service Group was going to salvage it. So now we will have to go to Licata with our paperwork filled out for the parts we need. White made Warrant Officer the day before yesterday. He is at the rest camp and does not know anything about it yet. He is supposed to come home tomorrow."

Bob now had some time again to update Toots on the local events and colorful characters. "Master Sergeant White finally made warrant officer in our squadron. He will fill the vacancy that I made. As yet Stracke has not made it but he should in a short time. One of the pilots in our squadron brought home four canaries the other day so now we should have some singing. Another fellow took a cage away from a peddler and proceeded to turn all the birds loose. After he turned them loose he paid for them all. I think about half the population of this island has been deported from the states at one time or another. Every place you go you find someone that spent some time in the States. I have talked to a lot of them that belonged to Capone's gang and many others."

---

§ Flt Off Elmer Jackson, Flt Off James Loffredo and TSgt. Dalton Cull were lost on July 11, 1943.

## Italian Weather

By October 2, the Group was getting a taste of the rough weather that was bogging down combat operations in Italy. "This morning at about 0600 the wind started blowing. It was the worst wind storm that I have seen. It got under my air mattress and would lift me right off my bunk. Then after the wind had torn down half the tents, it started raining. The rain beat right through the tents that were left standing. All our beds got good and wet and everything was a mess. The Mess Tent blew down and the men had to put it up again in the mud and rain.

"At about 1000 they decided to get the airplanes off the field (Mazzara airfield) before they sank out of sight. We flew them all over to Castelvetrano airport where they have a hard-surfaced runway. No. 4 had to stay here because of maintenance. Other Groups were moving there, too. The field was full of airplanes. The crews took rations and their bedrolls because they are going to operate out of there until this field dries off. This field is a mess and the runway is cut up terribly from our ships taking off in the mud. It ended up raining most of the day so we stayed in the Engineering Tent and played poker."

The rainy weather continued into the next day and made the field almost unusable. "No. 8 landed here today and when he pulled off the runway the ship went down like a rock. It took us all morning to get it out again."

The conditions were so bad Bob complained to Toots in one of his next letters. "The rain kept us in the Engineering tent all day yesterday. The mud here is terrible. We will now be able to compare the sand with the mud. Had to wring the water out of my blankets yesterday morning. The wind blew so hard that is beat the rain right through and my bunk got plenty wet. It is a miserable feeling. Last Sunday we flew to the cemetery where Singer and the other 22 are buried. They have all been moved to an American Cemetery that is being built. It is in a beautiful location. It is on the side of a hill overlooking a beautiful valley. When it is finished it will be a nice cemetery."

*Left and Center:* Mess facilities in Sicily; *Right*: Living quarters - not greatly improved from North Africa. (Mike Ingrisano collection)

By the 5th the weather had cleared enough for normal operations to resume. "All our ships started landing here again today because the field has dried enough to hold them up. About all of them needed maintenance, so it looks like we will have plenty of work to keep us busy tomorrow. Tonight at 1700 they had a stage show for us. They had Anna Lee, Adolph Menjou and a cast of special service men. It was a pretty good show but we have had a lot better ones. But beggars can't be choosers."

Unfortunately, the heavy weather cycle started again on the 7th, almost a carbon copy of the days before. "It started raining again today so we had to fly the ships over to Castelvetrano airport again. This time, we sent TSgt Crook, Sgt Robinson, Sgt McDonald and Sgt Johnson along to take care of minor maintenance. Every time they need parts for the ships over there, they call us her and we haul them over in the jeep." The next day, "Sgts Love, Comitz, Brown and I drove over. After the maintenance was completed we drove on into town and had an Italian dinner."

The situation at Mazzara had still not cleared up by October 9. "It has not been raining since our ships left so it is now dry enough for them to land here again. So all of our ships are coming back here tonight. It looks very much like rain though and it would not surprise me if we had to get up in the night and get the ships off the field again. It gets to be old stuff moving every time it rains. The first ship to land here today pulled off the runway and we don't have it out of the mud yet."

Bob described the miserable conditions to Toots. "Just place yourself out on someone's farm and set up camp and then let it start raining and you have it. No roads here but the ones you make yourself and then when it rains, the bottoms go out and you are in a mess. So, the people back there don't know there is a war on. I wish to hell they were over here where they could eat these rations then they would not have time to piss about the ration points they have to use."

## Maintenance Difficulties

October 10 saw a rash of mishaps in the unit that the maintenance crews had to address. "Last night No. 4's crew chief Sgt McPherson came back from Palermo on another ship and brought his aileron and wing tip along. The pilot of No. 4, Captain Roberts, ran through a fence and damaged the aileron beyond repair. The wing tip has a slit in it but we can repair that. This makes three damaged ailerons this week.

"This morning we have a spark plug change in Nos. 8 and 11 and a 100-hour inspection on No. 7. Also we have to pull up the floor on No. 9 and have a look at the cables because an Englishman shot a flare pistol inside the airplane the night before and it burned two holes in the floor."

To cap off the excitement, "I called MSgt Lafflen about an aileron this morning and he told me that compared to his unit, I did not have much trouble at all. It seems one of his ships ran through a flock of Spitfire fighters and now the C-47 needs two new wings and elevators. Also they had one of their ships shot down in Italy yesterday but they don't know yet how bad it is damaged. Also the ship that went to the rest camp in Morocco to take men down and bring back Col Garland and the rest of the men has been missing over a week now."

But the rest of the day was a series of ups and downs. "This afternoon, just after lunch, No. 53 came in with Col Garland and the rest of the men, so that takes a lot off of our minds. I flew to Gela in No. 6 to take glider pilots to their new station. On the way back we stopped at Agrigento and Sciacca trying to find an aileron, but without success. I got to make one landing and one takeoff. When I returned I found that No. 10 had broken two actuating arms on the flaps."

Bob ended that eventful day with a letter to Toots. "We had a movie here last night named *Million Dollar Baby* it was not the best I have see but it will do.¶ We won't gripe as long as we get to see anything at all in the line of movies. The mail has been coming through much better now. I hope it continues the same. It makes me feel good to get mail from you. A kid is grazing his sheep out in front of the Engineering tent this morning. They have pretty large flocks of sheep. Since White made Warrant Officer I have been taking it a lot easier. Love is Line Chief now and he is plenty good so he takes a lot of my worries away. I can depend on him more than anyone else. It sure helps to be able to get to your tent and know that everything is going to be taken care of."

Aircraft repairs continued as other men had an opportunity to take some leave at the rest camp in Morocco. "Sent Miliman on the ration run to Palermo with No. 4's wingtip which we have just finished repairing. He is going to try to find an aileron. Major McMenamin was over and he has just returned from the rest camp which he tells us they are going to close because it is getting too cold."

---

¶ One of the stars of the film was future U.S. President Ronald Reagan.

## New Commanders

October 13, 1943 was a change-of-command day for the 316th. The 36th Squadron Commander, Lt Col Garland was transferred to higher headquarters. His replacement as 36th Squadron Commander was Captain Jim Farris. "Col Garland gave us a farewell speech last night and he could hardly keep from crying. I know just how I would feel if I had to leave the Squadron.

"We now have all the wing tips and ailerons repaired and all of our ships are flying again. It's 0945 and Col Garland just took off for his new home and we all feel a little let down from losing him after all these months with us. I have been serving with him ever since he graduated from flying school."

Bob wrote to Toots about the changes in the unit. He was careful not to make any judgments on how things would work under the new leadership since Jim Farris had been his roommate for many months. "Colonel Garland has been transferred and now Captain Farris is squadron commander. I will tell you in the near future how it is going to work out. Our whole squadron flew low over his tent this morning before he left. He gave us a talk last night and he sure does hate to leave. Stracke made Warrant Officer yesterday and he is very pleased about it. I am sure glad he made it because he sure deserves it if anyone does because he sure knows his job. He has been doing very well over here. We are about to finish winterizing our tent so we can keep half way warm this winter. About the time we finish they will move us and then we will have to do it all over again. I am having you a scarf made out of a German equipment parachute. They don't use such hot material in their equipment chutes so don't expect too much. It is bright red so you can wear it around home."

With new leadership in place, the Squadron had a fairly steady list of logistics missions to carry out, but these did not come with the excitement of the airborne operations. "Yesterday (October 14) Love, Comitz, Lt Oliver and I went to Castelvetrano for parts so we had chow there before we came on home. We have been hauling supplies up to the front in Italy for the past month just like we used to do in Libya.

"We have two mice that live in our tent with us and we have named them Pete (the small one) and Mike (the big one.) They chase each other all the time and the only time that we can scare them is when we make a movement towards them. We can holler at them all the time and they just look at us. In the morning just after it gets light they start really going to town to wake us up."

Bob's tent-mates tried to engineer a solution to the mouse problem. "Last night Farris and Coursen came back from town and brought a cat with them to catch Pete and Mike, but by the size of the cat I don't know who is going to chase who out of the tent."

"Today we had an excuse to go to town after supplies, so Hershberg, McDaniels and I went in and had an Italian lunch. I also had my uniform pressed to go to a dance. It was sponsored by the Group officers. A woman that was born in the States made the arrangements for the women. Most all of them brought someone from their family along. I had a very good time, but one bad feature it rained all night and the roads were terrible. The jeep top kept off very little water."

Bob wrote Toots about the dance, but reminded her she was the only girl for him. "McMenamin invited me to a dance Saturday night. The Group officers hired an orchestra and they rented a big shack. No, don't worry about me. I danced two dances and that is all. Furthermore to relieve your mind each of the girls had some piece of her family along. We now have three inches of mud in our Engineering tent. It seems that we did not trench in front of the door so the water decided to move in during the night and now we have a lovely mess. Mac and I drove into town yesterday and had some spaghetti and eggs to eat. Also a very small quantity of wine for the stomach's sake."

Naples had been captured by the Allies in early October. They advanced slightly north of the city, but became bogged down in Italy's mountainous terrain. The front became more or less static, and there were no further airborne operations planned for the Mediterranean theater. By mid October it looked like the 316th would stay in place for a while so Bob and the men prepared for the colder months ahead in Sicily. "The grapes are all gone but we have some of the largest pomegranates and persimmons. Since it is getting much

colder, we put up a new Engineering tent that we can make weather proof. It is a pretty fancy job and we have it fixed up very nice. We have come a long way from eight barrels and a piece of canvas in the desert to our present set up with all our files and equipment. One of our engineering clerk's dogs had five pups night before last. We don't know if they belong to one of those Italian dogs or an African job. That makes our roster two large dogs and five pups in engineering now. Col Garland came in to visit us day before yesterday and he is very pleased with his new job. Bones should be a master sergeant by the first of November. He will then be a flight chief. By White making warrant officer that made another vacancy so we are going to promote Crook."

The lousy Italian weather carried through almost the whole month of October. "We are bringing our ships back from Castelvetrano this morning. Nos. 10 and 11 are due for 100 hour inspections. When they were bringing in the ships, No.3 got stuck clear up to the axle. This makes the third time in the past week for No. 3."

## The 36th Moves to Castelvetrano

Perhaps due to the substandard operating conditions at Mazzara, the 316th Group made the decision to move the 36th TCS from Mazzara to Castelvetrano to join the other Squadrons and consolidate the Group on a single airfield with a hard-surfaced runway. The 316th was sharing the field with the C-47s of the 314th TCG. The threat of air attack had also diminished so there was not as great a need to keep the Group aircraft dispersed over several different flying fields. The move would also greatly simplify the logistics of keeping the C-47s flying, as Bob and his men would soon happily discover.

The living conditions at Castelvetrano were also a significant step up from their previous locations. Bob's diary for October 24 records the move: "Today we moved to Castelvetrano airport. The men are moving into what used to be a hospital. It is very new and it is going to make nice quarters. The buildings are built close together and are long. Each barracks houses 80 men and there are 5 of them. We rented a hotel up in the town but that has only room for 28 officers, so the rest of them have to live in the barracks. I am in a room with four other officers. The room is a very large one and we have a large mirror. The only bad feature about the whole thing is the water which is only turned on from 0700 to 1330. During that time I am never there so we still have to keep water in cans to wash with."

Once settled at yet another base for yet another unknown duration, Bob wrote to Toots. "Today we are just finishing putting up the Engineering tent and starting operations. Yesterday it rained and we had to put up tents in the rain and mud and I tell you it was anything but pleasant, in fact it was just one big mess. We are living in a small town close to the field and so we have a roof over our heads once again. In almost twelve months we have only been in buildings six weeks. So living in buildings is quite a treat to us. But as usual, Engineering is in tents, but it is not bad because we fix things up pretty nice. Our water situation is like always, we have to haul it to where we live. The officers are living in a hotel if you want to call it that and we have to haul water there also."

This particular move was less disruptive than previous similar events, particularly since the 36th was moving from a remote location to a more fully-equipped field. So the settling in process was much easier and faster. But the end of October, the Squadron had completed the move. "We were all given today off (October 30). So in the afternoon Doc Boyer, Captain Chiros and I took our cleaning in to town and then went to eat. We shot the bull with an Italian family for about two hours and then came back to the hotel. In the evening they had a dance which was nothing to talk about so I went to bed."

## 316th Has Its Own Insignia

Before retiring on evening, Bob had a chance to write to Toots and give her some news about an important event for the Group – they had now designed their own distinctive insignia. "We just finished having an inspection and Col Fleet complemented us on our neat appearance. We now have a Group insignia. It has

nothing that ever connects it with Troop Carriers, but it is still our insignia and we are all having it painted on our jackets. The picture [drawing in letter] is just a faint idea of what it looks like. At the bottom it says, *Nobilis est ira Leonis*. Noble is as the Lion's wrath (we think.) The artist who paints them for us has been given many awards for his paintings in Venice. You should see some of his paintings. It is almost unbelievable because they are so natural. From a distance they look like photographs. I am having a large one made for you. If he fixes it like I told him to it should be a beauty. You should be able to use it for a table piece if you want. If there is anything you would want me to have painted for you just write and tell me. Well, since Marcum is overseas the war should be over in a short time, in fact I should start packing my bags. The stuff will sure fly now. I want to hear his stories when I get home. They should be the best in the west." Bob included a small sketch of the insignia on his letter of October 30. The blue and yellow insignia also appears on the front cover of the 316th yearbook that was produced at the end of the war.

The sketch of the new insignia Bob sent in his letter to Toots.

The next day was a sad one for Bob. "I received a letter from Toots telling me that Lagana had been killed in New Guinea on 20th September 1943. I can tell you now it just about knocked me off my feet. He was my best friend and I know how much I will miss him in the future." MSgt Charlie Lagana was still with a C-47 unit in the Pacific. Charlie and Bob had been friends ever since their days as junior enlisted at Patterson Field. They endured the long deployment to Alaska together and looked out for one another. The war had separated them and they ended up serving half a world away from each other. So in addition to the friends Bob lost from the 316th, he had now lost another one in the Pacific Theater.

Bob wrote back to Toots that he was still in shock over the news. "We were off yesterday so we walked around town and looked the place over. It is a pretty nice town but they don't have anything to sell because they just don't have anything themselves. Your letter about Lagana knocked me off my feet and I am still not over it. I can hardly make myself believe it. I liked him best of all and was looking forward to seeing him after this mess is all over."

The early part of November 43 was quiet for the 36th. The men were coping well with operations from their new location. "All of our ships but No. 12 are flying today. We have been servicing them with fuel from 5 gallon cans, so tonight when they all come in we will have one sweet job. We service them with oil from one quart cans so that is just as bad. We transferred No. 10 this morning. The crew is taking it to Oran to turn it over to the depot. Group sent us a new C-47A to replace it. The new ship is a honey with all the new gadgets on it. We are going to let Fenimore crew it."

Thankfully mail delivery was now back on a regular schedule so Bob could correspond more regularly with home. "Received a letter from that lame-brain brother of yours [Omer Fultz] yesterday. He seems to be doing OK. I see by his letter that he is now a PFC. He will soon be a big shot. We get mail just about every day now. It makes one feel so good to received mail. We still have our mud and I think she is here to stay of the winter, but it is not near as bad as that last place. Where we were last used to be a lake bed and when it rained, well, draw your own conclusions. I won't be eligible for captain until December 10 and that just means they can put your name in, but as far as anything being done about it is something else. Ratings like that don't grow on trees."

To pass the time, Bob and the others had been playing a lot of poker and Bob himself had won almost $200 over the course of week – serious money for 1943. Captain Hamilton left the outfit and so Bob changed

rooms to take his place and he described his new set-up to Toots, including efforts to build an Officer's Club. "Hamilton moved out so Major Wannamaker had me move in with him. There are only two of us to a room and we are fixing it up real nice. It is the best deal we have had yet, but we are afraid it is too terribly, terribly good to last. Both the officers and enlisted men are going to have a club. Each man put in ten dollars to buy furniture and we are getting some very nice looking furniture. We have a radio and a record player and reading lamp. We also have a very good looking bar and we each take our turns tending it. Everyone is much happier here than anyplace they have been yet. In fact, this is half like living again. If we could go out and find a decent restaurant everything would be OK. But here you go out and eat spaghetti and eggs or else you don't go out and eat."

To supplement their canned rations, the men tried to get out and bag some local game whenever possible. A group from the 36th decided to go off on a hunt on November 8, and Bob was one of them. "It rains about every other day and this morning was one of the days it rained. I was up at 0430 to go duck hunting. MSgt Toach is in the 314th TCG here at the field and he used to be in the 5th Transport Sq. back at Patterson Field. He made all the arrangements for the hunting trip. We picked up three Italians and their dogs and started out. There were four of us Americans. The Italians got the shotguns for us and the ammunition. The ammo is very hard to find so most of them reload their own shells. These Italian shotgun shells are loaded so light that the shot can just about make it out the end of the barrel.

"We started out in two jeeps and had to drive about ten miles. All the Italians were telling us about how good their dogs were. After we arrived at the ponds and swamp we turned the dogs loose. You have to wait until the ducks fly over the land to shoot. There is no way to get one out of the water if you shoot one down. We killed one duck and he fell into the swamp where we could not get to it. We began to make cracks about the dogs and talking about how it would be if we had a couple of our hounds over here. The Italians never knew what we were talking about unless we told Comitz to tell them. We laughed like we were silly and it was pouring down with rain the whole time. We had a lot of fun even if we didn't get anything.

"At noon we went to a barn to eat. The Italians had brought bread, cheese and wine along. We thought the cheese was going to get up and walk off and leave us. All in all I laughed so much during the day that I was weak."

By mid November the 316th Group was fast approaching their one-year anniversary of overseas deployment. They had endured frequent moves and been saddened by losses but their recent missions had switched from combat to routine resupply and logistics flights. There was a lot more war ahead for them, but for now they took the opportunity to relax and blow off some steam. "We are getting ready to give a big party on Monday 15th November to celebrate one year overseas. They bought three steers yesterday to barbecue. We have a cook here in Engineering that used to run pit barbecues in Kentucky so he is going to run this one. We dug the pit this morning and we are now all ready."

In his November 11 letter to Toots, Bob reflected on his year overseas and on how the war had changed relationships among people. "I am still not over that duck hunting deal. I laughed until I thought I could not stand up. We all had a sweet time. The Italians kept raving about their dogs but we still don't believe that they can smell their own bottom. We couldn't say anything about them and they could not tell what we were saying. I would like to show them some good hounds like we have back in the states. Major McMenamin and I have bought an Italian radio and it is a very good one. It is small and has a very good short wave reception. I have it on now. You know how much enjoyment that I get out of radio, well, I have not changed.

"I now am listening to a German broadcast. It is a very funny thing over here about broadcasting; the Germans always have the best programs. Every night they have one and always play all American pieces. The whole program is in English. They announce every once in a while the following, 'This program is coming to you through the courtesy of your enemy.' As you sit here and listen to them, you think about how foolish it is to be killing each other. You see the Italian soldiers everywhere and you stop and think that if you had seen them two months ago you would have killed them and now they are your friends. It will probably be the same with the Germans when they are beaten. Put us all in a room and take our clothes off and no one would know who to try to kill."

## Group Anniversary

November 15th dawned as a day of celebrations – not only for the Group anniversary, but for individuals as well. "Captain Farris made Major today. Also, the men in Engineering beat the daylights out of me because I am 29 years old. We had our barbecue tonight and it was a great success. The meat was very good and everyone had plenty to drink. Most of the Group officers came down because no one else was having a party. The enlisted men came over to the officer's bar and the officers went to the enlisted men's bar and drank. It was the best party ever given in the Group. Pappy Street and TSgt Wallace and some others that never drink got stinking drunk and I laughed until I hurt all over. I got to help put Street to bed and that did me more good than anything. I will never let the old fart live it down. Most of us had tears in our eyes when we drank to the men that were missing. You don't know what a feeling that is. Colonel Fleet was very much pleased with what we have done towards fixing everything up. Mac and I were together most all the time and we talked a lot about what we used to do at Patterson Field and also about all the fellows that we knew there and what they are doing now."

The next day was dedicated to recovery. "Everyone is off and it is a good thing because I don't believe many of them would have made it anyway. At group staff meeting Col Fleet told the rest of the squadrons to come down to the 36th and see what we have done for the enlisted men. Major McMenamin and MSgt Santomissino were here to have Group inspection. They said everything was satisfactory. Talk about mass production, they told us the Group had a flag flying for two weeks before somebody noticed that it only had 42 stars, there was one row of stars missing!

"We have new crews coming in every day now so it looks like some of the older men are going to get to go home. All enlisted men and officers that have over 800 flying hours have taken a physical in anticipation of going home. Kawohl was relieved as crew chief of No. 8. He has been showing less and less interest in his ship so I just had to relieve him."

The fighting in Italy had become stalled just north of Naples by mid-November 1943. This position would become known as the "Winter Line" and the front would be static there for many weeks. Stationary armies require far fewer supplies than armies on the move (as in North Africa) so there was little need for airlift resupply operations. German planes had been pulled back up the Italian peninsula or all the way back to Germany, so the threat of air attacks in Sicily was greatly diminished. All this combined to make the Troop Carrier units still in Sicily feel that the war had now passed them by.

Bob described to Toots some of the ways the men were keeping themselves busy on the quiet days at Castelvetrano. "It is still plenty damp here and I have a honey of a cold. It takes so long to get over one after you get it. The radio that we have is wonderful. I get more out of it than anything else. I remodeled my pistol holster because too many men out here on the line have been out-drawing me. I am now pretty good; I can draw the pistol and cock it all in one motion. We are getting our place here at engineering fixed up like big time. We have a small Engineering office fixed up now also good work shops and supply tent. Everything is in tents but we have it fixed as good as if we had it in buildings. What makes it feel even better is to know that most of the equipment is home-made. The inspector from command told us yesterday we had the neatest and best setup in the command and that alone is worth working the past year for.

"We try to do good work and you know how I am about having everything in its place and clean, well, I have beaten it into most of my men's heads also and now it is starting to pay dividends. Crook is looking better every day. You would not believe how his health has changed since he stopped flying. We can no longer get off the post because of food shortage with the natives. So I won't have any spaghetti for a while. Everyplace I go anymore I run into someone I know. By the looks of things, there must not be very many men left back in the States."

**Departing Comrades**

A number of 36th men were ordered home on November 19 but Bob Uhrig was not one of them, despite the fact that he had been overseas for a year now and had not even been able to get home on leave. "Today orders came through to send home Captain Coursen, Lt Leach, Lt McDaniels, Lt Arben, Lt Riggs and Lt Davis. Some of the other pilots were downhearted about not getting to go. Out of the original 48 officers, there are only 11 of us left in the Squadron and that makes you stop and think."

That night a small celebration was held for the departing men. "I was bartender tonight and Pace was on with me. We started making 'Planters Punches' and they sold like mad. As the evening wore on, I made them stronger and stronger and by 2200 I had everyone plastered, including Pace, my helper. Every time we made a new batch of punch we would sample it, but Pace would sample each batch about three times, so I lost him early in the evening."

The weekly inspection was held the next day, with some excitement. Afterwards, Bob and his comrades reflected on their year overseas and Bob recorded these poignant thoughts in his diary. "We had our usual inspection today. The men were given the rest of the morning off. During the arms inspection Sgt Fetter shot off his submachine gun. He had loaded a clip and did not know it. We all went to work in the afternoon and Sgt Love and I held an inspection of Engineering and found everything OK. Then we went to town to inspect our new Parachute Department. We have part of a school building where we pack and dry our parachutes. We use the hall for packing.

"From 1800 to 0100 I sat with the fellows who are going home and shot the bull. We drank to all the places we have been stationed since coming over here. But like always, the talk goes back to the desert. There is something about the desert that gets under your skin and everyone will tell you the same thing. We were talking about being homesick for the desert and it is the truth. When you talk and think of all the places out there you have little chills run up and down your spine. You can sit at night and stare into space and if your thoughts are not about home they will be about some lonely deserted spot out in Libya. We were also talking about sickness in the desert – most of the time there was hardly anyone on sick call yet here there is a lineup every morning.

"We were reminiscing about every place that we could compare with some place in the States but Libya is a country with no comparison. At times during the evening we had tears in our eyes because we hate to separate after all we have been through together. We also drank to the fellows that will be left behind forever. I will always be able to see the cemetery on the side of the hill near Gela where most of our boys are buried.

"I forgot to tell about two of the men coming back that were shot down and captured in Libya last December 24th. The pilot was dead before they hit the ground and the Germans captured the rest of the crew.** The copilot (Wells), crew chief (Van Gorden) and one paratrooper escaped two months ago and have been that long working their way back through the lines. Out of the ones that escaped, only the copilot and paratrooper are the only ones that have made their way back to Allied lines. They told us the Germans treated and fed them well but the Italians were terrible."

Bob wrote much of the same feelings to Toots. "There are eight of our officers on their way home and when they arrive there will be a couple of them call you and may stop by to see you. But I doubt the latter because they all live too far away. Before they left we drank to all the places we have been stationed and also to the ones that will always be here. But most of the talk always goes back to the Libyan Desert. In fact at times I get homesick to go back. There was less than nothing there but everyone liked it. After this war there will always be a place in our hearts for the desert and the times and men we knew. Someday I want to take you through it and see if you get the same feeling. No one ever knew what it was to be sick. Before the men left they all had tears in their eyes. They wanted to go home but still they did not want to leave the 36th Squadron."

The following day: "The fellows left for home. Major Farris flew them to Algiers this morning. Everyone was off today except the alert crew and the men that are flying. Crook, Street, Comitz, Miliman, two Italians, one dog and I went duck hunting again this morning at 0430. As per usual we did not bring anything home

** This was the C-47 of the 37th flown by SSgt Roberson. He was killed in the crash but the rest of the crew (SSgt Wells, Sgt Van Gorden and Cpl Lowry) were taken prisoner by the Germans.

but we had the time of our lives. I knocked the 'landing gear' out from under one but it kept on flying. We had enough rifles and ammunition for a small war and that is what it sounded like. We had plain old target practice most of the time. We cooked our chow with gasoline and sand [Benghazi Burner]."

**Rain and Mud Continue**

On November 22 there was a tragedy at Castelvetrano that reminded the men it was still a dangerous place even without enemy fire. This particular incident involved a C-47 from the 314th TCG, and Bob was a witness: "The weather was bad all day but this afternoon at about 1600 the fog started moving in and we had a lot of ships that had to come and land in it. One of the 32nd Squadron airplanes did not fare so well. He tried to come in through the fog and flew into the highway hitting a cart and killing the man and the horse. The entire crew (three men) was killed and the ship burned. It was torn apart more than any transport I have ever seen. The engines were thrown about 300 yards from the crash site."

Bob and some of his men had to travel across Sicily on November 24 to recover some 36th Squadron aircraft that had been forced down. "Received word last night that No. 5 had a flat tire at Catania and No. 00 had one at Naples, Italy. So this morning we loaded two wheel and tire assemblies on No. 7 and Crook, Brown, Poppleton, McConnell and myself went along to see about changing them. When we arrived at Catania, we found No. 5 stuck as well as having a flat tire. We had to wait on a crane to lift it up so we decided to leave two men behind then flew on to Naples. Just before we left, an English DC-3 ran off the taxi strip, the left wheel went down and the propeller hit the ground tearing off the nose section which cut the fuselage in half just behind the pilot's seat.

"We flew past Mt. Etna and I took a picture of the old gal. Then Lt Welter let me take over the controls and we flew through the Straits of Messina and on towards Naples. We ran into some instrument weather but Welter let me fly her on through. We flew over the beach where the American troops invaded Italy and where so many men were lost. This is the same place the paratroopers were dropped. It must have been slaughter to land there because it is completely surrounded by mountains. All along the beach were destroyed landing boats and farther out are some sunken transports. We flew past Mt. Vesuvius which is smoking all the time and lights up the sky every night. It is a sight to see.

"At the Naples airport the Germans dynamited each building so the roofs would cave in. There was also an underground aircraft factory here which they destroyed. Naples is only 40 miles from the front so there are lots of pursuits operating out of here. While we were waiting to leave we saw two squadrons of A-36 dive bombers†† take off on a mission with two 500 lb bombs, one under each wing. They drop their bombs and then strafe. Naples gets bombed about every other night. I got some apples there, the first I have had and they were very good. I tied the legs of my flying coveralls shut and filled them with apples. Italy, from what I have seen of it, is just like Sicily. Everything close to Naples is green and the country is very pretty. But the people there are going hungry just like every place else. They pick through the garbage cans all the time."

Bob's Thanksgiving letter to Toots on November 25 contained some homesickness between the lines. "Well for Thanksgiving we had fried chicken and it was plenty good. It was cooked brown and was good and done. We had breakfast from 0830 to 1000 and dinner from 1400 to 1600. Early in the evening, Street, Crook, Miliman, Hershberg, Fenimore and I took the command car and started looking for the old ruins which are close by here, but we never did find them. In the evening we had a dance at the hotel and the girls and all their relatives were there. I went to bed (alone) very early and listened to the radio and thought about what you all were doing at home and wishing all the time that I was there with you. Had an inspection this morning and Colonel Fleet was there to inspect and everything went off OK. Then he came out here to Engineering to inspect and he said everything was satisfactory. We have a different inspection about every other day. Having new crews replace the ones we lost and also the ones we sent home makes me feel lonely. I will be so glad when we get somewhere that the people speak our language. It gets to be old stuff after a year of trying to make people understand you."

---

†† The A-36 was an early dive-bomber version of the famous North American P-51 Mustang fighter.

On November 26 there was an incident in a nearby town and Bob was called on to help the Squadron Commander. "Major Farris came into my room last night while I was shaving and told me to put on my gun and come with him. We went to a whorehouse to get two of our officers who were causing a lot of trouble. It was quite a place but how anyone could ever even force themselves to go to such a dive is beyond me. We took them to the barracks and placed them under arrest in quarters.

"We still have to carry guns because every day and night there are some American troops killed in Sicily. The natives say they are getting less to eat now than when the Germans were here. Bread has gone from 1.5 cents a loaf to 60 cents. These people have to live on bread and spaghetti, when they can get it. It is terrible to see people starving. Old men and kids are begging for anything at all to eat. They will eat garbage or anything else that they can get. This is not all the people. Some of them are not so terribly bad off.

"I have just been listening to some good dance music. It is a Saturday night program given for the British and America troops by the Berlin Broadcasting Station. The Germans always have the best programs. They have another program on which they announce that, 'This program is coming to you through the courtesy of your enemy.'"

The 316th was still coping with the rainy, muddy conditions at the end of November. The majority of the repair work done by Bob's Engineering teams involved not battle damage, but aircraft that had run off the hard surfaces and become buried in the mud. "We got No. 2 stuck this morning. The pilot landed downwind and the ship got away from him and he ran into a soft spot and the landing gear went in so far that it bent the propeller and wing tip. It will take at least a day and a half to get it out. Two ships left for Cairo today. Sure wish I was going along. It would be something like going home. Some of our ships flew up close to France today and they had fighter cover. There were in sight of German territory for at least a half hour's flying. I had turkey for supper tonight. It was a little late for Thanksgiving."

**Training for War (Again)**

The 316th was experiencing a fairly long period of minimal activity, with only training missions to keep the crews occupied. With extra time on their hands, the men began to get restless and homesick. This affected Bob too, particularly since he was now seeing his long time comrades rotate home after a year overseas, and he was still stuck in Sicily with no prospect of returning to the States soon since he was not a combat crew member. The rotation policy did not apply to "ground personnel" and they were expected to stay overseas for "the duration."

Part of the training regimen now included towing troop-carrying gliders – either the American WACO CG-4 or the British Horsa. These gliders would be used in the forthcoming invasion of France so the Troop Carrier crews needed practice towing and releasing them accurately. "Today (November 29) we started getting gliders in again and also the personnel. The only good part is that they are not transferred to our unit. They are only on 'Detached Service' to us for one month for training purposes.

An American WACO glider (towed by a C-47), taking off from an airfield in Sicily on a training flight. (NARA)

"Got No. 2 out of the mud today and now we can start working on it. The Italians shot up a jeep last night and threw a bomb. Boy, how these people live I do not know. I don't see what they live on. I pity the kids something terrible but I can't keep them all. I only wish that I could. War is just hell on innocent people.

"For the first time I get the real lowdown [information] tonight. There are only 11 of us original officers left in the Squadron. I either want to go home or go back to the desert. I guess I am just homesick. I think about seeing Toots all the time and also just how I am going to act. I love her so terribly much. I want to hold her in my arms."

The next day, the first official word of the Group move to the United Kingdom was leaked out to the officers. The 316th would now be part of the force for the invasion of France. "Went to a Staff Meeting at Group Engineering today and got the lowdown on the future move. We are to get everything ready to move to England. It looks like I am never going to get home. I keep telling the men that I like it over here just to keep them from getting down in the dumps. I think I have most of them believing that I really do like it over here and it helps them a lot."

Although not a 'West Pointer,' Bob Uhrig instinctively knew a thing or two about leadership. He understood the importance of morale of the unit. He knew how to keep men focused on their jobs and looked after them on a daily basis. His prior service as an enlisted man provided him insight into the troops' mindset and what kind of leadership style they would respond to. His unique perspective allowed him to get the most out of his men and keep them motivated even in difficult conditions and surroundings.

Bob gave Toots a tiny hint of the upcoming move in his next letter. "Tonight there were 30 bags of mail that came in for the Squadron but they have not sorted it out yet. A ship just returned from Cairo and I now have you one ounce of Channel #5. You are supposed to be able to cut it but you run a test on it first and see what you think. Saw McMenamin today and he sure thinks that he is going to get to come home. He should hit it before long. We sure wish they would stop bombing Germany when the good radio programs are on because they are about the only ones that have anything worthwhile. All their stations went off the air tonight. We can always tell when they are catching it by how many stations are off the air and tonight they are all off. There is so little to write about anymore. Besides, business is picking up again and I may not be able to write to you every other day."

December 1 was payday, an important day in the life of every G.I. "Most of the men were paid last night but anyone that had longevity pay coming will not get paid until December 3 because the vouchers did not get to the Finance Office in time.

"I stopped by to see Pall, the Italian who takes us hunting. His daughter had given him four handkerchiefs with handiwork on them, for me to send to Toots. They tried to get us to eat with them but we had just finished eating at our barracks so I told them we would be back tomorrow night for dinner. When these people like you, there is not much they won't do for you."

The next day, maintenance and flying continued, with more concentration on glider operations. With the news of the impending move to the UK, preparations for the upcoming invasion now received priority. "We received two engines for No. 8 today so we will start changing them tomorrow. We are doing all local flying now and tomorrow we start towing these cussed gliders of which we have ten.

"Tonight at 1700 Comitz and I went to Pall's house for supper. We had wine, spaghetti and eggs. We looked at my pictures I took along. They were bound to keep the pictures of Toots and myself. Then Pall played the accordion for us. Like Comitz and I said afterwards, memories like these will live in our hearts forever. I would not take anything for the fond memories that I have. No matter how rich or poor one may become, these memories of things and friends will remain in our hearts forever. But every time I listed to the music, my thoughts always go to Toots, the men who have gone home and that little cemetery in southern Sicily. I just can't get those men off my mind. There is now a curfew for the civilians at 2100 because of all the shootings in the area."

Bob's next letter to Toots was about some of the new additions to the Squadron. "I have not told you about our dog situation for a long time so I will try and bring you up to date. Vino (wine) had six pups and after they got their eyes open men in the other squadron started asking for them. Warrant Officer McCormick

and MSgt Toach both from another group picked one out. Both these men used to be with me at Patterson Field and we visit each other about every day. A few days after the pups had their eyes open Suzie (another one of our dogs) killed all but one and it happened to be the one McCormick and Toach picked out. So we kept putting them off, and yesterday McCormick and Toach brought a lawyer over and we held a trial as to whether we had to give her up or not. We were all wearing guns to we won. So today we promoted her to private first class (PFC) and had her signature in ink on her paw and put on paper. She got in her first flying time today and she liked it very much. We have named her S-4 (it means all of engineering). Comitz and I ate at the man's house who takes us hunting tonight. He is always asking us to come and eat. Comitz does all the talking for me. His girl made four handkerchiefs for me to send to you. I sure think it was swell of her."

**VIPs at Castelvetrano**

December 8 was a red-letter day for American forces in Sicily and at Castelvetrano in particular. The Commander-in-Chief, President Franklin Roosevelt paid a visit to the base, his only stop in Sicily. Escorted by a dozen P-38 fighters, the aircraft touched down at 1400 at the airfield. He was on his way back to the U.S. after attending the Tehran and Cairo Conferences with Churchill and Stalin and stopped at Castelvetrano for a review of the troops. He was met by U.S. Generals Mark Clark and George Patton. Bob saw the spectacle first-hand. "President Roosevelt was here on the field today and also most all the Generals in this theater. The 314th Group and the 36th Squadron held a review for him. This will be the only stop that he will make in Sicily. The sky was full of airplanes all day long." Roosevelt took a jeep tour of the airfield and presented medals to General Clark and others for their actions at the battle of Salerno.[39]

*Left:* Generals Mark Clark, Hap Arnold and George Patton await the arrival of President Roosevelt at Castelvetrano. (NARA)

*Right:* FDR with General Eisenhower at the airfield. (NARA)

Left: FDR reviews the troops at Castelvetrano. The 316th TCG C-47s are just visible in the background. The officer next to General Clark (l) is Col Reuben Tucker, CO of the 504th PIR. (NARA)

The following week, Bob and his men had their hands full with an unending series of maintenance tasks on the C-47s. "I did not feel too good last night and Lt Wanamaker and Capt Conroy proceeded to get drunk on cognac and then started singing to me. They said I looked cold so they stacked everything in the room on my bunk including four chairs and a bed roll that weighs about 75 pounds. I put up with it until 0030 then I went up to the next floor and went to sleep in Conroy's bunk.

"We had to work today because we had too many ships out. For the past week it seems like everything went wrong and every ship that landed had something wrong with it. Some of our ships returned from Bari, Italy today and the crews told us that when the Germans bombed the place the other night they sent 18 aircraft over. Crook came back to work today. He has been off with a bad throat. We have three ships out for engine change. We are getting some bad gasoline now. Some of our ships have been cutting out on takeoff and when they are pulling gliders that is no good. Out of ten gliders we have two flying. They sure are a washout. It takes as much maintenance on them as it does on our airplanes."

On December 15, the 36th lost some more experienced men to Stateside rotations. These senior sergeants were especially critical losses to Bob as he needed as many "old hands" as possible to keep the C-47s in the air. Their places would be taken by replacements who would have to quickly get up to speed on Squadron operations and prepare for the move to England. "Our crew chiefs that have in their 850 hours are going home. They are Boggs, Winnie, Middendorf, Kawohl and Ridgeway. Boggs and Middendorf are leaving this morning. The other three are at a rest camp on the Island of Capri and have to be flown back here before they can leave.

"We received a bunch of junk (fuel tanks and fittings for cabin fuel tanks) from Egypt yesterday. So now we will have to start mounting them to see if we have all the parts. We found some parts that we took off the ships when we first came over. Little did we think then that we would ever have to use them again. Every day it begins to look more like we are going to England. At least all the rumors point that way. The way things are now I would just as well stay here."

Medical services were still in short supply at Castelvetrano, so Bob was forced to seek out a local civilian dentist. "This afternoon I went to town to a Sicilian dentist to have him look at my teeth and see how many fillings I need. He told me that I only need three fillings and it was hard for me to believe because I thought I had a lot more than three bad teeth. I had him clean my teeth and he put temporary fillings in two of the bad teeth and so I have to go back on December 20 and have them finished."

The glider training operations continued into mid-December, but this type of hauling was hard on the C-47s which were not purpose-built for them. "Last night we had steak for supper and it was very tender, believe it or not. We have been getting some good meals here of late. We are still pulling gliders and they are tearing our airplanes apart. We have had engine changes on Nos. 7, 8, and 12, and are now working on Nos. 2 and 9. This glider business is no good.

"First Sergeant Johnson had been sick for over a week now and he does not look the least bit well yet. In fact, he does not seem to be getting any better. They now have vegetables for sale around here. We buy our share of green onions and peppers. Everything is green and the grass is growing every day. It is cold enough for a sweater and leather jacket but not too cold for things to grow. It is just like early spring in Ohio, where everything is green but still cold. It rains here about every other day and it stays very damp. I don't see how these people keep from freezing to death because they don't have any heat in these stone houses. It is a wonder they don't die from the dampness."

Although far from combat, life in Sicily was still dangerous for the Americans and locals alike. "The civilians here are still not getting any food. We still have soldiers being killed at night. You can hear shots every night. Farmers hide what little grain they have and the gangsters then find it and hijack them. This is supposed to be the toughest part of Sicily and it sure looks the part. I would not trust one of them two feet from me.

"They raise plenty of oranges and tangerines. They are not the best in the world but they are better than nothing. Both are more sour than ones grown in the States. The Red Cross comes in about once a month with coffee and doughnuts for the enlisted men. They were here this afternoon (December 17) so we let the men

go in at 1430. We have been having at least one flat tire on the aircraft each day. We won't have enough tires if this keeps up.

"We have fixed up six gliders now so they will fly but we had to use parts off the other four, so now we are going to transfer the four bad ones to the 322nd Service Group and let them try to fix them up. At least they will be off our hands."

**Another Christmas Away from Home**

On December 20 the 316th Group had a party for the ground personnel who came to Egypt by boat, marking their first year overseas. The dance was held at the Officers' Club which was decorated for Christmas.

Although the holidays were almost upon them, Bob and his men continued the fast tempo of maintenance and repair operations. "Love, Quisenberry and I went to Group this morning to see about getting the cabin fuel tanks* and glider tow releases. We are repairing a wing tip on No. 00 and checking the valves on No. 8. When we were putting the slow time on the engines for No. 2, it sprang an oil leak and we lost most of the oil before we could land. It is a clear day here today and the sun is shining for the first time in over a week. It rains a little most every day. But then once in a while it will pour down for about two days and flood everything.

"The new Service Group, the 322nd is the worst yet. We can fix more here in the Squadron with hardly any equipment than they can with all their equipment. We are still on a training program just like we were in the States, but with 90% new pilots. I guess we can use the training. The new pilots need taxi instructions more than anything because they run into everything on this island.

"One sees more screwy things over here. The other morning we saw a man leading a pig which had a collar on. You also see people riding cows. They sure do load these two-wheel carts down over here. They gather up every little scrap of wood to burn and they eat less than nothing."

In no time it was Christmas Eve. "We got caught up with most of our work today so we could have tomorrow off. But after we were finished, Group called and said that we had to fly tonight and tomorrow morning. Everyone is about to kill all Group personnel after a stunt like that. After all, we are just on a training program."

After the work was done there was some time for Christmas Eve celebrations and reflections on their time in the desert. "In the evening all the desert rats from last Christmas got together in the Enlisted Men's' Bar and had a few rounds of drinks. There are very few of us left anymore. We talked about old times and what we were doing last Christmas Eve. It was a lot different last year because we did not know what a building or a bar looked like. After we sang a few songs, Love, Fennimore, Johnson and I went to an Italian family's house to celebrate. We drank and sang until quite late and then we all went to our quarters and to bed. I still like the desert the best."

Christmas Day brought with it a hectic flying schedule which was followed by a rash of maintenance write-ups. "Merry Christmas, everybody. After last night's flying and this morning's flying the status is as follows: No 2. Out – brakes; No.3 - stuck in the mud; No. 11 – flat tire; No. 6 – rough engine; No. 00 – hydraulic leak. All this because Group wanted us to fly. Now the men have to work on Christmas Day on about half of the airplanes. We quit at noon and let the rest of them sit. The 314th Group quit on Thursday night (December 23) and is not going back to work until Sunday morning (December 26).

"We bought and gathered up a lot of chow for the Giaffeo family in town; turkey and all. So today, Love, Pappy Street, Fenimore and I went there for Christmas dinner. We drank enough wine to get a buzz on. We stayed all afternoon and evening. In the evening they had some friends come in and we danced until about 2200 and then went home. All in all we had a pretty nice Christmas. All I could think about all day long was being home, and when you are doing that you can't enjoy yourself too much."

Bob's Christmas letter to Toots was sad and hopeful at the same time. "Well another Xmas and I was not home. That makes about three if I remember correct. We had a much better Xmas last year even if we were

* The cabin fuel tanks would extend the range of the C-47s to allow them to fly around Spain, Portugal and France on their way to the UK without the need to stop for gas.

in the middle of nowhere."

On December 29 the Group hosted some senior officer visitors. "Everyone was here from the 52nd Troop Carrier Wing today to have an inspection of the 314th Group and our Squadron. The General (Clark) and all his staff were here. Col Berger (Group Commander) inspected Engineering and found everything OK, at least he said it was.

"The weather is getting colder and colder here each day. It has not frozen here yet, but I have just about frozen to death. It is so terribly, terribly damp and I think that is what makes it seem so terribly cold. These stone buildings don't help much either.

"Another ship got stuck today and also another flat tire. Pall, the Italian hunter, was here this afternoon and brought us some tangerines and oranges. He brings them to us about every three days. He won't take any money for them because he says that we are his friends. No. 10 just landed and they said they went up to 21,000 feet in altitude. It is the newest ship we have so that is why they could get that high. Lt Pace, Wharton, Lee and Fenimore were in the ship."

**Bleak Midwinter**

New Year's was quiet at Castelvetrano as the men began to get more anxious about the date for the upcoming move. A number of other Army units had already been shipped out of Sicily to England in preparation for the upcoming invasion. Events in the local area made the situation even more stressful. "We just about have all our cabin tanks in now and most of our glider tow releases back in and operating. The sun has been shining for the past two days but she has clouded up again and it looks like it is going to start raining any minute.

"They still continue to kill about two people in the town every night. It is lucky for us that most of them are civilian police. I think half the population here has guns and hand grenades hid away. Most of those killed are from grenades. You can hear shooting every night in town. I think these people have killed each other ever since there was an island here. Yesterday a sheep herder took his sheep into an orange grove against the owner's order and when it ended up there were two sheep herders dead and one 'orange man' shot but not killed. We have things like that every day. To me it don't look like these people care one bit about killing each other. Quite a few of our men have had pot shots taken at them and a jeep full had a grenade thrown at them.

"All the people cook with olive branches. A bundle that you can put one arm around sells for 20 cents a bunch. Some people that have enough money use charcoal. Each kitchen has about five places to cook on and one oven. Just about half of the smoke comes out in the kitchen and then on out the window. In the evening, they take a bowl of the ashes and put it in the living room to keep the chill off the room and keep their feet warm.

"This morning we went out to No. 2 and where the freight door latches into the floor, there is a bean plant growing about six inches out of the floor. We are trying to preserve it and see how tall it will grow. S-4 (the Squadron dog) is growing every day and is very cute. She is so fuzzy that you can just barely see her eyes. She gets a bath every Saturday night and when she drips out she looks like a ball of angora yarn. She just about runs this place here now. We feed her part of our canned rations each day and she has not complained. She also eats like a horse."

By early January Bob and his men had just about finished preparing the C-47s for their move to the UK. Other than preparing their own gear for the move, they had done all they could to ready the aircraft. "We are just about to finish the engine change on No. 1. This is the third day since we started the change. Also Comitz is painting a new name on the airplane. It is 'Ali Baba & His Forty Thieves.' It is the nickname for Engineering. We also have the same name on our Engineering jeep. Stracke was here this afternoon looking for some parts. We only have about two sets of cabin tanks to install yet.

"We have more fun together here in Engineering. The things we do and say would kill you. I have to laugh about every five minutes when we all compare notes about what we are going to do when we get home. If

you don't think some of the stories are killers, think again. We are building a smokestack out of five gallon cans so this stove of ours will have plenty of draft. Love, Tucker, Milliman and Wallace made one for their tent and it had so much draft that it took half the papers in their tent up the stack.

"Just now Butler the cook is making some beans and everyone is arguing over who eats first. We are always cooking chow here. We take it off of the U-Rations[†]. U-Rations have everything from toilet paper to salt in them. It is a ration for five men for one day. It is the best ration that I have ever seen. Our dog S-4 is getting pretty big now and she has a new box. She has moved into the file cabinet with our Technical Orders.

On January 7: "Major McMenamin left for England this morning to arrange things for us. We finished the engine change last night on No. 1 and Lt Quisenberry and I flew it for 15 minutes and the only thing wrong was the fuel pressure was a little too high. This afternoon we put the rest of the slow time on No. 11 so it is now ready to fly.

"Col Fleet inspected us today and awarded Air Medals to the crews that were on the last mission to Salerno. An airplane from the 314th dropped a glider tow rope through the wing of No. 5 the day before yesterday and broke two ribs and two stringers in the wing so it will be out for repairs for about a week."

In the week that followed there was more work on the engines, wing repairs and drawing up lists of equipment and parts that would be needed to take along on the move. But by mid January, there was still no firm news on a departure date. "We got an engine cowl flap cylinder from the 37th Squadron today. This makes the fourth one that we have been able to get since we have been overseas. We are putting a door on the engineering tent so that we can keep the cold out. We have a fancy stove in there. The stove is made from two wide five gallon fuel cans and the stove pipe is made out of the long type of five gallon cans. When we open the dampener, the stove just sits and pants. We have to watch our paperwork or it will be sucked up the stove pipe. We burn gasoline in it and it gives off all the heat we can stand.

"Milliman came back from early chow and said that the mess was very poor, so we gathered up some cigarettes and went out and traded them for eggs. We get five eggs for one package of cigarettes. Then we opened a U-Ration and made coffee, spaghetti and meat balls, eggs and pork sausages. Six of us stayed here and ate. White and Crook took a can of green beans and a can of beets and took the labels off of them then told the Italian it was canned meat for which he gave them two bottles of wine. So we had wine with our meal.

"I went up with George Quisenberry and we were towing two gliders together. Two of those gliders sure make the engines moan. It takes 45 inches of mercury[‡] to get the ship off the ground and to climb it takes 35 inches of mercury and 2400 revs per minute and even then you only make 100 miles per hour airspeed."

In one of Bob's letters home he reveals some of the resentment that the soldiers felt against the so-called "draft dodgers" that were able to stay in the States and avoid the service. This resentment became more pronounced as the war went on. "In this place a person gets their first cold and then they never get rid of it. This damp climate is the best ever for colds. One thing, 90% of the other men here have the sniffles. How does it feel to be such a big shot and still making as much money as you did before the war started? What I said in the last letter still goes. Quit if they can't pay you more. They get it for their whole batch of draft dodgers. Every time I think about it I just about blow my top. Do you know that since the war started I have been away from you for 17 months? To keep myself in shape, I run from the hotel to the mess hall each morning at 6:30. It is about five city blocks. You can be driving along in town and you will have to stop and wait until some local farmer drives his cows, sheep or pigs across the street. Go out to the hog pen and put a collar on one of our hogs and then try to lead him around, well, you see a lot of it here. You even see an old hog following some Italian through town like a dog would follow a person.

"Did you ever get the Group insignia that I had painted on a piece of leather? It is painted on the back of my jacket that I used to wear at home. I wore it out over here so I had the back cut out and the picture painted

---

† U.S. Army Unit Rations "U-Rations" contained numerous canned, dehydrated or dried foodstuffs, as well as components from other rations and other non-food items. They were intended to be used where basic cooking facilities were available. The first were '5-in-1s', later followed by '10- in-1s'.

‡ Piston engine aircraft like the C-47 with constant-speed propellers used inches of mercury as a way to measure manifold pressure in the engine which gives a relative measurement of engine power.

on it. I sent it the same time I sent the silk ones. I have also sent you Channel #5 and four handkerchiefs. They were made for you by a little Italian girl. She is the daughter of the man that takes us hunting and she wanted me to send them to you."

The quiet days continued uninterrupted. The men of the 316th were getting restless after many months overseas. "By this morning (January 21), we were to have all our ships ready to go on a parachute mission, but now it has been called off. We don't know for how long.

"Castelvetrano sits on a 600 foot hill and the sea is approximately three miles to the south. The low land between the town and the sea is beautiful. It is covered with orange trees and all types of green vegetation. When we drive to work in the mornings the sun is just coming up and there are always little patches of fog here and there in the valley down to the sea. The view is spectacular.

"Every morning I see a man and his goat delivering milk. Of course this is done by more than one man and goat, but I will tell about the man that I see each morning. The man has a tin can and the goat stops in front of a door and the man squats down behind and milks a can full then pours it into a container for his customers. Then the goat takes off and runs to the next customer's door and stops. The first time I saw this procedure I thought the goat was running away. What is so funny is that the goat knows which door she is supposed to stop at.

"If we did not have a lot of wrecked enemy airplanes and destruction everywhere we would not know there was a war on. For the past three months we have had it pretty nice except for this training program. We got in three new pilots yesterday. They have never flow a transport. They just got out of flight school last November so you see what they are sending us.

"Pappy Street and I rode over to Group and then Lt Mulder and Warrant Officer Santomissino took us up to the City in the Clouds (Erice), which is on top of a 2900 foot mountain. The mountain is on the north side of Trapani. The road going up is just one hairpin turn after another. It is a beautiful drive up. From the top the view is beautiful. Once can look almost straight down on the city of Trapani. On the far side of the city you can see the harbor and the salt flats.

"Erice itself is beautiful and above all it is very clean. The streets are very narrow and paved with stone. There are two castles on top and they both sit on the edges of cliffs. The cliffs are about 1000 feet tall. The vegetation on top is like ours back in the States. The city gets its name from being blanketed by the clouds during the rainy season."

*Left and Right*: Views of Erice, Sicly, January 1944. (Mike Ingrisano collection)

On January 24 there was another mishap at Castelvetrano. This time it involved one of the WACO gliders being used for training. "Last night one of our gliders cracked up and broke the pilot's left leg and the copilot's right leg. The glider was a total wreck and I don't see how anyone got out alive.

"I had dinner yesterday at my Italian friend's house and it was very good. First we had spaghetti and it had ground almonds and cauliflower in it. It was very rich and you could not eat much. Then we had turkey, fried potatoes (second time in 14 months) and fried cauliflower. Of course we also had wine all the time. We had Italian bread which I like very much. Then, we had an egg omelet. I want to make some when I get home. They take the whites of eggs and beat them. Then they add the yokes, grated cheese, diced onions and peppers. They put plenty of olive oil in the skillet and after it is hot they pour the egg batter in. As it cooks they keep rolling it up. When it is finished it is in a roll and is very fluffy. I never did like omelets until I tried these. After we put this away, we had fruit, nuts and coffee. I was so stuffed that I was in misery."

The next day Bob continued his diary entries on the interesting aspects of Sicilian life. "These Italian people are sure the stuff. I find something new to write about each day, but by the time I get to write about them I forget most of the things. They have flat irons that they heat by filling them with charcoal.

"Our dog, S-4, was killed this morning. She fell off the weapons carrier (truck) and the rear wheels ran over her. Now we only have Duchess left because I shot Vino the day before yesterday. Vino had been sick for over a month and had not eaten anything for the past week so we decided to put her down. Duchess has had relations with a small black dog that is about her size. So in about six weeks we should have some new offspring. These American soldiers ruin all the young women. We are hoping that Duchess does not turn out like some of the other Sicilian girls. The venereal disease rate on the island is very high. If you don't think so, just step out of line and you end up with everything in the book.

"I went to the Group Engineering meeting today. We have one every Tuesday. We also got two new pilots in the Squadron today. One of them is a captain so that does not make any of our pilots very happy. Love and Tucker fixed an exhaust whistle on our jeep so that the Italians will get out of our way. The truck driver for the Engineering truck liked it so much, last night he stole a whistle off an Italian steam roller which was parked. They have just put it on the truck and it is plenty loud."

**V-Mail**

Bob's most recent letter from Toots on January 25 was a "V-Mail." V-Mail was a new process in 1943 that involved the microfilming of specially designed letter sheets. Instead of using valuable cargo space to ship whole letters overseas, microfilmed copies were sent in their stead. The microfilm was developed at the overseas location and the letters printed out and delivered. Unfortunately, the printed copies were only one-quarter the size of the original letter which made them difficult to read. "Received a V-Mail letter from you last night dated Jan 8, so it made it in pretty fair time. Also I received all of the comic sheets from the newspaper. The inspectors and line chief are this minute opening them and putting them in rotation. Just now there is a big fight going on because one of the men started reading one a week ahead. You know you have your name on one of the engines of one of our airplanes. I hope you don't fall off sometime while we are flying. The name is in black letters (TOOTS). Love made a new sign for in front of Engineering. It is (FUHRER'S DEN). I now have two names – Ali Baba and The Fuhrer."

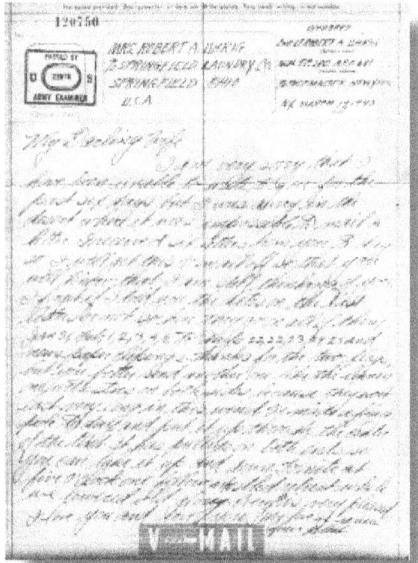

One of the V-Mail letters Bob wrote to Toots along with the original envelope. Actual size of the letter was about 4 inches by 5 inches.

January 29 was another Group inspection day, so Bob took time to write home while he awaited the inspection party. "It is Saturday morning and I am dressed up waiting for the inspection. Do I ever look pretty; you should see how handsome I am. We are having some beautiful days now here and they remind me of spring back home. This island is very pretty with the exception of the towns which are not so much to look at. Everything is so green; the trees are coming out in bloom. The other night we were in the room and each one of us voted as to what we would take if we could have all we wanted. Well this is how it came out: milk was first and beer was second. Things that you people never even think of twice back there we want. It is such simple things that we want to eat and drink but we never get. We can't stop at the corner and pick up a few of these and those. Also stop and get a glass of milk and some ice cream. You get broken of the habit of eating between meals over here. Well, the inspection is over and I am now out on the line and I don't have my pretty clothes on anymore."

On February 1 Bob showed his mettle by taking a ride in one of the WACO gliders on the field. This was not for the faint of heart. Since the glider was constructed of steel tubing and canvas, many soldiers and even some airmen would not set foot inside one, considering it close to suicide. But Bob's experience was a pleasant one. "Today I went up for my first glider ride. We went up to 10,000 feet and I got to fly it most of the time. It was lots of fun."

The next day, the 36th Squadron received one of the new C-47s equipped with the ground mapping radar to aid in navigation. Bob estimated that, "the radar equipment on it is worth half a million dollars. They are getting us ready to move again and I hate to leave Sicily."

*Left and Right:* A formation of 316th TCG C-47s over Sicily. All aircraft now have the standard "Stars and Bars" blue and white US insignia on the wings and fuselage.
(Mike Ingrisano collection)

As soon as the new aircraft arrived, it was gone again on February 3. "We had our new ship leave for England this morning and we don't know yet how soon the rest of us will go. I hate to leave here more than any place else. My heart is very heavy these days. I will miss the two-wheel carts rolling down the streets in the early hours of the morning, and also the men and boys going along the streets singing. It sounds good to hear singing in the streets at night and in the early morning. All in all I don't want to leave Sicily."

That week, Bob's expert crews accomplished a change-out of both of the Pratt and Whitney R-1830 engines on a C-47 in record time. "Yesterday they told us that we had to change engines in No. 3 and have it finished by tonight (February 5). I said that it was impossible unless we worked all night. The best time we every made before was 17 hours. We took McWhirter, McConnell, Berryman and Little for one engine and Shanta, Useman, Wagoner, Nordness and J.D. Robinson for the other. From the time McWhirter touched the cowling until the engine was ready to run-up it took him 7 hours and 45 minutes. I still can hardly believe it is true but I saw it with my own eyes.

"We pulled the ship into the revetment at 0830 and had the new engine back in and run-up before dark. They even had to take the new engine out of the engine box and get the pickling fluid out of it. The worst part of all is that it rained and sleeted most of the time. They had no protection from the weather and a very limited amount of tools. I think it is even a record for units in the States where they have all the equipment to work with. As I said before, for changing an engine in the field is such a short time just doesn't seem

possible. Shanta had a little more trouble and it took him 8 hours and 15 minutes. Even this is a record for any Squadron in this Group."

Continuously bad weather in Sicily cut back on flying operations, so Bob and the boys took in a movie on February 6 and witnessed more local color. "We have had plenty of rain and sleet the past few days. Went to an Italian movie tonight and it was something to write home about. We had a box seat where we could look down on the lower floor seats and see all the goings on. The Italians crack nuts and whistle and talk out loud during the show. While the reels are being changed they climb over the seat, spit, and get into fights. To look down on the action is like seeing a three-ring circus. Very few of the men take off their hats during the show. Most of them wear very large caps and they have the bills pulled down over their eyes. All in all they are a sight to see. Upstairs in the balcony it is a little better, but one still gets walked on and the noise is just as bad.

"As cold as it is here, we still see people going around barefooted because they don't have shoes to wear. It is a pitiful sight to see. No one gets enough bread to eat. The ration is very small and it is also very cheap (2.5 lira or 2.5 cents per loaf.) But if they want more, they must get it from the black market where it costs 45 cents a loaf. This is quite a difference in price. Only the people with money can eat enough.

The following Tuesday, February 8, was another Group Engineering meeting but, "there was very little doing. A lot of the officers' promotions came in today, but mine did not. It was submitted on the 14th of December. I don't know if it is lost or what is wrong. If it was not going to be approved it should have been sent back by now. I would like to make it before I leave or it will be messed up for about three more months. We have all of our ships ready to move and we are now just waiting on the orders."

Preparations for the move continued, including receiving replacements for the enlisted men who had rotated to the States. "We have some new mechanics again. Before long we will have a whole new Squadron. We heard that there is very little to drink in England, so Fenimore and I are loading up with wine to take along. The sun is shining today so maybe it will warm up. I am getting in quite a bit of flying. They let me fly on most of the test flights we have.

"I can't get over the shoes that the women wear here in Sicily. They have such high heels that they can walk on their tip toes. Their shoes are not made like ours. The soles are all cork so you can picture how they look. They have very good looking shoes and I like them much better than the American shoes. I think they are much better in wet and cold weather because their feet are off the ground and the cork acts like an insulation and helps to keep their feet dry and warm."

**The 316th Leaves Sicily**

Bob and his men received their movement orders on February 14, 1944. They spent the balance of the day loading the C-47s with their equipment and personal belongings in preparation for the long overwater flight to England. The following day Bob said a sad goodbye to his local Italian friends and neighbors. "Told my Italian teacher, Fifi, that I was going to leave and her and the family felt very sad, but not half as bad as I did about having to leave. They have treated me royally for the past three months. They have made me feel like I was home. If I had to be overseas I would like to have spent all my time in Sicily.

"I stayed all night and this morning when Love came after me to leave, they all cried and so I could not keep from it myself. After the war I would like to send Fifi to America to live with her aunts. She would like very much to come to America. You feel like you want to do something for people like that because they try to do everything in the world for you. There is just Fifi, her brother John and her mother living together. They lived in Casablanca before the war. When the war started in Africa their father moved them to Sicily and now they have not heard from him in eight months."

Eleven C-47s of the 36th Squadron departed Sicily in formation on February 15. They were bound for Tunisia, Algeria and then on to Gibraltar, their final stop before the long leg to their new base in Cottesmore, England. They were the last of the 316th Group to depart as the aircraft of the other three squadrons had already left in the three prior days. Bob was fortunate to be with the Air Echelon again, so he was spared

the long boat trip from Sicily to the UK. He flew to his new duty station on board the aircraft piloted by his bartender buddy, Lt Robert Pace.[40] "We left Sicily this morning at 0900 and I watched out of the astrodome until I could not see Castelvetrano anymore. I have felt low the rest of the day."

**Long Hop to the United Kingdom**

The first leg to Tunis was a quick one. "We landed here at Tunis at 1000. Fenimore, Street, Love, Crook, Tucker and I fried up a batch of eggs." Unfortunately they were weathered-over in Tunis and would have to wait until the next day to continue the trip. "They have quarters here on the field but we have to have guards on the airplanes so Fenimore (crew chief) and I are going to sleep on the ship. We have our bunks all made up. Most of our old gang are going to sleep in the ships. We had the radio on tonight and Fenimore ran a piece of safety wire from the aircraft master switch to his bunk so he could turn off the juice for the radio without having to get out of bed. So we had some good music. Also tonight Crook, Love, Tucker, Street and a few more fellows came over and we had a good bull session. It has been raining all day."

February 16 dawned and Bob noted in his diary it, "makes 15 months overseas." The inclement weather continued so the 36th was stuck in Tunis for a little longer. "It rained almost all night and it was like sleeping in a barn with a tin roof. We did not get up this morning until 0930 and then we fried some eggs and made toast. I heated up some water to shave and wash up. Then we re-arranged the baggage in the airplane so we would have more room to play poker. Then we went over to an old German glider and used it as a latrine."

There was not much to do except wander around the airfield. "We all went to the Mess today for the first time. It is very good and they tell us they always have fresh butter. As yet we don't know how long we are going to have to stay here before we go on to Oran. This afternoon Pappy Street and I put all the names of the places we had been on our life vests. We also fixed up our short snorter bills§. There is a poker game going on in the back of the ship. It is now 1650 and will soon be time to eat again. We have been getting very good food here. It rained again today and the wind has blown hard all day and it has stayed pretty cold. In fact a little too cold."

The next day the 36th was able to complete the flight to their intermediate destination, the British colony of Gibraltar, with one stop in Oran for fuel. "This morning they got us up and told us that we were leaving at 0800. We took off in No. 10 at 0805 and we landed at Oran at 1345. Our Squadron Commander, Major Farris, told us that as soon as we were serviced we would take off again. No. 6 had carburetor trouble so we had to leave them behind.

"The field at Oran has changed 100 percent. In fact, no one would know that there had ever been a war there, even near the field. The runways are all like new, all buildings have new roofs on them and been painted. It looks like a municipal airport back home. We took off from Oran at 1500 and landed at Gibraltar at 1650."

**The Rock**

"When we first sighted the big rock it was a real thrill." Bob wrote, "It was still about 20 miles away but you could see it in the haze. We had to circle once before coming in for a landing. The runway looks just like a landing strip on an aircraft carrier. At both ends it extends out into the water. Everyone made just a fair landing.

"After we landed, Major Farris went to the Weather Office to check the weather over the Atlantic. When he came back he told us that we would stay overnight. He also told us he was going to put the town of Garrison City¶ on limits until 1200 tomorrow. We were also told that the pubs stayed open until 2100, but there was a curfew at 2300.

"They have huts here and each man was assigned a hut and a bunk. We ate at a British Mess which was

§ Short snorter was a dollar bill or other local currency that friends traveling together would sign creating a "keepsake of your buddy's signatures."
¶ That portion of Gibraltar south of the airfield which was controlled by British Forces.

very good. Then every man made a line drive for the showers. The shower facility had both hot and cold water, but it is salt water so one can get very little lather, but at least you feel better. Then everyone, that is almost everyone, got dressed to go to town which is only about a 15 minute walk. On the way to town you pass more barbed wire entanglements than there are in all of Africa; tank traps and guns both large and small. After you reach the town you must pass through a large water gate that is about 50 feet thick.

"The city is not very large and it is built on the south west side of the rock. That is on the tapered side of the rock. The tapered side features the Straits and the sheer-drop side faces Spain. This is the part that you see in all the pictures. The runway is just to the north of it and runs very close to the base of it. All of us were very surprised when we walked through the town. There are hundreds of pubs serving all types of whiskey and beer. We could hardly believe our eyes when we saw all the stores and merchandise on all the shelves! There is no blackout there so you can picture the sight at night.

"Our money had to be changed into Gibraltar money before we could buy anything. After we exchanged it we had a few scotch and sodas and a few beers. But this time it was 2100 so we could not drink any longer. So we went window shopping instead and went to bed about 2230."

The next day, February 18, the weather was still too bad to attempt the leg to Britain, so the 36th had another day to walk around Gibraltar and enjoy some aspects of civilization that had been missing from their Sicily experience. "We were up this morning at 0800 and had breakfast at 0830. Then Street, Fenimore, Love, Tucker and I went to town. We all got shaves, haircuts and shampoos. After this we went shopping. Most of the stores have Spanish and Chinese merchandise. All of us bought a few souvenirs and plenty of fruit from the fruit stands. This place does not seem to be lacking in anything at all. Still, of course this is like Panama because all of the boats passing through. There are two very large docks and there are over 100 boats in the harbor just now.

"We returned to the field and walked over to the main border gate. This is where all the people and traffic must pass when going to and from Spain. We wanted to put one foot through the gate so that we could say that we had been in Spain, but we were told that Spanish territory was still 200 yards down the road. These 200 yards are neutral and policed by Spain which is also neutral. It is cleared of everything but barbed wire entanglements, tank traps and pill boxes. Just on the other side of the gate there are Spanish police and soldiers. The soldiers all have German equipment – from uniform to guns to helmets. We were told that most of Spain is pro-Nazi.

"Most of the civilians that work on the Rock are Spanish and they are searched by both the British and the Spanish. Cars and horse and carts drive back and forth all the time but they still have to be shaken down each time. There is a road that runs from the Spanish frontier to the Rock. The road cuts across the middle of the runway. Every time a ship takes off or lands, the road must be closed. The road has to cut across the runway because the runway runs from one side of the Rock to the other (east-west). The Rock itself is 1400 feet high and the longest distance in any direction is two miles. This is the most fortified piece of ground in the world. At night there are powerful lights that light up the whole harbor and airfield. The only time the lights are turned off is when an aircraft has to take off.

"The runway is 5400 feet long and is built of stone and tar macadam. The parking strip is not very large so they can only handle one squadron of ships at time.* The British have one squadron of Spitfires based there and two mixed squadrons of bombers. Most of the bombers are used for submarine patrol. They have about four B-24 bombers here that have large searchlights on them. These are used to bomb German U-Boats at night when the subs are on the surface charging their batteries. The B-24s use radar equipment to locate the sub and when they are in range they switch on the lights and search the sub out and attack it.

"This evening when the Major checked the weather he found that it was still too bad to fly so we are remaining another day. Fenimore and I went to bed very early because we wanted to get up early the next morning and ride around the Rock."

On February 19 Bob and some of his men ventured out to see more of the small British colony. "Fenimore and I were up this morning at 0730 and we ran around without a shirt getting up Crook, Street, Love and

* The ramp size constraints at Gibraltar are likely the reason for the staggered departure of the four squadrons of C-47s from Sicily.

Tucker and then we came back to the ship and finished dressing. Then we all went to breakfast. After this we caught a ride into town. We all bought a few small items and then we hired a taxi and rode around the streets. We stopped for a short time at the Dry Docks. These are very large and each one is in use. The seaside docks are enormous and even battleships can pull up to the docks without the aid of tugs. The harbor itself is beautiful and very large.

"We then drove on up the Rock. The taxis are only permitted to go about half way up but the drive was worthwhile because all the buildings are very pretty and they are stuck up against the rock whenever there is enough flat space to build a house. Even the shrubbery is nice. Where we had to turn around we were able to get a good view of the Straits and we could also see Spanish Morocco on the other side. Each place you look you can see a gun of some type and they are not very far apart.

"At noon we went into a pub and started drinking beer. Some American sailors came in and we talked to them. After a while the Americans were making all the noise and the English were just sitting around looking dumb. There were three Spanish girls that would dance about every five minutes. There was a five piece orchestra that played in the pub all the time. We drank here until 1315 and then we took a taxi back to the field. We got together this evening before chow and had a few scotch and sodas.

"It is now 2100 and Love, Tucker, Street, Crook, Fenimore and I are sitting in No. 10 listening to the radio and waiting to hear if we are going to take off for England tonight. Update – the flight has just been called off so we will be here for another day. Everyone is pleased because the weather here is very good, the chow is good, the beer is good and there is plenty to see. In fact, I would like to stay here a long time because I don't want to go up to England and freeze to death.

"We are still here in the ship making wisecracks. The 'staff' as we call ourselves have just voted on what we are going to do tomorrow. We call Fenimore 'the orderly' and we always vote him down on whatever he wants to do. I only wish I could write down all the sayings I have heard during this war. Someone is always doing or saying something that just about kills you. Love has been trying to gas us out of the ship all evening. Each time we holler 'GAS' and the later we sound the 'All Clear' signal."

It would be February 21 before the weather had cleared enough for the 36th Squadron aircraft to make the last leg of their transit to Britain. In the meantime, Bob and the men all agreed there were much worse places to be stranded. Their stay in Gibraltar turned out to be a nice bit of leave that they were not able to enjoy in Sicily. "The winds were still too high last night also, so we are still here on the Rock. Love, Tucker, Crook and I hired a boat this morning and had a sightseeing trip through the harbor and bay. The owner showed us all the boats that had been torpedoed.

"Then we went into town. I had dinner with Major Farris and then we had drinks at an Officers and Civilians club which had an all-girl orchestra. They are much better than most of the bands in the States. The drummer is just out of this world. They were all Spanish girls and all very pretty.

"I was talking to an RAF pilot who is stationed here on the Rock on submarine patrol. He told me the Germans were hauling submarines over land and putting them in the Med. He also told me that the RAF had sunk three subs in the few days we have been here on the Rock. He told me that if a German sub reaches the Straits, he shuts his motors off and lets the current take him through. They tell us now the weather is going to be OK tonight so we are making plans to take off.

"We took off for England at 2230. We are to fly out to sea about 100 to 150 miles so that no enemy fighters can pick us up. Then we fly up the coast of Portugal until we approach the coast of France, then we head west again to clear the French coast by a wide margin."

After a long pleasant stop in Gibraltar, Bob and his men were looking forward to getting some sleep on the long, and hopefully uneventful, flight to Britain. But this was not to be. In fact, they almost didn't get there at all.

316th TCG C-47s in Gibraltar on their way to the United Kingdom in February 1944.
The aircraft are still sporting the RAF "fin flash" insignia on the tail.
(Mike Ingrisano collection)

# Chapter 9

## Crossroads

*You will enter the continent of Europe and, in conjunction with other United Nations, undertake operations aimed at the heart of Germany and the destruction of her armed forces.*
Directive from the Combined Chiefs of Staff to Supreme Commander, Allied Expeditionary Forces; February 12, 1944

It happened somewhere out over the Atlantic in the middle of the night. Short of enemy action, it was the worst thing an aircrew could face – an engine fire. Bob was awakened by the crisis in the making. "I was asleep in the back when it all happened. We were flying along when the right engine spit and backfired and shot flame out about three feet. The pilot thought the engine was on fire and shut it off. We were flying at 9000 feet, just above the clouds and ice. The pilot told the crew to prepare the ship for a water landing. The crew chief ran by and told me to go up front and see what the trouble was. The copilot got out of his seat and let me sit down. I asked the pilot what happened and he told me. While doing this, he was circling a break in the clouds and he told me that with all the weight aboard he was unable to maintain his altitude.

"By this time, we were in the edge of the clouds and were picking up ice which made things even worse. The pilot asked me what I thought and I told him we could not stay up long with one engine unless we got rid of weight and quick. I also said we should start up the bad engine again and see what happens. So we did just that and the flame shot out as before and although it sounded like a tractor engine, it did not vibrate enough to be dangerous. With this extra power we climbed back out of the clouds again. Then we started talking about what to do. At 0230 in the morning, over the ocean, with a bad engine and an overloaded airplane is just no good. We were only half way to England so it would be just as bad to turn back as it would to go on.

"Our only way out was to land in Portugal and be interned.† But we knew that we would lose all our personal equipment and things that we could never again replace, ever. The navigator told us we had half an hour flying time left along the coast of Portugal. The pilot asked me how long the engine would run like it was. I told him it was one of two things – either an exhaust stack had blown off or a spark plug core had blown out. I told him if that it was a bad exhaust stack and we still had five more hours of flying ahead of us, there was a danger of burning into an intake pipe and setting the ship on fire. We decided to fly on until we had to turn out to sea again. If it was no worse at that time, we would go on.

"It did not get much worse as time went on so we decided to try and make it. Altogether we flew 11 hours and 45 minutes and that is a lot of time in the air. When someone would open a window to throw out a cigarette, everyone would jump because the noise the engine was making sounded worse. At least we were able to stay warm because No. 10 has hot air heat instead of water heaters."

Thanks to the combined efforts of everyone on board the crippled C-47, they all made it to Britain unharmed. "When we landed here in England we found out that we were the second one to land. During the next hour most of the others came in, except for Nos. 2, 5, 6, and 9. In the afternoon, Nos. 5, 6 and 9 came in. They had run low on fuel and had to land in the south of England and refuel. We heard that No. 2 had landed in Ireland.

"Nos. 1, 5, 6, and 10 needed maintenance so we had to work this afternoon. Oh yes, it was a spark plug core that had blown out in No. 10's engine. Everyone in the other ships about froze to death on the trip last night. About half of them were lost once or twice during the night and some of them iced up pretty bad. Everyone who landed here was low on gasoline. We received our first bottled Coke in 15 months today. Some of the men went to a WAAF‡ dance and they said it was very good."

---

† Allied airmen who landed in "neutral" European countries (Portugal, Spain, Sweden, Switzerland) were "interned," meaning they would be kept locked up for the rest of the war, but were not technically considered POWs.
‡ Women's Auxiliary of the British Royal Air Force. Similar to US Army WACs.

The airplanes would be quickly repaired, but the 316th was still at an intermediate location. It would be two more days at this "undisclosed airfield" before they arrived at Cottesmore. "Today (February 23) at noon they told us that everybody but the pilot, copilot and crew chief had to leave tonight by train to go to our new station. The reason is to make more room for the new Groups coming over.

"I was coming back from lunch today with a few other men when I heard someone holler 'Uhrig!' When I looked around I saw Major Petty. He was the copilot with Krebs on the airplane when we flew to Alaska in 1941 and 1942. He is now a major and is the commander of a new transport squadron that just arrived here today. After he finished lunch he came back to shoot the bull, so Captain Etter, Major Petty, Sgt Fenimore and I started drinking champagne to celebrate.

"Major Petty spent nine months in the South Pacific§. He has the Air Medal and the Distinguished Flying Cross with one oak leaf cluster. His airplane was down on a coral reef for eight days with 18 casualties. When the Navy sent in two flying boats to pick them up on the eighth day it was too rough for them to take off so they drifted for another 24 hours. The Navy sent a destroyer to try to pick them up and the two flying boats ran together and one of them sank so they had to use life rafts. It took the destroyer another 12 hours to pick them all up. He said that about half the patients they used to haul out of Guadalcanal were crazy. I had to take him to his quarters before I caught the train. He said, 'Uhrig, this is the second time that you have gotten me drunk.' The first time was on our way to Alaska. I sure did enjoy seeing him. We all caught the train at 1830 and it was 2030 before we pulled out of the station."

The following day was a long train ride to their new and final location in what the men considered to be "typical" British weather. "The train ride was sure something. When we arrived at Derby at 0300 they sent us to the British YMCA to get tea and toast. At Derby, we had a three-hour wait so a lot of us shaved and cleaned up at the YMCA. During the night there were card games, men sleeping in the aisles and men drinking and doing almost everything that could be done.

"When we changed trains at Nottingham, it was just getting light and the fog was so thick that you could see only about ten feet. This was our first taste of England's fog. On our way to Oakham where we were to get off the train, we got out of the fog and were able to see what the country looked like. The villages were beautiful and each one had its own little church with a very tall steeple. All the buildings are built out of brick. The farms are very neat and you never see anything just lying around. We arrived at Oakham at 0830 and there were trucks to pick us up and bring us to the field here at Cottesmore. We were about dead from the long train ride."

**New Home at Cottesmore**

The 36th Squadron then began to settle into the new permanent base at RAF Cottesmore. The field was located northeast of Leicester in north-central England. The runway was finished in the late 1930s for use by the RAF, but was turned over to the USAAF in September 1943. It was to be the home for all four squadrons of the 316th for the rest of the European war.

The 316th and its four squadrons would now be part of the IX Troop Carrier Command which was activated in October 1943. A troop carrier wing and three TC groups were assigned to the command in October and November 1943. Evolution of the command was accelerated in February 1944 by assignment of a second troop carrier wing and two additional groups. March witnessed expansion on a sufficiently large scale to build up the command to full organizational strength by the assignment of a third wing and nine groups.[41]

For the men, the adjustments to their new environs in the UK were going to be a little tougher than their other moves. Gone were the sunshine, dry climate, sand storms and open spaces of the Mediterranean region. These were replaced by something the men had not really seen since leaving the States – winter weather with ice and snow. At the end of February, "It snowed most of last night and this morning. The ground has about three to four inches of snow already and it is still snowing.

"We have to work today (Sunday) because we don't have all the cabin fuel tanks out of the airplanes yet

§ Major Cecil E. Petty, Commander of the 92nd TCS, 439 TCG, ditched a C-47 on the Huon Reef, north of New Caledonia in October 1942. Full story of the action is described in *Young*, pp. 56-57.

and we have a lot of other work to catch up on. And we only have 34 men in Engineering. That is counting everyone, Supply and all the other sub-departments. We do have our bicycles now and Engineering has received five of them. They are English made and they pedal very easily.

"In the afternoon we all had to go to the General's reception and tea. At the reception Pace and I met two English gentlemen who live in Oakham and they invited us to come and visit them. These two and one other were the only civilians at the reception. Major McMenamin was the president of the reception. He is also in charge of the tea parties in the future. He told me that if I ever wrote home about it he would kill me.

"It has snowed most of the day and the trees and bushes are covered with it. Everything is beautiful. We had quite a few snowball fights today. I was shaving tonight and Welter let me have about ten pounds of snow right in the face. Pace and I would watch and every time someone would go to the toilet, we would give them a 'snow bath.' Later Pace and I went to the NAAFI¶ tonight to drink beer with the enlisted men. Kestner told me that they had made out my promotion papers to send them in again. My last ones must have been lost because they never came back approved or disapproved."

The 316th now struggled to launch missions in this English weather. The Troop Carrier crews were getting their first taste of the difficult operating conditions endured by American fighter and bomber groups that had been operating in the UK for almost two years. "It froze last night and is much colder this morning. They wanted to send out four of our ships this morning but we were unable to do so because of the ice that was under the snow on the wings. We sent out all our men to sweep the snow off. The sun came out at about 1000 and the ice began to melt. The ships took off this afternoon. Three of them went to Northern Ireland and one took crews to get new ships.

"We are going to have three new C-47s. None of them have more than 90 hours on them. Two of them have glider pick-up equipment on them. They are used to pick up gliders from the ground and on the fly. Of course no one cares much for that idea. As yet no one knows how it works and the crews don't care to find out. There is a winch inside those airplanes that must weigh a ton. If they keep putting more equipment on these ships we won't be able to haul anything at all."

The last day of February (leap year in 1944), the men finally had a chance to get out and see the local area around the base. "Last night a lot of us went to the Red Cross club which is about a city block from here. It is very nice with a lounge, ping-pong and pool tables and all sorts of games. They also sell coffee, tea, cakes and Coca Colas. Just across from our barracks there is an English YMCA that has about the same thing. It is wonderful to be able to go out in the evenings and buy coffee and doughnuts.

"The three ships that started for Northern Ireland yesterday had to turn back because of the weather. This morning they were going to try it again but the airplanes were covered with frost so they still could not take off. They are going to try it again this afternoon. It was very cold last night and it has not warmed up very much this morning. No. 2 finally came in late yesterday and the ship is OK, and also the crew. They were unable to tell us anything that happened or where they were because they had signed a pledge. As soon as they landed, they let them change clothes and then put them on a train for London where they had to report to Troop Carrier Command."

Bob's first letter home from the UK on March 1 was necessarily secretive. "Well here I am but I can't tell you where it is. I got news from home from Major McMenamin because he had been getting mail from home. Mac said that I had better never write home about last Sunday but here it goes. He was president of the General's tea. We had a reception for the General and Mac had to take care of all the details but the tea was the last straw. All of us are getting into a terrible frame of mind. We have been over here almost 16 months and we hear about strikes back home and the wages that everyone is making. We hear so much about rationing and then when we get new pilots and crews we ask them all about it and they all tell us the same thing that no one in the States is going without any more than they ever did. All this don't go down so good. Major Petty who was co-pilot when we flew to Alaska has just come over here. He was in the South Pacific for nine months. He said after he was home a short time he put in for overseas again. He told us that things at home make you so mad that you can't stand it."

¶ Navy, Army and Air Force Institute. The British equivalent of the American Post Exchange (PX), canteen or recreation hall. Found on almost all bases housing UK Forces.

That same day the 36th had their first minor mishap in Britain. "Last night we went to the Red Cross and had our coffee and doughnuts. Lt Frost taxied No. 8 into No. 13 just before dark so now we have to put a new wing tip and aileron on No. 8. No. 13 was not hurt bad enough to change wing tips."

Although the work was mounting in preparation for the invasion, the quality of life for the 316th was considerably better than it had been in Sicily or North Africa. For once they were operating from a field fully outfitted for military aircraft. "This field is called the Randolph** of England and it is a very beautiful field with all brick buildings. There are four large hangars here. The paratroopers [82nd Airborne] have one, the RAF has one, the Service Group has one and our Group [316th] has one. The offices in the hangar have steam heat. Engineering has one and Technical Supply has the other. It is a little difficult for all the squadrons trying to work out of one hangar. All the men want to move out on the line so that when the Ground Echelon arrives we will have to go back into tents out on the line anyway.

"There is one theater on base which has a movie every other night. Both the enlisted men and officers living quarters are very good. We have steam heat in every room. There are three officers to a room and it is very home-like. We have a good shower and bath with hot and cold water. We even have a wash room for washing clothes and a drying room. Lt Quisenberry and I built a table for our barracks room. We now have our radio sets on it and also a table lamp. We also built a set of shelves for our toilet articles."

On March 2, Bob took a short trip to another Troop Carrier base in England to scrounge some spare parts. The number of Troop Carrier groups in Britain was expanding considerably to handle the airborne requirements for the invasion. "Flew to Ramsbury this morning to get some parts but as usual there were none. We picked up six new jeep trailers to bring back. Also heard that Marcum and Guiles are over here somewhere and I was also asking where their Group was but no one knew. Our base restriction is up tonight so I guess that everyone will go into town. There are trucks on base that take the men to all the towns in the area."

The next night, Bob and a small group ventured out to see the local area. "Quisenberry, Welter, Etter, Stracke, Love, McWhirter and I got the Engineering weapons carrier†† and went to a town called Leicester which has a population of about 200,000. We went to a dance and I just sat and listened to the music and watched them dance. There are just as many women here as we have been hearing about, but the good looking ones are few and far between."

And the following day Bob was able to visit another local town. "Went to a town called Grantham. It is not too hot a town. This morning we had a review for Col Fleet. It was so cold that everyone froze. We have a ship going back to Sicily today. I would give almost anything to be going along.

"We had a large dance at the Officer's Club. It was the best Army dance that I have ever seen. We had good punch, good music and there were quite a few women to dance with. They even had Scotch which is very hard to get anymore."

*Left and Right* Scenes from the dances at Cottesmore. (AFHRA)

** Probably a reference to Randolph Field in San Antonio, TX, which was a showplace of the Army Air Forces and known as the "West Point of the Air."
†† Dodge trucks, US Army designation WC-52, were know as "Weapons Carriers."

A Pass In Review ceremony was held on March 5 for Troop Carrier Command and decorations were presented to a number of 316th men including Bob's buddy Jim Farris, for his actions over Sicily. "We had a review here this morning for a General. Our squadron commander and my ex-roommate, Major Farris received the Distinguished Flying Cross (DFC.) He sure did deserve it but it took them long enough to get it to him. We have almost caught up on our work so I don't have to be running down here every night. We have more fun in the barracks than I ever have from going out. The few that stay in the barracks each night always get together for a session and I laugh until my sides are sore."

*Above*: The 316th Pass In Review ceremonies held at Cottesmore. (AFHRA)

*Right:* Major Jim Farris receives the Distinguished Flying Cross on March 5, 1944. He would be killed two months later in a mid-air collision that occurred duing a pre-D-Day exercise. (AFHRA)

**First Losses in the UK**

Unfortunately with the awards and camaraderie came equal amounts of tragedy when the Squadron lost another C-47 and crew, but for the first time due to an accident and not combat action. "This afternoon we received some very bad news. No. 7 crashed and killed the crew. They were Lt Munger, Lt Oiler and Sgt Poppleton. This is the first fatal accident we have had since the Squadron was formed in February 1942. All the news so far is that they flew into a hill." Later on Bob would learn that the aircraft flew into the ground during some bad weather.

On Tuesday, March 7, the Squadron "held services for the crew members that were killed in No. 7." Bob also received some letters from the wife he had not seen in well over a year. "Also received some mail from Toots today. One letter was dated February 29. If the mail keeps up like this, we will be able to get news before it is too old to get any good out of it. Toots sent me some pictures of the room at home and they are just about to drive me mad. Some of them were of the bar, bed, furniture and my Toots. I just keep looking them over and over. One of them was of Toots in bed, the little rat! She has changed a lot. She looks older and prettier than when I left, if that is possible. I feel like stealing a ship and flying home to Toots.

"We are closing in the Major's jeep with canvas sides so he won't freeze to death driving around here. He is going to let me borrow it once in a while. Our transportation is very hard to get because it is all kept together in one area. The Group controls the transportation and we have to call every time that we need to go somewhere."

On March 10, Bob was finally able to write to Toots and tell her about his harrowing trip from Gibraltar to England. It was of particular interest to her since it was "her" engine that almost conked out. "It makes me nervous that some of those 4F guys at home might like your looks too. So it tickles you to know we have your name on the engine of one of our airplanes (*TOOTS*). Well, at 2:30 in the morning out over the ocean you caught fire and tried to quit so we had to shut you off. We were at 11,000 feet and when you quit we started down because we were loaded too heavy to stay up on one engine. We prepared to make a crash landing when the fire went out so we decided to try to start you again. Well you started but how you ran we won't talk about because it was terrible. So we just sat up there for the next six hours watching the sparks fly and waiting for you to quit, but you spit and sputtered and kept running. We were glad to see land and get on it again."

**Witnesses to the Air War Over Germany**

As the men of the 316th grew accustomed to their new surroundings they quickly became aware of the immense air war already going on around them. The strategic bombardment campaign against Germany and other targets in occupied Europe had been underway for many months. The USAAF bombed by day and the RAF by night. The USAAF bomber fleets were being escorted further and further into Germany by fighter groups equipped with aircraft such as the P-51 Mustang that could reach all the way to Berlin. The first daylight raid on the German capital was conducted by the U.S. 8th Air Force on March 4, 1944. Bob Uhrig was witness to the enormous streams of bombers flying overhead on their way to targets on the continent. "Three days ago American bombers started raiding Berlin in daylight. The first day they lost 86 aircraft out of 850. The second day they lost 68 out of 850. This morning they started going over at 0600 and at 0800 they were still going over. This must be the largest raid of all.

"We had a review this morning (March 12) and Generals Clark and Williams‡‡ were there. My roommate Lt Quisenberry was awarded the DFC. There have been only five of them awarded in the entire Group since we have been overseas. He is sitting here now looking at it. Manning was here tonight and we shot the bull. There was another dance which we went to, but Manning and I talked instead of dancing."

Lt George Quisenberry receives the Distinguished Flying Cross from Brigadier General Harold Clark, 52nd Troop Carrier Wing. (AFHRA)

The next day, "I got Captain Etter and we flew Manning back to his home field which is about 25 miles from here. In the evening, Capt Etter, Lts Quisenberry, Welter, Gwyn, Mulder, Sullivan and I sat here in the room and drank Coca Cola and wine until 2200."

Bob's diary on March 17 makes mention of preparing for a mission for the past few days, but not knowing if it was "the real thing or not." It might have been a practice airborne mission or an airdrop of supplies or special operations agents on the Continent. In any event, he wrote, "It is now 2230 and they just told us it was called off. We were to take off at 0045 tonight and get home at 0400."

Bob also saw first-hand the carnage of the strategic bombing campaign. "A bomber that was lost landed here a couple of nights ago. He had five dead on board and everyone was wounded. You should see how some of these bombers come back. I don't see how they stand up. We should all take our hats off to the bomber crews because they are taking a terrible beating on these Berlin raids. They have to fight their way over and back. But I guess it is worth the loss because they are really making Germany hurt.

"A P-47 landed here on a return trip from Berlin and he told us that they flew so low that they had to pull

---

‡‡ Brig Gen Harold Clark, 52nd Troop Carrier Wing and Maj Gen Paul Williams, Commander of 9th Troop Carrier Command.

up to go over the towns. They saw people riding bicycles along the roads and they would wave to them before they knew that the planes were enemy."

The loneliness of separation was starting to get to Bob as he wrote in his next letter home. "Last night we had a dance here at the Officer's Club and there were 150 girls here from a college which is about 40 miles from here. I did not dance a single dance. They have started calling me an old man. But women just don't even appeal to me because I am always thinking of you. You have something on me and I don't know what it is. You have it all over the women that I have seen since I left you."

*Above:* The 316th's own band (l) and orchestra (r) play for the dances at the Officer's Club. (AFHRA)

"Crook is sick again but he is not in bed this time. He does not look at all well. His head hurts all the time. We get a lot of good programs on our radio here. We can at least tell what the people are saying, that is, sometimes. But we still can't pick up the German broadcasts."

On March 20, Bob was able to take some time to travel across England to seek out Toots' brother Omer who was also in the USAAF. "I am sitting here alone tonight. Quisenberry went on leave to London and Welter went to a town nearby. I got a ship yesterday and we flew all over the neck of the country looking for Omer. We had very little to go on because his unit is on detached service all over the island but not one Sgt got the wise idea about looking on their promotion list from their headquarters and there, big as life was Omer Fultz. So then they looked up where he is. Now if he doesn't move soon I am going to fly down and see him. He is about 150 miles from me.

"Just don't think anything about it when I blow up in a letter. They used to try and make us do the impossible with no men back in Africa before the invasion. I never told you before but a couple of time I took stock in myself and I thought I was going crazy. More than once I almost went to pieces. You don't know how we lived and what we had to put up with. Tell Mrs. Singer I have pictures of his grave and the plot and grave number and I will give it to her when I come home or can send it. Tell her it is a beautiful place and very well kept. It is an all American Cemetery." [Bob was referring to TSgt George Singer of the 36th, killed in action on July 11, 1943 over Sicily.]

Bob said goodbye to a long-time roommate on March 22, with the departure of Lt Quisenberry. "Just came back from the Engineering office. It is 2130 and McMenamin and I had a long session. I am listening to a rebroadcast of Bob Hope's show and it is very good. Things are beginning to turn a little green even if it is still cold. It is just about like early spring back home. They are very strict on blackout regulations. They have blackout screens that you place in the windows each night. Here in the barracks we have drapes that we pull in front of the windows. Quisenberry moved out of the room because he is going to a special school. Lt Sullivan moved in with us by our own request. He joined the squadron about 12 months ago. It does not look like I am going to get any time to go over and see Omer because we are so busy."

To meet the absolute necessity for expert navigation in airborne operations, a Pathfinder School was established by HQ IX TCC on 1 March 1, 1944. Intensive training was given to carefully selected pathfinder crews in the use of all aids which enable a navigator to pinpoint his position in normal, night, or instrument

weather.[42] George Quisenberry was on his way to become a charter member of the Pathfinders unit in the UK and was assigned to the school and Pathfinder group for the duration of the war.[43]

By March 26, "We were pretty well caught up with our work, so we took this morning off and also part of the afternoon. But tonight there is going to be a parachute jump, so we will have to work. In the afternoon, Captain Chiros and I rode our bicycles to a small town named Market Overton which is about two miles from the field. We had a few drinks of beer in a pub before it closed at 1400 (Sunday hours). The beer here is terrible stuff. The bottled beer is the only kind that even tastes like beer. Then we went to the edge of town and stopped at a small church and cemetery.

"The church was built by the Romans in 1200. It has been rebuilt but the original arches and walls still stand. They tell us that most of the churches are of this type and origin. We also saw the Roman baptismal bowl and a Roman casket. The casket was made from stone and it was built in the shape of a body. We then started looking at the graves in the cemetery. We found one dated 1695 and quite a number in 1700. The tombstones here are much different from the ones in the States. These here are plain and stand about four feet tall and three feet wide. They are only about three inches thick and are not marble, but stone. The town is about the size of Donnelsville. Every little village is a clean and neat. I noticed that every window in every house just shines."

The 36th Squadron received two more C-47 aircraft on March 29, bringing the squadron total up to 16 aircraft. The next day Bob went on another junket to find spare parts and at the same time tried to track down his brother-in-law. "I got a ship and two pilots today and we flew down to Aldermaston which is about 100 miles from here. I wanted to find Omer. Also, we need some brake shoes and there being a Troop Carrier Group there, we thought we might get a couple.

"I located Omer in the 40th Mobile Communications Squadron (Detachment C). We talked about home and the places we have been and seen. I showed him a bunch of my junk – guns and pictures. It was sure a thrill to see him. We had a very good time. He is much heavier than he was when I left the States. And yes, we also got seven brake shoes, so I was able to kill two birds with one stone.

"A new B-17 stayed here at Cottesmore last night. The ship and crew have only been over here two weeks and they have been on five raids already. They put you to work just as soon as you arrive now."

*Left*: Toots' brother, Omer Fultz.

*Right*: An airman at Cottesmore with a B-17G in the background. Markings indicate it was assigned to the 401st Bomb Group at Deenethorpe, about 20 miles from Cottesmore. (Mike Ingrisano collection)

On his return from Aldermaston, Bob quickly got a letter off to Toots to let her know he had met up with her brother. "We have just caught up with our work so we took today off. It was a wonderful feeling to lie in bed this morning and not have to get up. It is the first time for so long that I cannot remember the last time. We took a ship and flew down to see Omer day before yesterday. He is in a small detachment and that made him hard to find. I opened a door on a radio truck to ask about his whereabouts and there he sat running the radio. We both about passed out. He was relieved in about 15 minutes and then we went to his tent. I told him it would do him good to live in a tent. I also told him that he would get used to it after twelve months. His tent sits in a large grove of big pine tree with a large lake at the foot of the hill. It is a much prettier part of

the country than were we are living. Best of all he has to cut his own wood. We compared notes and pictures for almost two hours before I had to leave. I want you to tell Opal not to worry about him because the highest he ever gets off the ground is when he climbs up into the radio truck. I got a big kick out of watching him take code and sending it. I called him a dit-dot artist - that is what we call all radio operators. We are going to try to get together again in about two weeks.

"The chow here is wonderful. We get fresh meat at least once a day and fresh butter every meal. Soldiers that have been in England all the time don't know what it feels like to go hungry and do without water. Last night I washed all my own clothes. I have done it for so long that it is nothing new. But the last place almost spoiled us because it was no trouble at all to get some lady to do all your laundry and have it back to you the next day, if you had the soap to give them."

The Group held another award ceremony and review on April 6, this time for General Royce. At that event, Major Jim Farris received what was apparently his second DFC, "for bringing his ship back from the Sicilian mission." There was also some continued reshuffling of the Group aircraft as Troop Carrier Command filled out all the units in preparation for the invasion. "We are going to transfer Nos. 2 and 5 today. This will leave us with only one original ship, No. 6. This will leave us with 17 transports and one Cub.§§ I have been up flying in the Cub once already.

"In Leicester the day before yesterday I bought a pair of high top boots and I am having them cut down to 11 inches. I also found some potted flowers and we now have them in the Engineering Office window."

A few days later, Bob had an enjoyable reunion with an old friend. "Lt Colonel Washburn (Group HQ) called me today and told me Colonel Krebs¶¶ was coming to the Group dance. Col Krebs was the pilot when I went to Alaska, and he also taught Toots to fly. So this evening I went to the dance and met up with him. He is just like always, he does not drink, smoke, or dance. He is a group commander and also base commander. He tried to talk me in to coming over to his outfit. It was one of the biggest thrills I have had in a long time. I always thought more of him than any officer I have ever known. We came over to the room and I showed him some of my photos and other odds and ends. Then we shot the bull until 0300 and then returned to the dance. There we met Major Petty (copilot to Alaska), now a Squadron Commander [92nd TCS], so the three of us got together."

Sunday April 9 was Easter so it was a quiet day at Cottesmore. "No work today except to change a tire. MSgt Crook and four men are the only ones working today. I slept until almost noon. Then I shined my new boots before going to chow. In the afternoon, Capt Chiros, Capt McCullough, Lt Sullivan, Lt Cunningham and I went for a bicycle ride. We rode to Market Overton again and had a few beers then we went to the little church to see how they had it fixed up for Easter. It was beautiful inside with flowers everywhere. It started to rain just as we started for home and we all got wet. By 1700 it had cleared up so we all decided to go to Oakham to the movies. We got back about 2300."

The next day it was back to regular work. "This morning we transferred four English Horsa gliders and are getting in exchange American WACO CG-4As instead. Lt Pace came into the office and said he felt like flying so I told him that we could always test hop something. We flew over to see Krebs. I took my diary along to let him read it. He took us up to his quarters and we talked for an hour, and then Pace and I flew home. Four of our ships went to Northern Ireland today. Major Farris would not let me go because it was an overnight trip."

In his next letter home, Bob continued his descriptions of the quality of life at their new location. "Yes, a pub is a bar. None of them are fixed up like the ones at home. Everything is plain. But they are cute because they are so different, with the fireplace and all. Had real honest-to-goodness wieners day before yesterday. I ate them until I could hardly walk. Just think we get fresh butter almost every meal and fresh meat once a day. But as yet (17 months) no fresh milk have I had. A fellow (new one) at our table the other day complained <u>about the food</u>. When we finished with him he thought he had killed someone. He will think the next time

§§ Piper "Cubs", known in the USAAF as L-4 liaison aircraft were sometimes assigned to combat units to allow the pilots to get additional flight time when not on primary missions.
¶¶ Colonel Frank X. Krebs, at that time Commander of the 440th Troop Carrier Group

before saying anything about the food. The fellows that went through all the hardships in the desert just can't stand something like that. That is what I am afraid of when I come home, the first person I hear gripe about something will have me to whip. Have you read *They Flew Through Sand* and *The Libyan Log*? If not, try and find them. I am going to try and send you an Eighth Army book."

**Invasion Preparations**

To keep the men sharp, Troop Carrier Command decided to have a "practice deployment" on April 11. This was probably in preparation for moving aircraft from the UK to France as soon as possible after the invasion. Plans were already underway to construct temporary airfields right behind the beach areas of Normandy as soon as the area was secure. Although most of these fields would be used for fighter aircraft supporting the ground troops, it was envisioned that transport aircraft would be required to bring in supplies and evacuate the wounded, as the 316th had done in North Africa. But Bob figured the 316th was already well prepared for short notice moves. "They told us this morning that we have to make a practice move. As if we have not moved enough! The Wing Commander tried to get our Wing out of it because he said we have had enough practice. We are going to move everything away from the field for 24 hours.

"We are now on a large glider training program. We are pulling both English Horsas and American CG-4As. The sun shone today about half the day. It is much warmer now. We have just been talking for a while about the things we used to do in the desert. I don't think we will ever forget them. We still get a lot of laughs."

In one of Bob's letters to Toots he described the quaint yet extremely dangerous practice of USAAF men – washing their clothes in high octane aviation gasoline instead of soap and water. "I went over back of the mess hall yesterday and watched them stack their straw. They do not use a bailer. They use a conveyor and there are three men on top of the stack that shape the stack. I have to admit that everything they do is done well. These blackout screens are a lot of bother. You have to put them in all the windows each night whether the lights are going to be on or not. This afternoon I washed one pair of pink [tan] and one pair of green pants in gasoline. Then tonight I will have them pressed. We have a man in the back of the barracks that has two electric irons so he does our pressing." Since Bob made no mention of explosions in the laundry, the process seemed to have been successful this time around.

On April 14, Bob recorded that the Squadron transitions were continuing. "We are still getting in new ships and crews. We now have 16 ships. We transferred two of our old ones out. No. 6 is the only original ship that we have. It has 1437 hours on it. Old Willie Coursen would like to know that.

"Last night we had an air raid without any bombs. It was almost like old times but no show like we used to have. There is supposed to be a Group Formation tonight but it is raining and we hope they call it off. Our mail is still pretty good. Most letters get here in ten days to two weeks. I even received two flight caps in the mail from Toots a couple of days ago."

Bob continued his descriptions of the British living conditions for Toots. "We have some work to do so we are working tonight so we can be off tomorrow. We should be finished by 2100. Lt Pace and I have to test hop two of them tonight. I am chief test pilot (don't I wish I were.) Don't fret about the Red Cross entertainers because they are only for the enlisted men. We can't pick up many German stations here so we don't get much good music. There is an American station in London that puts on some good programs. They also broadcast Bob Hope and other shows. They use a lot of Italian prisoners here to work on the farms. They are planting potatoes here now. That is one thing that these English have plenty of. If you want to see gasoline rationing, you should be here. You might see two cars a day. No one has driven their car for four years. They cannot even use them to go to work. They have to ride the trains, buses or bicycles."

Sunday, April 16 was, "a day of rest which is what I used it for. I got up at 1145, ate and was back in bed by 1330. Afterwards I lost all my change in a crap game that was going in high gear in our room. In the evening I got out a $12 bottle of Scotch that came from Ireland and Lt Phillips, Lt Welter, Lt Sullivan, Capt McCullough and myself proceeded to drink it. The whole conversation for the evening was about hunting and fishing."

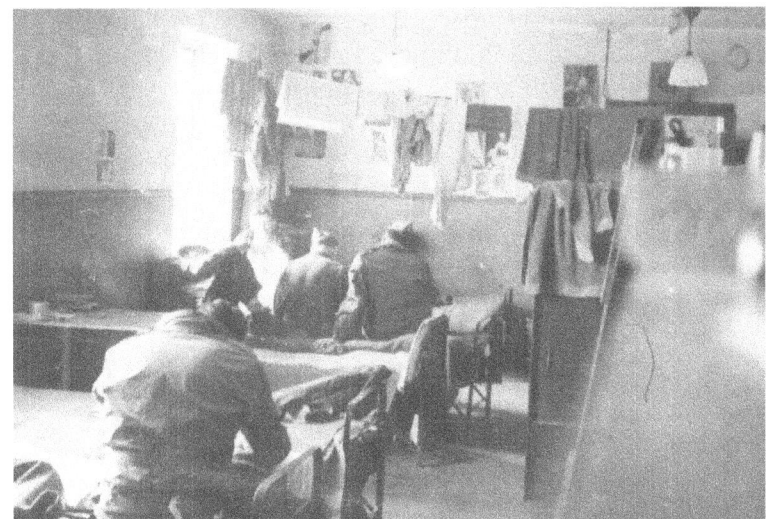

Yanks living in Britain.

*Left:* The ubiquitous British bicycles - familiar to every American airman.

*Right:* Enlisted Men's Quarters at Cottesmore. (Mike Ingrisano collection)

Training for the invasion now continued in earnest. All the Airborne divisions and Troop Carrier Groups were now based at various locations in Britain and they began a series of exercises which grew more complex as the date for the invasion drew nearer. The lessons learned in North Africa and Sicily about training were being put into practice. All the units were brought up to or over strength in personnel and equipment. "Today we have 11 ships to fly and tonight there are 14 flying in a Group formation. We have been doing more formation flying here than any place that we have been before. The pilots are getting pretty tired of it. Every fourth day we tow gliders from 0800 to 2100.

"We have so many new pilots that I don't know half of them. It has been pretty bad weather all day but we are still going to fly tonight. We now have 18 ships and are supposed to get two more. We have four flights now instead of three, so we made Wallace the new Flight Chief. I still have this cussed cold and I cough about half the night."

On April 18, the Squadron took on yet another C-47. "We received another airplane today which brings the total to 19 transports. I don't know what we are going to do for crew chiefs.

"Later today Sgt Fenimore, Sgt Street and I went into Leicester. First we had lunch at the Grand Hotel then we went to the Bruin shoe store to have my shoes fitted. After that we went to the Odeon Cinema and saw *The Cat and the Canary* and *The Mystery of Morgan's Creek*. Then we had dinner at 1900 and drank some scotch. We got home about midnight."

The 316th Group staged a practice drop for British Airborne forces on April 21. Bob commented that it was the first time they had dropped British paratroopers and that, "It was a lot of extra work getting the ships equipped for it."

## Captain's Bars

April 22, 1944 was a big day for Bob Uhrig – he was officially promoted to Captain. The Group Commander, Col Burton Fleet and Major Jim Farris pinned on Bob's new Captain's bars. The promotion was actually effective on April 15. Bob wrote home to Toots to give her the good news, along with some other updates on his Squadron mates. "Well I am now a big juicy captain. It is the biggest thrill that I have had for a long time. Colonel Fleet and Major Farris put the bars on me Saturday night. I can hardly get over it. Has Major McMenamin told Peg that he is going to be operated on for his appendix? I was up to see him and Lt Quisenberry yesterday at the 303rd Hospital.*** Mac is going to be operated on this morning. They have been bothering him so he decided to get them taken out while we are at such a good place. The hospital here is as good as any in the States, that is, field hospitals. It is located in a grove of trees and there is a castle close by where the staff lives. The hospital is one series of Nissen huts connected by brick corridors. They are built

*** 303rd Station Hospital, located on the grounds of Lilford Hall estate near RAF Molesworth where the U.S. 303rd Bombardment Group was based.

on the grounds of the castle. It is the best location I have ever seen for a hospital. Everything is so peaceful and quiet with plenty of beautiful landscaping. I think that I would like to go there for a while. It is about 35 miles from here. It is no wonder the British are in love with their country. No rushing here and there like we do in the States. You just feel like laying back and having the world go by.

"The trip to the hospital is through rolling hills with green meadows and small wooded areas. The streams and lakes are very clear. They even cut the grass away from the sides of the road like we trim our sidewalks at home. All the buildings being of stone or brick, their age does not bother them and they all look nice. The pheasants were thick all the way. You could look out just about any time and see four or five. There are many castles in this country and the landscaping around each one is a work of art. They all have lakes bordered by trees and shrubs and everything is well kept."

The 303rd Station Hospital near RAF Molesworth. The original estate (castle) is in the background. (US Army Signal Corps)

Bob continued his description of Britain in his next letter and introduced his new roommate, Richard Etter. "He is a captain in our outfit and we roomed together for some time. In one of my letters I told you about him. It must have gotten lost. He is a big Dutchman and we talk about how we are going to run Germany after the war. We are still at the same place. They just made the change but we remain the same. The weather here is beautiful. It is just like lazy springtime at home. The grass is ready to be cut and the trees are coming out in bud. Another thing is that it doesn't get dark until 10:30 pm and it gets light at 4:30 am. They tell us that in the middle of the summer there are only about three hours of darkness. We are told that they have chilly days in the middle of summer. The weather goes from good to bad in five minutes just like Ohio."

A big package from home arrived on April 28. "I received a gabardine trench coat, pen and pencil set and toilet articles from Toots. I shall write this day starting with my new pen. The trench coat is a beauty. It cost Toots $62. Everyone is raving about it.

"They are still flying every ship that is marked 'in-commission.' It is about to kill us. We are so short on help and we now have 20 C-47s. We have less than three men to a ship and it is just too much work. On top of that we fly every other night and pull gliders every third day. These English Horsa gliders are pulling the guts out of our engines. We had an inspection today by the Wing inspectors today and they seemed well pleased, so I am happy.

"We have had some beautiful weather the past few days. The sun has been shining and it has been very warm, just like the lazy spring weather at home. The grass is green and all the trees are coming out. You take one look at the nice green grass and you want to lie down and roll in it. It is 2230 and it is still light outside. I don't know what we will do later in the evenings. We will have to use the blackout curtains to keep the light out instead of in. Major McMenamin came back from the hospital last night and he brought his appendix with him. The kept putting the operation off and he got to feeling better so he came back."

By early May in the UK the days were very long as Bob had observed. There were only a few hours of darkness which is one of the reasons the invasion was originally planned for that time. Once the paratroops were dropped, they would only have to fight for a few hours before daylight and the start of the seaborne invasion. The 316th and other Troop Carrier Groups were spending most of their time in night practice missions since a night drop in Normandy was the main part of the airborne invasion plan.

On May 4 the 36th Squadron lost the last two of their original complement of C-47/C-53 aircraft. "We

transferred our last two ships today, Nos. 6 and 12. Now we don't have any of our original ships left. They are still bombing Germany both night and day. There is just a steady stream of bombers going over all the time. Pappy Street is in the Base Hospital. Doc Boyer put him in to try to cure his cough. He has had a cold ever since Sicily and keeps coughing all the time."

A few days later, Bob was able to spend some more time with his old friend Col Frank Krebs. "Krebs flew up to see me today. He arrived about noon so we had dinner together at our mess. We had fried chicken. Then he wanted to see some more of my photos. So we looked at photos and shot the bull. Then I showed him how we had made everything in Engineering. He told me that he was going to bring his engineering officer up and show him how we had made everything. He is group commander of a Troop Carrier group just like ours."

One of the traditions that developed for American units in the UK (particularly Air Force units) was to "adopt" British children orphaned by the war or the Blitz on London. The 36th followed that tradition on May 10. "Sgt Johnson went to London today and adopted two kids for the Squadron. To support them, we used the money that was made from our bar in Sicily. We adopted a boy, 8, and a girl, 3. We were able to give them $800. That makes enough to keep them until they are big enough to take care of themselves. Next month they are coming to visit us for one day.

The size of the American Air Force in Britain continued to grow. By the end of March 1944, 65 of the 161 units which were to comprise the IX Troop Carrier Command had been organized. By the end of April, three troop carrier wings and fourteen troop carrier groups had been allocated to the IX Troop Carrier Command. An indication of the growth in operational strength of the command preparing for the invasion is the increase of C-47 and C-53 type aircraft; from 353 in February to 1226 transports and 2500 gliders in May 1944. At that time, each troop carrier group was authorized 64 aircraft and a reserve of 25% totaling 80 aircraft. Two combat crews manned each aircraft.[44]

Bob saw this buildup and the impact firsthand. "We received four more ships today and now we have 22 DC-3s and one Cub. I don't know what we are going to do for crew chiefs if this keeps up. We don't have any spare men to work on the ships. Between KP [Kitchen Police], and Guard Details, there are no men left to do any maintenance.

"Tonight I developed two more rolls of film. All but two rolls have been developed. I have been worrying about the film for the past year because I was afraid it would spoil. All of the pictures have come out good, to my surprise. If I can get some bromide paper I can start enlarging and printing them."

An exhausted Bob Uhrig falls asleep in his quarters listening to the radio. His new gabardine trench coat hangs on the wall behind him. The months of mental and physical preparations for the invasion of France took their toll on all the men of the Group.

**Countdown to D-Day**

The intensive training for all the components of the airborne force continued. The paratroops and glidermen continued their ground exercises, while the Troop Carrier force practiced progressively larger formation flying missions. On May 11 there was a large combined exercise. "We are going to have a practice jump

tonight and each squadron is going to have 18 ships in the air instead of the usual 9. In total we will have three Wings in the air at one time."

One of the participants in that airborne practice operation, Corporal Richard Ladd of the 502nd PIR, 101st Airborne remembered it this way:

> "This was called Operation EAGLE. It was in early May and the only practice jump we had before D-Day. We were trucked to Greenham Common airfield and loaded into C-47s. We thought it was a night practice jump for us in the 101st, but found out it was really to give the AAF practice in formation flying (making the V of Vs) and dropping a lot of troops at night. We were dropped into the area held by the American 28th Division, about 20 miles from Hungerford. The planes circled for some time -- 2 hours -- before dropping us. Many injuries were sustained in that operation – out of the 2000 or so men of the 502nd who jumped that night, there were 108 broken bone injuries. This was many more than a normal jump operation. I was briefly 'captured' by 28th Division troops as part of the exercise, troops later 'escaped' and fell asleep in a hay stack. Eventually the 502nd returned to Hungerford by truck at the end of the exercise."[45]

In addition to the injuries sustained by the paratroopers in this exercise, there was another, more serious, tragedy. This one befell the 316th Group and cost the 36th Squadron a number of men, some of them close friends of Bob Uhrig. "Last night we got our 18 ships in the air. Our Group had 73 planes in one formation. There were over 1000 DCs [C-47s] in the air at once.

"On the way back from the Drop Zone, the front part of the formation had turned and started back when one of the ships pulled up into our Squadron formation. Nos. 1 and 2 on the first element saw him in time and went down and out of the way, but our lead ship did not see him coming up and they collided in the air. Our No. 3 ship had just dived down out of the way too, but when the two ships above collided it threw gasoline on it, but it burned out without any damage. The other two ships (Nos. 1 and 2) exploded and went down together. Some of the pilots that saw it said that the two ships blew into small pieces and there was fire everywhere."

One of the two C-47s lost was from the 36th Squadron, and the other was from the 44th Squadron. "In our ship we lost the Group Commander, Colonel Fleet, our Squadron Commander, Major Jim Farris, Group Chaplain Captain Richart, co-pilot Spencer, navigator Lt Still, crew chief Tomchek and radio operator Sgt Elliott, as well as a number of members of the paratrooper staff [82nd Airborne.] I had lived with Major Farris from the time we came overseas until we moved to Castelvetrano, Sicily. We are not in very good spirits today."

This tragic blow hit the 316th very hard. They had lost two of their most experienced air commanders on the eve of the largest airborne operation ever undertaken. On May 16 Bob wrote home, still a bit shaken from the tragedy a few days before. Unfortunately, due to security concerns, he could not share the details of what had happened. "We now have a little cold weather here now. It good and warm and then it turns cold before you can turn around. Well this makes 18 months overseas and there is still nothing that makes it look like coming home. I often wonder just how long it will be before I come home. Of course we all wonder that all the time. There is just not much to tell you. It is getting hard to find things to write about. It cost me over $30 when I made captain because I had to buy the drinks and then I bought a barrel of beer for Engineering. We are plenty busy and we have very little time to do anything. I have been wanting to look this country over but as yet I have been unable to do so."

Although Bob was torn by a desire to get home and the need to do his duty, he, like the rest of the 316th, knew that their most important missions were still to come. It was now late May, 1944. The invasion was "on."

# Chapter 10

# Invasion

*We were there right on the button. It was a great delivery.*
Lt Col Benjamin Vandervoort, Commanding Officer, 2nd Battalion, 505th PIR.

The airborne phase of the Normandy Invasion was originally planned to begin on June 4, 1944. The paratroops would load up in the Troop Carrier C-47s on the night of the 4th and be dropped in France just after midnight - about six hours prior to the start of the seaborne invasion. Glider operations would begin at first light on the following day. A forecast for bad weather on the 4th led to the postponement of the invasion for 24 hours. The decision was made by the Supreme Commander, General Eisenhower and proved to be a correct one as the weather on the 5th was much better, although far from perfect. It was judged by all the land, sea and air commanders as the best that could be hoped for.

**Assault Planning**

The general timing of the assault was to be determined in accordance with considerations of weather, availability of resources and co-ordination with a Russian spring offensive. Selection of the day would be determined by the conditions required for H Hour*. H Hour, in turn, would be chosen based on advantageous light conditions for airborne operations and tidal conditions on the landing beaches.

After considerable analysis of previous landing operations in the Mediterranean, the Allied senior staff decided that the landings on the coast of France should take place as soon as possible after first light so that, while maximum use could be made of darkness in covering the approach of the invasion fleet, the preparatory air bombardment and naval gunfire could be delivered in daylight. This fire support was considered essential to the success of the landings and at least an hour of daylight was needed for accurate observation of the effects of the supporting bombardment. Adjustments in the landing times would be necessary to account for suitable tidal conditions.

The landing plan was flexible enough to account for the possibility of a day-to-day postponement in case of bad weather. There was a narrow range of acceptable H Hour times from 30 minutes after daylight to one hour and a half. The required conditions of daylight and tide were confined to three days during each 14 day period. However, there was an additional requirement for sufficient moonlight to allow for successful paratrooper operations. This narrowed the available times to only three days over a whole month. Using this formula, the invasion planners set the first target dates for the invasion of 5, 6 and 7 June.

A centerpiece of the invasion plan was the employment of 23,000 British and American airborne forces, both paratroop and gliderborne, to hold the flanks of the 60 mile invasion front. They would begin arriving the night before the assault troops would hit the beaches. However, British Air Chief Marshall Leigh-Mallory, Chief of the Allied Expeditionary Air Forces, was convinced that the American portion (14,000 troops) of the airborne operations would be a disaster, based on the unsuccessful operations in North Africa, Sicily and Italy. He predicted 70 percent losses in glider forces and at least 50 percent in paratroop strength even before the paratroopers hit the ground. Leigh-Mallory warned of the "futile slaughter" of two divisions of airborne soldiers. In spite of these dire predictions Eisenhower retained the airborne portion of the invasion plan.

Every means at hand had to be employed to make the invasion a success – it was an operation that could only be undertaken once. Britain's moral and material capital was nearly exhausted. Failure, to say nothing of disaster, would have had incalculably negative consequences for the war effort. The U.S. might recover from a physical defeat on the beaches, but morale would be dealt a serious blow.

---

* H Hour is an arbitrary military term denoting the opening time of an amphibious landing operation. All times for the operation are expressed in terms of H Hour, e.g. "H+30" means 30 minutes after landing operations have commenced.

## The Weather Factor

Eisenhower knew that the invasion plan was being launched with ground forces that were not overwhelmingly powerful. The operation was feasible only because of Allied air superiority. If the Allies did not have that air superiority, the landings were too risky. Airborne troops could not be dropped as needed, and the flanks of the American and British landing areas could not be secured. High- and medium-altitude bombers could not find their targets – the imposing concrete bunkers on the French coast that housed heavy artillery pieces that could shell the invasion fleet or pin down the troops on the beaches. Air support was also vital in the days just following the initial assault to keep the Germans from moving mobile reserve forces into the landing areas. Air power was essential; there could be no invasion without air superiority. Good weather was required for the airpower to be effective.

The Troop Carrier Groups in the UK had been practicing for weeks in the techniques vital to success of an airborne operation: night operations, close formation flying, and above all, precision navigation. The Pathfinder force originally organized in Sicily had been increased in size to handle the drop zones for the two full American airborne divisions designated to make the main assault: the veteran 82$^{nd}$ and the newly-arrived 101$^{st}$ Airborne (Screaming Eagles.) American airlift forces had now advanced by leaps and bounds in a few short years. They were now key to successful combat operations.

What also worried all the Allied commanders was secrecy and security. The troops designated to make the initial assault were confined to areas all over southern England by mid-May. The 82$^{nd}$ Airborne troops who were married up with the 316$^{th}$ were billeted close by so that they could easily be loaded up in to the aircraft. Once enclosed in these pre-invasion areas, none were allowed to leave and there was no contact between the troops and the outside world. The plans for the landings were revealed and the invasion forces were briefed on their part of the big operation.

June 5, 1944: Mike Ingrisano stands in the paratroop door of his 37th TCS aircraft; fully loaded for the D-Day mission and freshly painted with the black and white "invasion stripes." (Mike Ingrisano collection)

Now that the plan was unmasked to the bulk of the combat troops, there was a serious concern that the secret of the date and location of the landings must be kept. The longer the troops were confined, the greater the chance that some details of the impending operation would leak out. In addition, the morale and training of the troops would suffer the longer they were kept penned up. For reasons of security and effectiveness it was most advantageous to move them through the staging areas, embark them on board aircraft and ships and get them to the battle zone as soon as possible. There was also the possibility that the large concentrations of troops could be discovered by German spies or (more likely) by aerial reconnaissance.

The 316$^{th}$'s partner in the upcoming operation was the veteran 505$^{th}$ Parachute Infantry Regiment (PIR).

The regiment was split between two airfields, Spanhoe and Cottesmore.

Sgt Bill Tucker of the 505th recorded the events as his unit was moved into position for the operation. "The whole regiment was transferred by truck to the airfields. The Third Battalion was quartered in a huge hangar at Cottesmore with one Company distributed to each corner. By some magnificent effort, each man had been provided with a cot so it was fairly comfortable. They still didn't know where they were going and there was all kinds of speculation that it would be some place like Norway or Holland, etc. The four or five days they spent there were rather pleasant, particularly in view of the Air Force chow."[46]

Eventually a total of 2095 paratroopers of the 505th were ready for the operation. To maintain security all the D-Day airfields throughout Britain were sealed off by the Military Police. "The only movements between them were the two [505th] Chaplains and their escorts to hold religious services."[47]

All this was in play as Bob Uhrig and the other members of the 316th "sweated out" the first airborne missions on the night of 5-6 June. The long months of preparation were over and it was time for the 316th to go to into combat once again. The overall operational plan for the 316th was this: 72 C-47s would be used to carry the 2nd and 3rd Battalions of the 505th PIR and some members of the 456th Parachute Artillery (82nd Airborne). The jumpmaster in the leading aircraft was Lt Col Benjamin Vandervoort.† The drop zones for this force were just slightly northwest of Ste Mere Eglise. Final briefings were held at 2045 on June 5, and the first takeoff was at 2300, followed by the rest of the aircraft from all four squadrons in 10 second intervals. The last plane, from the 36th Squadron, took off at 2320 hours. The rest of the 505th and the 307th Engineers would be carried by the Group at Spanhoe.

Leading all the C-47 of the IX Troop Carrier Command were 20 Pathfinder C-47s. The other Groups, arranged in the "V of Vs" formation, flew south over England until they reached the English Channel. Here, in a single stream, they followed the navigation signals provided by Allied ships in the Channel until they reached the French coast. They flew straight across the Cotentin peninsula of Normandy from west to east, and after their drops continued out over the invasion beaches and back to England. Of the 1276 troopers aboard the 316th aircraft, all but two jumped and no Group aircraft were lost.[48]

Lieutenant James Coyle, a paratrooper of the 505th described the scene on the flight to France in a 316th C-47. "On the night of 5th June we took off from Cottesmore airfield where we had been isolated under strict security for several days. I was the jumpmaster in our C-47 which carried 18 paratroopers…I had a beautiful view of the English Channel as we crossed. The door of the plane had been removed, and as I sat by the open door I watched the moonlight shine on the waves. It was a peaceful prelude to a violent invasion… Shortly after we crossed the western coast of the Cherbourg Peninsula the anti-aircraft fire filled the sky. This continued at intervals in the distance until we reached Ste Mere Eglise…The plane's engines slowed down and I knew we would receive the green light to go soon, but I could not see the lights which were to be set up by the Pathfinders to mark the DZ. Suddenly we made a sharp left turn and I picked out the blue lights in a 'T' formation directly in front of us. Just at that moment the green light beside the door flashed on and giving the order to 'GO', I jumped."[49]

**Witness to History**

Bob Uhrig's first notes on the operation were sketchy since there was little information available in the early hours of June 6. "Our ships on the mission last night ran into some bad weather and had to break formation. We were all very glad to see they all came back. They all started coming back in at 0400. As yet we have not received any new on the radio. On their way back, our ships passed over hundreds of boats so the Navy should be giving them hell now. I hope and pray that it is a success. If not, it is going to be a long war and it has been long enough already for me. We all feel better again because we feel as if we are doing some good once again.

"They ran into very little trouble. The Jerries must have thought they were the bombers coming back and

† Vandervoort's name would later become a household word when he was portrayed by actor John Wayne in the film *The Longest Day*. One of the other men in the 505th to become famous in the film was Pvt John Steele, who landed on the church steeple in the center of Ste Mere Eglise.

went for cover because our ships were in and gone before the big fireworks started. They saw quite a bit of flak to the south. The damage sustained by the aircraft was all from small arms. No. 10: four holes, two in fuselage, two in elevator; No. 7: tire shot out and one shell went through the navigator's compartment; No. 01: shell through the fuselage.

"We are now repairing our ships and getting them ready for another mission. We are going to use 14 ships with American [paratrooper] equipment for re-supply and we are going to install British equipment in the other eight ships. You see we may have to use British paratroopers for reinforcements. There were both British and American paratroopers used last night but they were used in different places. Also gliders were used.‡ All our missions are to try and keep the Germans from rushing in reinforcements where our boats are to land on the beach."

Later in the day on June 6, reports of the operation began to filter in. "Just heard the news on our Wing. We lost one man. He was killed from small arms fire. No ships were lost out of our wing except one that a paratrooper blew up on the ground from fooling around with a hand grenade, but no one else was hurt." The incident Bob spoke of occurred at Spanhoe. Just before takeoff a grenade carried by one of the men from the 1st Battalion 505th exploded inside a C-47 of the 43rd TCS. It turned out that four men were killed and 15 were wounded. The one paratrooper who was not injured made his way to another C-47 and jumped with them.[50]

Bob continued his description of the D-Day mission. "We were hauling mines, TNT, 75mm artillery shells and ammunition, 75mm guns and small arms.§ The land mines and TNT are very dangerous to haul because a direct hit by a heavy shell and the whole works goes up. That is what happened in Sicily. We hauled the 505th [Parachute Infantry Regiment] and it was the same outfit that we hauled in Sicily. There are not many of the original left in it.

"It is now 0800 and the English have just announced the invasion. She is now on full blast. We must win now. I have had little sleep in the past 72 hours but if we can win I can keep going."

Towards the end of this "day of days" Bob continued his notes on the progress of the invasion, in spite of what must have been a tremendous workload. "It is now 2130 and all is still going well. There were fewer lives lost than had been anticipated. All day long we have been getting news of the invasion. We are going to resupply the paratroopers tomorrow morning. We are only using 13 ships in our squadron, and it is going to be a daylight mission. Our ships will leave here at 0335 and will make the drop at 0629. Churchill gave a speech today in which he said that the Troop Carrier Groups were the spearhead, the best planned and a decisive factor in the invasion. I have moved a cot down here to the Engineering Office because I have to spend night and day down here and that is the only way I can catch any sleep at all."

**316th Executes a Successful Mission**

Although overall accuracy among the Troop Carrier Groups was mixed, the 316th drops were mostly successful. "Out of 117 sticks of paratroopers, 31 sticks landed on the drop zone, 29 within one mile, 20 within two miles, and the rest anywhere from five to fourteen miles away...it had been a fine achievement in night navigation and steadiness under difficult conditions by the leading serial of 36 aircraft from 316 Group; they kept their tight and disciplined formation the whole way round and appeared back over Cottesmore at 5 a.m. in perfect order." In fact, the 3rd Battalion Commander, Lt Col Krause, landed on the exact spot where he had planned his battalion rendezvous. By 0600, the 505th had captured Ste Mere Eglise.[51]

One trooper of the 505th remembered the jump vividly. "All went well until arriving near the DZ 'O'; the C-47 did not slow up for the drop. Everyone in the Second Battalion agreed that it was the highest, fastest jump ever made. Eyeballs had to be screwed back into their sockets. The Second Battalion landed on or near

‡ Night glider operations were kept to a minimum because of the difficulty in landing them in the Normandy hedgerow county at night. One notable exception was the enormously successful British glider operation to capture Pegasus Bridge.

§ In addition to the paratroopers on board, the C-47s again used the pararacks underneath the fuselage to carry equipment bundles, including the components for 75mm pack howitzers.

164

the DZ. Except for one stick from F Company and they headed for the center of Ste Mere Eglise."[52]

The official U.S. Air Force history of the operation went even further in commending the 316th. "The drops at DZ [Drop Zone] 'O' were, taken together, the best at any zone by the IX TCC in NEPTUNE. Half of the troops dropped were assembled and ready for action within eight hours. Among the factors contributing to this were the success of the 316th and 315th Groups in climbing over the cloudbank and descending to the DZ without losing formation, the absence of intense enemy fire during the approach and the lighted 'T'. The last was an [navigation] aid not available on other zones."[53]

The paratroops of the 82nd had all been dropped in Normandy on the night of 5-6 June. Once they were in place, the missions of the Troop Carrier aircraft shifted to glider operations and resupply missions. The glider troops of the 82nd began arriving at first light on June 6. These resupply missions took the form of airdropping supplies to the American forces on the ground. The unsophisticated nature of these operations was for the crew chief to push equipment bundles out of the cargo door of the C-47 when they were over the drop zone. The paratroopers inland from the landing beaches had a tenuous hold on their positions and needed a steady stream of resupply and reinforcements.

Bob was able to get a quick letter off to Toots to tell her about this momentous day, knowing it would be many days before the letter would get to her. "I guess by now you know the big show is on. I hope and pray that it continues as good as it has all day and night. We moved the radio down to the Engineering Office now so we could hear the news. We listen to the English and then we shift over to the German stations. The Germans are sure giving us hell over the radio. The Berlin Bitch (the same one we used to listen to back in the desert) is still putting on the best radio programs. She said she would have intercourse with the first paratrooper that landed in Germany. I have had only four hours sleep in the past 48 and I can just see. I have moved my cot down here in the Engineering Office because I have to catch my sleep in minutes. We have to change something or do something every five minutes. No one minds working overtime when they can see things being done. We are all OK. We had a short summer here. That nice weather we had is gone. It is so cold now that we are about to freeze to death. Yes, it is just like Ohio weather, cold one day and hot the next. It is now 11 at night and we have just finished work and I have to be with the ships at 2:30 so I have to lie down for a few minutes because I am so tired. I may miss a few days writing if they keep it up."

*Left and Right*: 316th crews prepared for D-Day missions. Note side arms, field gear, steel helmets and infantry leggings. There was every chance their C-47s would be forced down in enemy territory. (Mike Ingrisano collection)

## 316th TCG Aircraft Damaged in D-Day Missions

This is the 44th TCS ship that collided on the runway with another C-47 on June 7. Bob Uhrig witnessed the mishap and attempted to rescue the crew. The pilot, Lt Calvin Heinlein, was killed. (AFHRA photos)

Other C-47s that sustained damage. (AFHRA photos)

## Missions for June 7

Bob picked up the D-Day story again in his diary on June 7. "Our ships took off this morning at 0335 on a resupply mission. During the take off, the 44th Squadron was taxiing down the runway and one of the pilots thought the ship in front of him was taking off so he started to take off. When he realized his mistake, it was too late, and he ran over the ship in front of him. Their wings interlocked and they spun around until their noses were together. The left half of the cockpit on the ship taking off was cut in half by the other ship's propeller. The pilot had both legs cut off by the propeller. One leg was cut off at the hip and the other below the knee.

"Sgt Love and I saw it happen so we rushed out. The pilot was a mess and he died in a short time. Both his legs were left in the cockpit. No one else on either ship was hurt. The mission ran into bad weather and could not hold formation, so Major David from Group brought the 45th Squadron home. They did not get to France.

"One of our 36th Squadron ships that got lost from the Squadron went on and he is at an emergency field. We don't know how bad he is shot up. Col Berger, our Group Commander is also down there. The rest of the ships got through but almost every ship was hit two or three times by small arms. We had four casualties in the Group from this mission. One crew chief was standing in the back door after they had thrown out the bundles and he saw a German aim at him and shoot him in the thigh. Most everyone could see the Germans shooting at them with rifles. All the ships were just above the trees after they had dropped their bundles.

"It is now 1130 and we are getting all our ships ready for British paratroopers tonight. I don't know when there is going to be a letup. We are all going around in a daze from lack of sleep. Last night I took enough time off to shower and shave but other than that, I have not had my clothes off.

"Lots of curious things happened last night. One navigator had the map shot out of his hand, a crew chief had the grips shot off his pistol and then the slug stayed in his holster. The flak suits saved a lot of crew members because there were quite a few hits on the flak suits.

"This afternoon we found that our No.11 ship had been hit 23 times and had just been able to make it back to another field in England. We will be unable to repair it so we transferred it to the Service Group. There were two other ships that landed there that belong to our Group. Our Group Commander, Col Berger, was flying one of them. He had one engine shot out and his hydraulic system, two fuel tanks, aileron control cable and tank selector cable all shot out. On his bad engine the propeller feathering pump was shot out so he was unable to feather the propeller. But that ended up being a godsend because by the propeller windmilling it kept up the fuel pressure for the good engine, which had lost its fuel pump. The aircraft went down to within three feet of the water before Col Berger could get it under control and get it to climb a little. The crew threw out all the equipment they could to lighten the load."

9th AF C-47s lined up for D-Day missions with British Horsa gliders. (NARA)

The "V of Vs": Standard formation of Troop Carrier C-47s carrying paratroopers. (NARA)

**A Break in the Action**

Bob's next diary entry was not until June 9. Presumably he was finally able to get a brief respite and some sleep on the 8th. "We have all the equipment loaded for the British mission and we are on the alert. We have been waiting for two days now. They have a landing strip built in France already and fighter aircraft are operating from it. Troop Carriers have started hauling out casualties yesterday. After this mission of ours is over we will start doing that also."

In a short note to Toots on June 17, he confirmed the participation of the 36th Squadron in the Normandy invasion. "Well I can tell you we took part in the invasion. I guess you read all about it. We were one of the first to drop troops in France. We were there six hours before the sea invasion started."

His next letter on June 19 warned her of a potential visit to Dayton by his Squadron mates. "There should be some of our fellows in to see you soon. They are my very best friends and I want them treated as such. I could not have hand-picked a better bunch to come home with, but as you know I am not a combat crew member so I must stay here. There should be five or six of them. Take them home and put them up for the night and if they can drink all the whiskey that is OK. They are the only ones left that have been together since the start and we have seen hell together and I want you to do everything in the world (within reason) for them. Each one of them picked the bottle that he was going to open. I have been to the village to a show *Tunisia Victory* and it was very good. If you get to see it, watch for a transport aircraft unloading and you will see the tail number 511. It was one of our ships. We all about passed out when we saw it. It is now 12 am and I have just been listening to German news broadcast. They are sure good. We like to listen to them because we get a lot of good laughs. Now they are playing dance music and I am so very lonely."

His diary entry for that day, the first in ten days, confirmed that the operational tempo had calmed down after the frantic pace of the invasion activities. "We have just started flying again. Ever since the invasion we have been on alert. The invasion is going quite well. Our troops have now cut across the Normandy [Cotentin] peninsula and have the Germans at Cherbourg trapped.

"The Group started putting out passes again this afternoon so all the men are happy. We are putting new litter brackets in our ships now. They are made of web straps and we can haul 24 wounded now instead of 18. They are starting to send men to the States on 30-day leaves. Of course it is combat crews. I guess they never think that the ground crews get sick of overseas duty. When this bunch goes it will be only Captain Chiros and I that have not been to the States out of the original Air Echelon. We are sending Pappy Street home. Everyone is supposed to have 800 flying hours overseas to be eligible, but Doc Boyer helped us out on Pappy Street."

# Flight to France

Bob got his long-awaited chance to go out on a mission on June 22. It would be a resupply mission to one of the improvised airstrips just behind the battle lines in France. "Today at noon, orders came in for us to send ten ships to France to haul ammunition over and wounded men back. I did not lose much time finding Major Roberts [Squadron Commander who had replaced Major Jim Farris]. He told me it was OK to go so I told Captain Etter that I was going along on his aircraft.

"We left at 1315 and flew to Greenham Common where we were to pick up the ammunition. Upon arriving they told us that they would be unable to load us in time to go to France before dark, so we had to remain there overnight. Col Berger got his load on board in time and flew on over to see how the landing strip was and then he returned. We went to the movies and then returned to the ships to sleep. We did not have enough blankets to go around so some of us had to sleep without them. I had plenty of clothes on (as usual) so I slept without a blanket.

"About 0300 there was an air raid alert. We all got up because we thought we might see some of the new pilotless ships that the Germans are sending over.¶ For some of the men this was their first air raid and they about killed one another running around. They acted like wild men. We all had a good laugh because it used to affect us the same way. Etter, Wakowack and I decided it was too cold to stand outside the ship so we climbed back inside. They were bombing a field about ten miles away and they were putting up plenty of ack-ack but I did not see any planes hit. Our ships had already been loaded. Half of them had 155mm artillery shells on and others had the gun cotton and fuses. It was a funny feeling laying there with all that TNT all around you with an air raid on. But it would do no good to be outside because if a ship had got hit, the explosion would get you unless you were plenty far away."

The next day the flight continued on to France. "We were up at 0400 and a truck took us to breakfast. The first five ships took off at 0430 and then we took off at 0530. Five ships take off at a time and each flight must take off an hour apart. This is to give the first five enough time to unload and load up again. The parking area at the landing strip will only take care of about 15 ships at a time. We were to have a fighter escort but they never did show up, so we went on. We were leading our flight of five ships.

"As we neared the coast we began to see hundreds of boats, both large and small. Then we flew around a small island with our wheels down. This is the signal of friendly aircraft. Then we proceeded on up the coast to the landing strip which is located within a city block of the beach. The beaches are covered with destroyed landing barges and the English Channel is alive with boats. I did not know there were so many boats in the world! I tell you they are there by the thousands and barrage balloons are also there in numbers, both on sea and on land. We had to follow a path between them. We found the airstrip, which is just a short strip, cleared for landing. When we landed the first five ships had not yet taken off. They were still waiting for the wounded to be loaded. We all helped unload the freight. There were sure glad to see the ammunition because they were running very short."

C-47s towing American WACO gliders over Utah Beach. The Germans have flooded major areas inland to limit the available glider landing zones. (NARA)

---

¶ The Germans had started launching pilotless V-1 rocket-bombs at England in mid June 1944. When they flew overhead, their pulse-jet engines made a distinctive sound which earned them the nickname "Buzz Bombs."

The timing of this mission came just after a major storm had blown through Normandy and wrecked a good deal of the logistics infrastructure on the beaches, including some of the remarkable Mulberry Harbor equipment. There was a significant interruption in the Allied supply lines as most of their equipment and follow-on forces had to wait offshore for the storm to pass. Once the weather had cleared, off-loading and air resupply could continue. "For four days they have been unable to unload equipment from the boats because of the rough sea. We were told that the rough weather had wrecked more boats than the Germans had during the invasion.

"Already there are four wrecked aircraft on the airstrip. A C-47 had ground-looped and hit a P-47 fighter. Another P-47 had made a belly landing. During the invasion, he had been trying to knock out two machine gun nests in the side of a hill at the far end of the strip and had run out of gas. So he landed nearby and hid in the bushes until our troops took the hill. Just before our first five ships landed, two German Focke-Wulf 190s had been over strafing. They had shot down one barrage balloon and damaged another. The balloons look strange hanging in the air with one end shot away and hanging down, and the other end still full of gas, pointing skyward.

"We could hear the Navy shelling the harbor at Cherbourg. Our troops are on the outskirts of town but as yet it has not fallen. There is an enormous convoy waiting offshore for the harbor to be taken. The convoy reaches farther than the naked eye can see. They must have boats here from all over the world. The Germans were sure ready for our gliders because they had poles about 10 feet high and about 20 to 30 feet apart planted in every field that a glider could possibly land in."** They are what killed so many of our glider borne troops.

"While we were there they brought in some German prisoners that had been captured the same morning. They were plenty hungry because they had been cut off from their supplies for over a week. The day before, an American paratrooper was waiting at the Dressing Station to get his wounds dressed and a German prisoner was talking about being a sniper. The American pulled out his knife and cut the German's guts out right in front of the Dressing Station. The reason was the American paratrooper had found 11 of his buddies hung in trees by their heels with their throats cut.

"The landing strip itself reminded me of the desert because of the dust and destruction. They had to send our five ships home without a load because there were more ships circling for a landing and they did not have any place to park them. All we brought back was a few empty blood bank containers. We arrived back at Cottesmore at 1120.

"When we got back we found that the men going home on leave were going to depart at 1330. The ones going home are Pappy Street, Fenimore, Quisenberry, Welter, Sullivan, Wheeler, McCullough, Nagel, Pace, Roberts and Miller. Captain Etter has moved in the room with me. It was lonely in the room tonight with both Welter and Sullivan gone. I want to get home to Toots so terribly bad. If I was going home I could not have hand-picked a better crew. I imagine they will just drink about all the whiskey that Toots has stored away when they get there."

On June 26 Bob wrote home and asked for his camera to be sent to him in England. Up until then he had been borrowing one from his buddies to take a number of photographs in North Africa and Sicily. "Here most of our packages come through so you can send the camera. Send me some filters for it. Webber will know what I want. Tell him to get me some film and I will bring him some good pictures home. I have over 700 now. Tell him I overexposed most of them in the desert. Those stones that you received are from the ancient city of Carthage. They are part of the floor and pillars. I was all through the ruins. It was in Africa. Get out your history book and you will know where it is."

**Short Leave in Scotland**

The following day Bob was able to take a short jaunt to Scotland. The trip demonstrated the flexibility and carrying capacity of the C-47. "Major McMenamin got one of our ships to go to Scotland, so he asked me to
** The proximity to Cherbourg and the glider landing zones indicates that the airfield where the 36th landed was on the Utah Beach side of the American invasion beaches. These anti-glider poles were nicknamed "Rommel's Asparagus."

go along. We flew to Prestwick†† and left off a nurse and Major Brodie, and then we passed over Glasgow and flew on to Edinburgh. We carried a jeep along so we had transportation to town. We stayed at the Royal British Hotel which is the best one in town. It is a very nice one. We drank Scotch most of the afternoon and then Lt Eustice and I went to a stage show which was very good. Then we went back to drinking Scotch again. If you are a resident of the hotel you can drink as long as you want. We got to bed about 0230."

How to get a jeep inside a C-47 - by hand. This demonstrates again the fantastic versatility of both the Douglas transport and the Army's 1/4 ton, 4x4 truck. (NARA)

The next day the group went about exploring Edinburgh. "Scotland is very beautiful with very heavy vegetation. Lots and lots of high stone fences. The buildings are also stone just like in England. But there are no thatched roofs there. It was sure funny to hear the people talk. I almost had to laugh in their face. The say 'a wee bit' and all the other Scottish sayings. I wanted to send Toots some Scotch plaid (tartan) but I did not have any coupons. Their plaids are of all colors and each combination of colors has its own name.

"We went to a visit a castle which sits high above the town. From the top you could see the harbor and the entire city. We took off at 1530 and stopped at Prestwick to pick up Major Brodie and then we came on home.

"Received a letter from Sullivan and Welter today. As yet they have not left for the States and are they ever pissed off! They have roll calls every day and they are being given lectures on sex hygiene, how to behave in Britain and have to do calisthenics. They still don't know how long it will be before they leave for home.

"I also received a wire [telegram] from Pappy Street and Fenimore. It was the craziest one I have ever received in my life. It was sent to 'Abe Uhrig' and the contents are as follows:

'Captain Robert Abe Uhrig, Chief Moggo the Slave Driver, Ali Baba and his 38 Thieves. We are in the mood to break both of your arms. Pappy is restless. I want my baby girl in Leicester. You stay away from her. No boat in sight. This is the dog. Toots is having it by mental telepathy. Love you best, The Two Thieves.'

Besides the cryptic note from his buddies still stranded in the UK, Bob also received a special letter from Toots that day. "The letter from Toots contained some color pictures of the house and two pictures of her, one in a suit and the other in a red and white slack suit. I think she is getting better looking each day and her shape is out of this world. I thought maybe she was getting fat, but by the looks of the pictures she is just right. The entire lot of pictures is the best I have ever seen. One cannot imagine how much difference color pictures can make things look. I love to get pictures from home even if they do make one a little homesick."

By early July the fighting had moved farther into France and there were far fewer missions for the Troop Carrier forces. For the time being no further airborne operations were planned but that would soon change. "We are doing very little flying these days. We are wondering what they are going to do with us. They don't need us here just now and even when we can start hauling things to the front, they can never use all the

---

†† Prestwick airport outside Glasgow was a major transit point for U.S. personnel going to and from the States and Europe.

Transports that they have here. Some think that this Group may get to go home and others think that we will go to India.

"We talked to two wounded paratroopers that our Group dropped and they told us that our Group put all the paratroopers on or close to the Drop Zone. All of us were pleased to hear it. They told us that some of the other Groups dropped their men into the sea or into the marshes. This is of course, just some of them."

On July 2, Bob wrote to Toots. Now that the excitement of the invasion had worn off, he was growing weary of England and spent more of his time thinking about home. "Bones is on the alert today and we have been shooting the bull for a long time. We always talk about what we are going to do when we get home. I want to come home so terribly bad that it hurts. This country is not for me. I just don't like it. The country is beautiful but when that is said it is all said. The women (most of them) have no morals at all. I guess women back home are the same way from all the stories we receive.

"You have to stand in line for everything. If you want to go someplace there is no transportation and hotel rooms are almost impossible to get. I don't like any of their systems for doing things. That is why I stay home until I can get a trip across country. All in all if we did not have such good living conditions I would rather be back in the desert. Besides, a pound ($4.00) goes just like a dollar bill and that is the truth. You have to get away once in a while or you would blow your top."

The 36th Squadron held a picnic on July 4th to celebrate Independence Day. Bob was not sure "how the English are going to take it, but we are going to celebrate anyway."

**Buzz Bomb Menace**

Although the threat of German bomber air raids was diminishing as the Allies pushed further into France, the V-1 "Buzz Bomb" menace to England was steadily increasing. These early cruise missiles did not have the range to reach Cottesmore, but were wreaking havoc on London and southern England. Bob recorded the experiences of one of his squadron mates who experienced it firsthand. On July 7: "Lt Lee just came back from London and he has been telling us about these flying bombs. We don't hear about them unless we talk to someone from London. The Germans are sending over about 500 a day.‡‡ They send them over about every 15 minutes so there is never an 'all clear' signal. Everyone says that it is twice as bad as the Blitz. People who were never in air raid shelters before are now living in them.

"The bombs do not make a large crater but the concussion is terrible and they destroy half a city block at a time. They are shooting down a few and the barrage balloons are getting a few. Some of them travel close to 400 mph and others travel just over 150 mph. There are two sizes now. One is 1000 pounds (V-1) and the other is 2000 pounds (V-2).

"People are evacuating London by the thousands. There is no hotel room or private room in either Leicester or Nottingham. Everything is taken up by the evacuees from London. Lt Lee says that the peoples' nerves are going bad. Everyone is jumpy and tired from lack of sleep.

"Lt Lee was in a bank when one hit outside. It blew all the windows out of the back, tore the revolving doors off and threw them across the bank. It brought down three buildings. Lee said the bricks came loose from the ceiling and the walls and the building caught fire. After the debris had cleared from the air, Lee looked around and found that he was the only person not injured. He had been hit on the leg by a brick but was not hurt. The banker that Lee had been talking to was hit on the head but the cut was not too bad so Lee ran outside.

"Just on the outside of the door there were three people blown up. Lee knew that they were dead so he looked around to help any persons that were just injured. Someone brought out a first aid kit and Lee started giving first aid to the injured. In a short time some doctors arrived and some firemen. Lee says the amount of damage done by one bomb was terrible. When Lee went back into the bank to get his hat he found that his insignia had been blown off it.

‡‡ This was a bit of an exaggeration, but not by much. From June 12 to July 21, the Germans launched over 4000 V-1s from the French and Dutch coasts, of which about 3000 actually reached the UK. The larger V-2s were launched at the UK starting in September 1944.

"Lee said you could not sleep at night because they kept sending flying bombs over. He said that you could hear them coming over and going off all night long and he said he just laid there and sweated all night. He left London the next day but people are moving out of London by the thousands so he had to stand in line for three and a half hours before he could get a train."

## Lili Marlene

On July 10, Bob wrote home asking for a copy of song he and the other "desert rats" knew very well from North Africa. He was now hearing it again in England. "Will you buy the record and sheet music of the German song that is now sweeping the country? We have whistled and sung it for the last 18 months. We never knew any words to it because it was sung in German. Now the English have recorded it in English. Always when we felt low in spirits we would whistle or hum it. The name is *Lili Marlene*. We used to love to hear the Germans sing it."

By mid July a quiet routine had settled on the men at Cottesmore. The Allies were closing in on Paris and the German Army in France was in full retreat. There were now more than enough Troop Carrier Groups in the UK to handle the supply and casualty evacuation missions. On July 15, the 316th held a dance, and "Although everyone proceeded to get drunk, no one had a fight and that is something new." The next day the 36th sent 12 C-47s to France to haul freight over and wounded on the return.

The paratroopers of the 82nd had finally been taken out of the line and returned to their bases in England where they took on hundreds of new replacements and recounted their stories of combat to their comrades in the 316th. "This morning we started jumping paratroopers again. They are all volunteers from the infantry and have only had one week's training. This is their first jump. Most all the paratroopers are back from France and they are training the new boys. Our paratroopers caught hell in France. Out of 150 men in each outfit [company], the most they came back with was 65. It has been very warm here for the past three days and it reminds me of late spring back home."

With a lot of time on his hands, Bob could now catch up on his letters to Toots. On July 24 he wrote: "The war news for the past few days has been good and I hope that Germany blows up before very much longer. I am doing this writing with my gas mask on. They make us wear them every so often and I don't like it and I don't think anyone else does either."

In return, Toots had managed to send him the music and lyrics for the German song he had grown so fond of. "Thanks for the song [*Lili Marlene*]. It seems so funny that they should just now get it in the states. The words are different than what they have here. In fact they are no good. The only part that is I like is 'Underneath the lamppost by the barrack gate.' So I am going to get the German words for it. Now the Italians, English and Americans all have their version of it. Even the British version is much better than the American. But after all it is the tune that counts and I think it is beautiful. I read the part about where you said it was the first time on the air and it was hard for us to believe. The Germans have sung it for us ever since we have been overseas and I hum it all the time. I don't know how the Americans are doing it, but here in England they have to put the royalties away for the German copywriters until after the war. This war is a funny business."

Bob responded with the words for the other versions. "I am sending you the English version of *Lili Marlene* which is much better than the American one. Our nice days are shot again and we are back to the good old typical English weather. Maybe it won't last too terribly long. We are still building equipment in case they move us out in the sticks again. Tisdale has built a small gasoline scooter for Engineering. You should see some of the things we build. I hope we never have to use them again."

Bob also wrote home describing the hijinks of his barracks mates. "Last night 2nd Lt [Earl] Shank came into our room and said he wanted something done about the way his two roommates sprang their rank on him and made him do all the work in the room. He took Etter and I down to his room and both his roommates had their bars on their pajamas. They just made 1st Lt about a week ago and have been making Shank do all the work.

"They are about ready to cut wheat here and I am going to try and help on some of these farms. The wheat here gets much taller and has much long heads than our wheat at home. I guess it is because of the rain. They call all grain here 'corn.' They do not grow any corn here. The farmers talk about corn and I thought maybe I was crazy so I said, 'I never see corn here.' So they told me that all grain was corn to them."

**Temporary Move to Greenham Common**

In early August there were some significant changes made to the Airborne and Troop Carrier forces. All US and UK airborne troops were now organized into the First Allied Airborne Army (FAAA). The US and UK Troop Carrier Groups were also attached to this organization for the purpose of engaging in large airborne operations. The first operation planned by the FAAA was TRANSFIGURE.

On August 7, the 316th deployed to Greenham Common, over 100 miles to the south of their permanent base at Cottesmore. The base had previously been used by the 438th TCG, but they had redeployed to the Mediterranean in early July to support the airborne operations and invasion of Southern France (ANVIL/DRAGOON.) The 316th would now take over their facilities temporarily. As the base was just west of London, it reduced the flying time necessary for the C-47s to reach France. This was especially critical when the aircraft were towing gliders, since that type of mission required a greater degree of flying skill and reduced the fuel efficiency of the C-47. "There has been very little to write about in the past couple of weeks. We have been making a lot of trips to France and we have been doing a lot of glider towing work. About every flight we have a couple of glider pilots cut loose from the tow ship because of tail flutter. Then we go out and use our glider pick-up ship and snatch them up and bring them back home again. I have ridden in both the glider and the tow ship.

"Today we moved to Greenham Common. We only had one day's notice and we only brought 35 men from Engineering. They told us not to bring any equipment or supplies because we are to use the equipment and supplies of the squadrons on this base. The Air Echelons of these squadrons have gone to Italy.

"Well, as usual, I loaded most of our equipment and supplies because I have heard that old story before and it is a good thing that I did because they do not have anything left here. We only have nine men left out of the squadron we are supposed to be working with."

The next day, Bob and his maintenance crews attempted to set up shop at Greenham and take stock of what they needed. "We are so short on help that we cannot operate. Also, the only transportation that we have is the two jeeps that we brought along. So this afternoon the Colonel sent word back for our trucks and some more personnel to be moved down here. We got a good laugh when the men landed here this evening because they thought they were going to live a life of ease at Cottesmore after we left. They had plenty of transportation and no work, so they were looking forward to a good time.

"That all changed for about 25 of them. Butler was sore at us for bringing him down. Although we have plenty of cooks we wanted him here with us. We called him and told him to bring us some food because they were giving us C-Rations here. When he arrived it looked like he brought the whole mess hall along. So in the evening we had beer and sandwiches here at Engineering. We have here Hershberg, Etter, Fennimore, Love, Tucker, Crook, Butler and I so we could shoot the bull just like the old times. We are living in Nissen huts and they are plenty crowded. Etter, Chiros, Ryan, Morton and I have half a hut to ourselves so we have plenty of room. The mess hall is half a mile from our barracks so that is not so good."

Glider training continued at Greenham Common in preparation for the TRANSFIGURE operation. The 316th also received a visit from the Supreme Commander. "I flew to Cottesmore today (August 11) to get some more equipment and supplies. It was good to get home and get something good to eat once again. We got another new ship today so that makes 22 ships in the Squadron again. We have fourteen 100-hour inspections to do so we are working both night and day. We have to pull inspections on every airplane that has over 75 hours since the last inspection. We are changing engines on No. 19 and we want to try and get two new engines for No. 10 before this big move goes off.

"General Eisenhower gave all Airborne and Troop Carrier Groups a talk the day before yesterday. All

Troop Carrier and English and American Airborne are now under one Command. We are no longer in the 9th Air Force. We are now in the 1st Air Army (sic)§§. They took General Brereton from Command of the 9th Air Force and put him in command of the new unit, so it must be going to be something big.

"They are still blasting London with buzz bombs and these towns are full of evacuees. Thousands are becoming homeless. It must be terrible. I brought back two quarts of gin from Cottesmore last night so it got pretty high in our room. Etter heaved on the floor before he could get outside so myself and Chiros had to clean it up. It was a scream to watch Etter and Chiros play cards. Both of them were about to fall off their chairs."

In Bob's next letter to Toots it was apparent that there was some discontent among the troops overseas regarding the behavior and statements by the folks back home. The feeling seemed to be that once the D-Day invasion had gone off successfully, the final conquest of Germany was a foregone conclusion. The men at the front knew that was not the case and that there could be many months or even years of hard fighting ahead. "For the past week we have been working every night and we are all getting plenty tired. You people all have this war won already; well you better all get that crap out of your heads. That is what hurts when you people back home start letting down. I have read a lot of papers from the states and one would think by the stories that the invasion was a push over, well just go over to France one time and take a look around and see for yourself. And then talk to a few of the fellows that were there on D-Day and you people back home will start using that head and begin thinking. Yes, we too hope it is over soon but we give the other fellow a lot more credit than you people do. Old Jerry taught us a long time ago to respect him. So you just tell those people to buckle down a little more and not quit until it is finished."

By August 14 it looked very much like there would be another airborne operation in France. The 316th and the other Groups were ready and poised for action, and there was a good chance Bob would finally get to go along on an operational mission. "Our training period is over again and we have the Real McCoy coming up. We received orders today to install two cabin tanks in each ship and we have to have it done by tomorrow noon. So that means working all night tonight.

"There are going to be three missions. I asked the Colonel if I could go on one of them and he gave me permission to go on the last one. I am going to fly as copilot with Lt Lee. Lee and I do all the test-hopping so Etter is going to let me fly as copilot. They are going to be very short of pilots. We have 27 ships now. The glider pilots are coming in by the dozens. We are going to pull two gliders behind each ship. It is going to be the biggest thing of its kind in history."

The next day the men were still on edge, waiting for the signal to go. "We are about dead today but we have to keep going until we are finished. We had coffee most of the night. Lee and I finished test-hopping the rest of the ships this morning. Lee is the craziest person I have ever seen. We carry five gallons of this piss-poor beer around on the back of the jeep where the spare gasoline can is supposed to be carried. He can drink more beer than anyone I know of. Tonight we lined the cockpit of the airplane with extra flak suits. The Germans always shoot at the pilot and copilot with small arms so we are going to try and out-fox them. We had a lot of good laughs while we were fixing it up. We are now finished with our work and ready for the missions."

## Operation DRAGOON

The airborne and seaborne operations in the Mediterranean kicked off on August 15 and Bob heard the news at Greenham Common. "Good news this evening. The Americans, English and French invaded Southern France. So far it is a success but as yet we have heard very little other news. The Germans are still retreating in France and the Americans are within sight of Paris. The activity on post here is terrific and when we are flying we can see convoy after convoy on the highways. This invasion of Southern France today was mostly an air show and we think that ours will be the same thing. It makes one feel good to see things happening and something good being done."

---
§§ First Allied Airborne Army.

There were additional Allied successes in France, not only in the south, but in the center, where Patton's forces overran the drop zones that were to be used by the 316th's glider troops. This forced the cancellation of the TRANSFIGURE airborne operation on August 17. "Yesterday they postponed the mission for another 24 hours. This evening we received orders to take out the cabin fuel tanks and everyone is pretty sore about it. They tell us the mission is off because Patton is running wild over there and has taken the ground that our airborne was to capture. Now after at last getting permission to go on a mission they up and cancel it."

On August 18 Bob wrote to Toots about the long days his men were enduring to get the airplanes ready for these standby-missions. "We have been working night and day. Last night is the first night's sleep we had had for the past three days and nights. Last night we got to bed at 1210 am and believe it or not that was a good night's sleep. For the past few days it has been like a nightmare – men would fall to sleep if they stopped working for a few minutes. If I were to come home now everyone would stare at my pretty white skin. There is no such thing as a tan in this country. I wish you could have seen me last summer in the desert. I was just plain black. You will have it all over me this season."

**Return to Cottesmore**

The false alarms and alerts ended for a time on August 24, when the 316th returned en mass to Cottesmore. The Group that had previously been based at Greenham Common was now returning from Italy. "We loaded all our equipment last night to move back home. We finished at about 0130. Yesterday the weather was bad and they needed blood in France so they asked for volunteers to fly it over. We sent two airplanes out of each Squadron. One ship from the 44th Squadron did not make it and killed all seven on board."¶¶

Due to the rapid advance of the Allied armies, resupply missions now became the priority for the Troop Carrier aircraft, and airborne operations had to wait until the supply situation would permit the aircraft to be used for paratroops or gliders. So there was heavy competition for the use of the C-47s. "Tonight we are back at Cottesmore and everyone is very happy. This place is like a country club compared to the other places we have been. We have been making a lot of trips to France. We are hauling ammunition and blood over and injured back."

 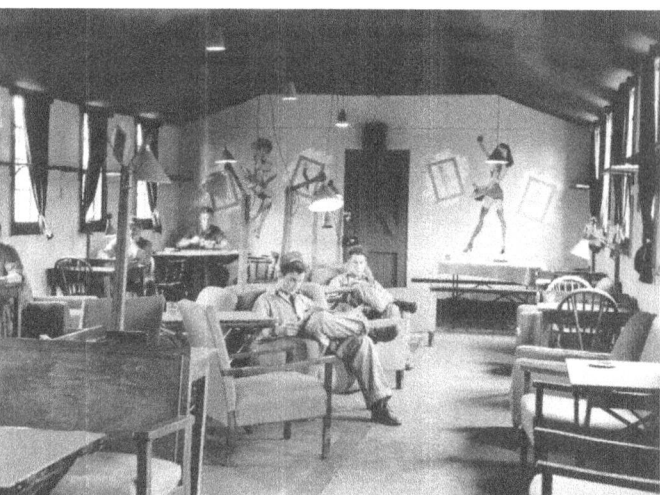

*Above*: The Enlisted Men's Club at Cottesmore where the mechanics and crew chiefs could relax - when they had the time. (Mike Ingrisano collection)

---

¶¶ According to *Ingrisano,* p. 146, the C-47 from the 44th crashed in bad weather on August 22 near Grove in southern England and all five crewmembers were killed.

Bobs' next letter home showed that he was clearly relived to be back at Cottesmore. "Well here we are back home again. It is next to going home to come back to this place. Now we have good food and quarters. This place is out of the world in comparison with the other places that we have been. Everyone is glad to be back again. This is the time of the year that they start getting that lovely fog and rain. When they get a fog over here, they get a good one. It makes the ones at home look like a clear day. We are just getting things going again since we came home. Everything was in a mess as usual when you move around."

Allied troops crossed into Belgium on September 1 and were pushing the remaining German troops out of France. But this headlong pursuit was about to come to an end and the gains of July and August would be replaced by stalemate in September; due in part to weather and terrain, but mostly to logistics. "In the last days of August the Allied armies had reached the stage of 'frantic supply.' The bulk of all supply was still coming in over the Normandy beaches and being carried by truck directly from coastal dumps and depots to the combat zone. The beach areas were congested. An insufficient number of transport vehicles, coupled with the long haul from the beaches to the front lines, made it impossible to move supplies forward in the quantities required for the daily maintenance of the armies. The stalemate in Normandy and the hand-to-mouth supply in August had prevented the stocking of supplies between the Normandy coast and the army dumps. None of the Allied armies had been able to build up large operational reserves."[54]

As a result of these difficulties, the only way to get supplies quickly to the front was with transport aircraft. Often this involved flying them into newly captured airfields as had been done in North Africa. There were also many cases when the aircraft had to land on unimproved dirt strips that had a minimum of navigation or air traffic control facilities. Unloading of the aircraft had to be done by whatever ground units were available or by the aircrew themselves. For a time the, the principle mission of the 316th became airlift of supplies and equipment.

But once back at Cottesmore, the action for the ground crews slowed down considerably by the end of August. "Crook and I were in town together yesterday. There is very little to do in these towns. Go to a show and then have a couple of beers is all one can do. Yes, the Colonel was right, this is a beautiful base. That is why we all sound off when they try to move us. We have the best living conditions in England. It is about time that we caught up.

"George is getting out of hand. He is our pet crow that we have had since he fell out of a tree. He steals everything in sight and then hides it. He flies into everybody's room and carries things off. We caught him hiding a watch under some dirty socks. The other morning we have been giving him some instrument flying. With this fog as thick as pea soup in the mornings he gets a lot of time. We threw him out the second story window when there is a heavy fog and so far he has spun in twice. He just can't use his instruments in this fog. He goes in and out of the mess hall anytime he wants. In fact he acts like he owns the place. You might be walking along and he will fly up and land on your shoulder. He can catch coins when you toss them to him and he immediately hides them."

The slow pace of activity for the Troop Carrier units would not last. The Allied command was determined to use the airborne troops for at least one more big mission. This one would not occur in France, but in Holland, and be the source of the phrase, "A bridge too far."

# Supplying Patton's Tanks
## Troop Carrier Units Fly Supplies Direct to the Battle Front

*Above:* Troop Carrier units were pressed into service to deliver supplies to the fast-moving Allied armies in the summer of 1944. "Assembly lines" were set up to bring the cargo direct to the fighting units. (NARA)

*Above:* Long lines of C-47s wait to unload. (NARA)

*Below:* The two most urgent needs were gasoline in five-gallon "Jerry Cans" (*left*) and ammunition (*right*) (NARA)

*Above:* Every available man, including the aircrew were needed to unload the aircraft quickly. (NARA)

# Chapter 11

# Liberation

*Almost from the day of its creation, this Allied Airborne Army
showed an astonishing faculty for devising missions that were never needed.*
General Omar Bradley

**W**hile Bob and the rest of the 316th was dealing with the ups and downs of deployment and cancelled operations, there were plans in the works for an even larger airborne operation. This one would be bigger than the D-Day drop and involve the whole of the First Allied Airborne Army.

On August 29 a conference was held with the major Allied airborne chiefs to brief them on the upcoming operation. It would be composed of troops from the 82nd, 101st, British and Polish Airborne forces. The plan was to land the paratroopers near Tournai, Belgium to seize and hold the crossings of the Escaut River to trap the German troops in Belgium and prevent their withdrawal into their Siegfried Line positions in Germany.[55]

More aircraft were added to the 36th Squadron fleet on September 1 in preparation for the operation, known as LINNET. "It is now 2200 and we are still working. This morning we received orders to get our ships ready for a parachute mission. This afternoon we received three new ships straight from the States. We now have a total of 29 ships. This makes over twice as many ships as we are supposed to have, and no more mechanics. They expect us to do the impossible I guess. I want a good long rest when this is over." The miracle of American mass production was now being felt on the battlefields of Europe, as more C-47s poured into the theater. How things had changed in three years – the 316th TCG alone now had almost as many transport aircraft as had existed in the USAAF in 1941.

The paratroops of the 82nd also arrived back at Cottesmore for the airborne drops planned for September 3. "It rained most of the night and everyone is plenty cold. They moved the paratroopers in yesterday and put them behind a barbed wire enclosure. They did this as a security measure. By the time the paratroopers arrive here, they have already been briefed for the mission, so they keep us away from all of them. This morning (Sept 2) they were all soaking wet and we could see them wringing their blankets out. They had slept in the rain and the mud all night. It is a sickening sight. The mud is about six inches deep. Today they finally put up tents, but they were already wet.

"I get to go as copilot on the second day of the mission. Lee and I are going to fly together. After the mission is over I will be able to get to write where we went (I hope) and also about how it was planned. Everyone is highly strung and on tension because the zero hour is getting close. This is to be the largest air show ever pulled. The flak is to be plenty rough.

The men waited all through the day on September 2 for the signal to go. "It is now 2300. There is a storm blowing and it is pouring down with rain. We have just finished tying down our gliders that we have marshaled along the side of the runway, ready for takeoff. The wind has wrecked about six gliders already. It has even blown one over on its back. It is a sickening mess here in the mud, the night before a big mission with everything going wrong. We are now to take off at 0540 in the morning."

The dawn on September 3 brought another postponement, but also some good news. "At about 0200 this morning we received a message telling us the mission has been called off for 24 hours. If I get cut out of another mission I am going to blow my top.

"We received the very good news this morning that our tanks are in Germany. The bombers have been going over this morning by the hundreds. There are to be three full Air Forces in the air* to knock out fighters and soften up the opposition for our troops, and also knock out the flak or at least part of it."

By the following day though, the prospects for the new mission were even dimmer. "We think that the mission is going to be cancelled. If it is it will be the second one that I have been beaten out of. We had to

* Referring to the U.S. 8th and 9th Air Forces and the British 2nd Tactical Air Force.

179

transfer three of our new airplanes to the 313th Group. They lost five of their ships and quite a few men in one blow. A crew chief was fooling with a land mine and it went off. All the ships close to it were loaded with land mines and the explosion set them all off."

In reality the missions were cancelled for the same reason the TRANSFIGURE operations were called off – the Allied ground forces were advancing so rapidly that they were almost upon the drop zones on the day set for the drop. This fact, plus the bad weather that Bob wrote about, forced the cancellation of yet another "big drop" in Europe.

**Plans for the Invasion of Holland**

The idea of a big drop was still alive. Again the airborne commanders met on September 10; this time the plan was to drop the Allied airborne troops deep in Holland to capture critical river and canal bridges. Once these were in Allied hands, the ground troops could move north into Holland and avoid the fixed defenses on the German border by swinging around behind their flank. The flat terrain of northern Germany beyond the Dutch city of Arnhem offered a favorable route for armored forces to drive into the industrial heart of Germany: the Ruhr valley.

The date picked for the initial assault was September 14. This was revised to September 17 because the Troop Carrier aircraft were heavily involved in ferrying gasoline to U.S. ground forces and could not be cut loose for airborne operations until that date. The plan for the operation now code-named MARKET GARDEN: "Entailed dropping three airborne divisions behind German lines in southern Holland to seize bridges over canals and rivers and hold them until the British Second Army could drive north from the Belgian-Dutch border over the bridges and turn east into the plains of northern Germany."[56]

Operation MARKET GARDEN would use the same airborne forces already earmarked for LINNET and another similar operation called COMET – the American 82nd and 101st, British 1st Airborne Division and a brigade of Polish Airborne troops. These paratroops and glider forces would be dropped in a long continuous line stretching over 60 miles. They were to hold the routes over which the British Second Army and its leading elements (XXX Corps) would pass.

From bottom to top, the operation shaped up like this: the 101st would drop near Eindhoven and seize two bridges at Son and Best and be the first linkup with the British XXX Corps. They would also link up with the 82nd to their north, which was assigned to capture four major river bridges and five canal bridges including a large bridge over the Waal River at Nijmegen. They would also have to hold an area of high ground near Groesbeek and block German forces expected to come from the east. Finally, the British Airborne and the Polish Brigade would capture the city of Arnhem and the road and rail bridges over the Rhine. Over 33,000 men were to go into battle by air – a feat never attempted before or since. It was expected that XXX Corps would be able to cover the distance from the Belgian border to Arnhem (over 60 miles) in three days.

The operation was extremely complex from an air-planner's perspective. Because of the size of the airborne forces involved and the number of available American and British Troop Carrier aircraft, it was not possible to carry and tow all the troops to the drop zones and landing zones on one mission. Each division to be dropped and landed had different requirements for equipment and timing. In order to assure the maximum accuracy in landing the troops, it was decided to launch the operation in daylight. This would simplify Troop Carrier navigation and formation flying, yet expose the C-47s to more anti-aircraft fire and the possibility of interception by German fighters.

The paratroopers were resigned to the fact that despite the risks, a daylight drop was the best plan. "One good element of the operation was that we were to jump in daylight at about 1300 hours on 17th September. I felt a little better about the daylight jump for two reasons: one, there was no room for nighttime navigational errors on this operation, and two, we were promised fighter escort all the way to the drop zone."[57]

The IX Troop Carrier Command decided that on the basis of the distance to the targets in Holland, the available hours of daylight, and the requirements to rest crews and service aircraft, that only one mission would be flown per day. This meant that the MARKET GARDEN airborne operation would be spread over

three complete days and the divisions involved would be delivered piecemeal until they were at full strength. This would stretch the resources of the Troop Carrier Groups to the limit. If aircraft were lost or out of action after the first or second day, it would make delivering the successive waves of troops even more difficult.

Once the plan was complete orders went out to the FAAA and Troop Carrier units involved, including the 316th TCG, which would be carrying their old comrades the 82nd Airborne Division. For the men in the Squadrons one operation had stretched into the next and the final destination mattered little to the Troop Carrier Groups. On September 9, Bob recorded that, "We have been on alert for the mission for the past four days and every day they put it off another 24 hours. I am going to get to go on this one. We have been ready for three missions now and two of them have been cancelled on us and we are now sweating out this one.

"Last night we had a large drink in our room and Lee kept me awake until 0400. I told him it was going to cost him because I was going to get him up at 0730 when I got up. They always sleep until about noon, so getting them up at that hour just about kills them. So this morning I went up after him and he had taken his sleeping bag and moved out. I looked all over for him but I never did find him. When I did see him this afternoon, he sure did give me the horse laugh.

"We are all restricted to the base, so Crook, Tucker, Babbitt, Butler and I decided that we would go hunting around here. This afternoon we bagged some game and Butler is going to make them for dinner tonight. We have some beer and we are going to have a feast. There are a lot of pheasant and partridges around here and we are planning a way to get them. We have decided to use box traps. We will thin out this game for the English."

In the midst of all this, there was an unexpected visitor to Cottesmore on September 11 that needed Bob's help. "One of the hottest English fighters stopped her last night and had engine trouble. No one knew anything about the engine including Lt Lee and myself. So Lee and I decided to fix it. We had to feel our way around but we had it fixed by 11:00 pm last night. We feel pretty proud of ourselves."

An RAF Hawker Typhoon or Tempest fighter departs Cottesmore. This may be the aircraft with engine trouble that Bob and his mates repaired.

The same day Bob also received some great news – his work in North Africa was being recognized with a special award and he sent the news to Toots. "As you should already know our group received the Presidential [Unit] Citation. Now comes the payoff. I just received orders through Cairo, Egypt, awarding me the 'Legion of Merit.' It is the fourth ranking medal in the Army. They put me in for it a year ago. It had to be approved in Washington and then sent back. But even before that, is had to be approved by Wing and Command. I received it for the work I did with the 8th Army in Libya. I guess all that work paid off after all. I do not have the medal as yet. They will award it to me during some inspection."

## MARKET GARDEN Kicks Off

The massive Allied operation went off as schedule on September 17 and after so many weeks of false

alarms; the 316th TCG was poised for action. The 316th share of the operation was to carry the 504th and 505th Parachute Infantry Regiments to their drop zones around Groesbeek. It would involve 87 of the Group aircraft, including 22 from the 36th TCS. "Yesterday they told up to prepare for another mission so we had to work late last night. Today we are going to drop American paratroopers and then the next three days we are going to tow gliders over. We are going to drop troops 100 miles inside German territory. This is to be the largest air army ever dropped. We have 50 fighter squadrons for air and ground protection. This is the first daylight drop we have ever made.

"Our Group will drop their troops in between two canals and their objective is to keep the Germans from blowing up two bridges. The rest of the Wing drops their troops around where we do. If our troops fail to save the bridges, the rest of the troops dropped will lose their value. Etter talked to the Colonel and I get to fly as copilot on the last mission."

At mid day, the Troop Carrier forces were launched on time for Holland. "It is now 1100 and all the ships got off OK. Now we just have to sit here and wait. This is the hard part, to sit and wonder how things are going. One paratrooper blew a hole in his leg when a detonator for a hand grenade went off in his pocket. Another blew his toe off when he was handling his .45 pistol. Most of these paratroopers we are hauling have jumped in Italy and this is why they have been given the toughest job."†

*Above and Left:*
82nd Airborne troops arrive at Cottesmore for MARKET GARDEN. 316th C-47s are in the background. (AFHRA)

---

† The 504th PIR lost so many men in Italy that it did not participate in the Normandy campaign. They were back at full strength for MARKET GARDEN and again distinguished themselves in action including a remarkable crossing of the Waal River to capture the Nijmegen Bridge.

# MARKET GARDEN Forces Depart from Cottesmore September 17, 1944

*Above*: The Red Cross provides coffee and doughnuts to the aircrew and paratroopers waiting to depart for Holland.

*Right*: C-47s and WACO gliders are moved into takeoff position.

*Below*: Final checks and pre-flight before the mission.

(AFHRA photos)

# MARKET GARDEN Forces Depart from Cottesmore

*Left:* Tightly packed C-47s await the takeoff signal. Glider tow ropes are attached to the tail of the aircraft. (AFHRA)

*Left:* C-47s and gliders are carefully lined up to allow for smooth takeoffs and tight formation spacing. (AFHRA)

*Left:* Brig Gen James Gavin, 82nd Airborne, suits up in his jump gear before boarding a 316th TCG C-47. (NARA)

*Right:* American troops in a British Horsa glider. (NARA)

Maps of the MARKET GARDEN flight routes
and drop zones in Holland.
(US Air Force)

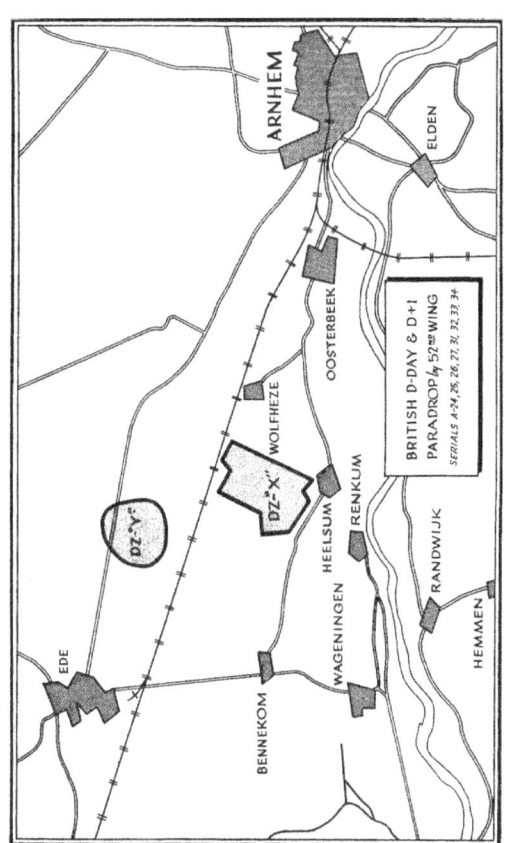

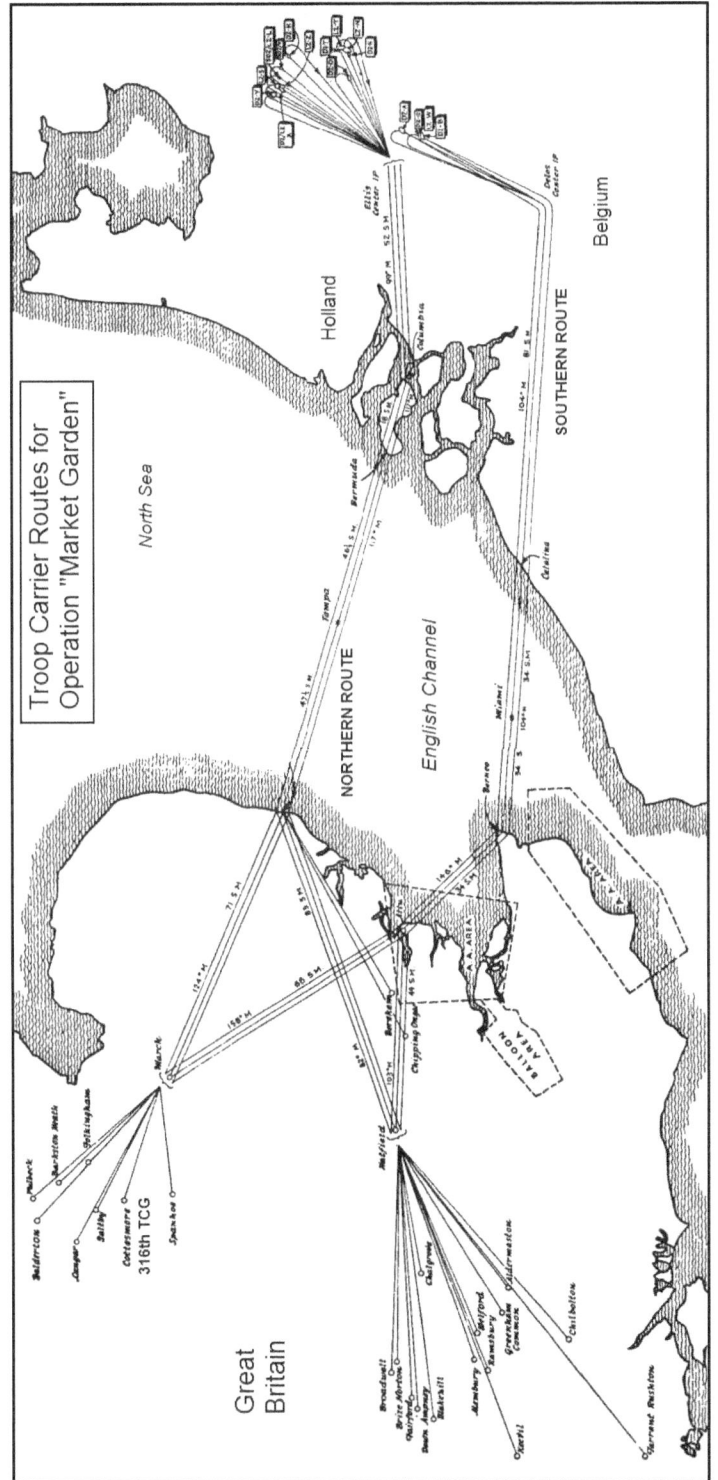

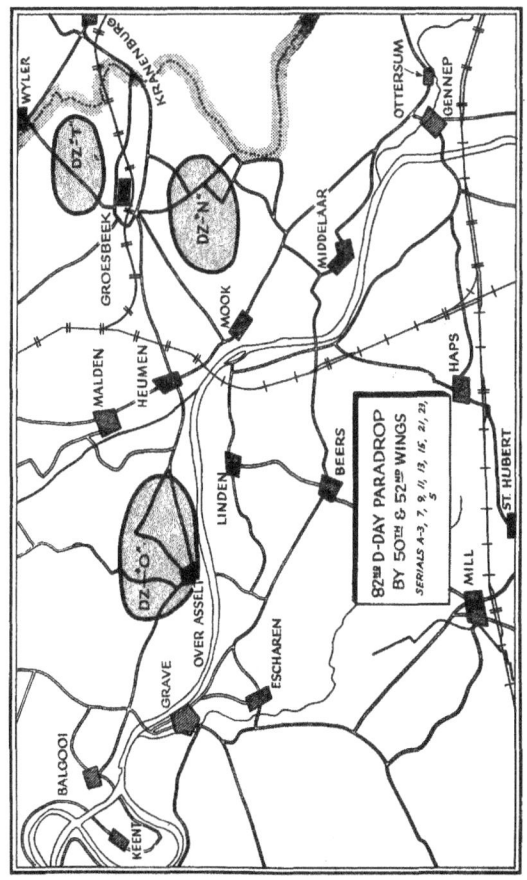

## Personal Accounts of MARKET GARDEN

As he had on the way to Normandy, 505th platoon leader 1Lt James Coyle recorded his impressions of the flight to the Drop Zone with the 316th: "Our flight to Holland was the most impressive sight I can remember from World War Two. There were over two thousand planes and gliders involved in the fly-in. I could see hundreds of them from my position next to the door of my plane. At one point, we flew under and parallel to a formation of gliders being towed by different types of British bombers that I had never seen before."[58]

By the end of the day on the 17th, reports were arriving on the success of the drop and Bob recorded the damage. "We only had two ships hit and they were not too bad. In total, 40 ships were lost out of the entire mission and one of them was our No. 19.‡ The pilots told us that there was plenty of flak and small arms fire. The crews said that every time a German gun opened up on them, three to six fighters dived down and strafed the position. There was excellent support from the fighters. One ship in the Group just in front of ours went down in flames, but some of the paratroopers were seen to get out before it hit the ground."

On September 18, there was good news and bad news. It was Toots' 29th birthday and although Bob wished he was "home to help her celebrate," there was plenty of work to do. The 36th Squadron ship that had been lost the day before was not accounted for. "Last night Lt Fulton, the pilot on No. 19, called and told us that his crew was OK. They had to ditch the ship in the English Channel. We were all pleased to hear that everyone was OK. The ship however, was lost. This morning we sent another ship over to pick up the crew but as yet they have not returned."

Once the crew returned to Cottesmore, they filled in the details. The incident was a testament to the toughness of the DC-3/C-47 design. "The crew from No. 19 has just returned and we now have the whole story. Another ship flew into them from the rear, knocking the rudder off. The top half of the vertical stabilizer caved-in the cabin and tore out a section of the navigator's compartment about four feet square which knocked off the right propeller and wing tip. When this happened they lost control of the ship and by the time they recovered, they were 100 feet above the water and were lost and flying alone. They were only flying at 105 miles per hour and were in the wrong corridor to return to England.

"Then the Germans opened up from both sides. They flew through this fire for half an hour and were hit time after time. The left wing was torn almost in two and the copilot's controls were shot away. The radio set was blown out and the cabin looked like a sieve. The crew cannot explain how no one was hit. Young [TSgt John Young, crew chief] told me it was worse than No. 8 in Sicily so it must have been terrible. A P-47 saw them about to go down so he flew in front of them and lowered his wheels so they followed him and he took them over to an air sea rescue boat and they landed close to the boat. They only had to spend five minutes in the water."

The full story on the lost 36th Squadron aircraft was that it was flying behind the normal group of 316th aircraft. Just after takeoff, one of the paratroopers had gone berserk onboard and the crew landed the aircraft and discharged him. This put them well behind the other 316th aircraft. They continued on their way to the drop zone where they were able to deliver their paratroopers without incident. After leaving the drop zone, however, Fulton's aircraft was hit in the tail by the other C-47 from another Group. After they were rescued, they had some minor injuries from the ditching and were taken to Martelsham Heath airdrome for the night.[59]

## The Drop was a Success

Some additional good news was that the 82nd paratroops dropped by the 316th had all landed on or very near their drop zones. It was a much more successful first day than Normandy. One officer from the 504th described it this way: "If ever we had made a perfect jump, this was it. All of my company landed exactly on the drop zone. I touched down within a few feet of the assembly point. Within one hour, my entire company was assembled with all equipment and no injuries and we had captured our objectives. None of our practice

‡ The actual number was 35 out of 1544 transport aircraft used in the first lift on September 17 (*Airborne Assault on Holland*, p. 12.), making the loss ratio less than 3%.

jumps had ever gone so well."§ The logic of a daylight drop was born out by the accuracy of all the drops in Holland.

The MARKET GARDEN missions continued on the 18th and Bob recorded the details: "We go on a glider mission today. The gliders are carrying heavy equipment to support the paratroopers and they are also hauling airborne troops. It is 1200 noon and three squadrons have already taken off. The 36th takes off at 1235. We are sending 80 gliders from this Group so if all the Groups all over England do the same you can tell the size of this mission. It is the largest airborne mission ever attempted. Yesterday the BBC broadcast about the invasion of Holland by airborne forces before our ships had even returned."

Unfortunately their luck on day two was not as good. Now that the Germans were fully alerted to an attack on Holland, they were more prepared for the arrival of more airborne serials. "On today's mission we all got a lot of ships shot up. No. 16 was seen to go down in flames but two parachutes were seen to come out so it looks like the crew chief and radio operator got out. Crews from some of our other ships that were flying close by said that the left engine and the entire cockpit were on fire. No. 14 was hit and damaged so badly that the crew had to bail out and one was seen to strike the tail. No. 1 went down and no one got out.

"There was no fighter protection and our outfit got lost and caught hell. They said it was almost as bad as Sicily. They were flying at 700 feet and could see the German gun crews run for their guns. Most of our ships were hit by flak which tore large holes in them. Also there was plenty of other small arms fire. The gliders caught hell and many of them were seen to go down out of control and crash. Also many of them hit land mines and blew up. The flak was so heavy that it is doubtful if many of the troops are still alive. It looked as if the Germans knew that we were coming.

"Today we lost Nos. 1, 14 and 16. No. 18 was shot up so bad that it could not get home and it had to land at Methwold. It had about 20 direct hits with 40mm shells which burst upon contact and made hundreds of small flak holes. Some of the crews saw hay stacks and buildings fold back and there would be one large and one small anti-aircraft gun in them. We are all wondering if the Germans machine-gunned the crews that bailed out.

"So far we have the following men missing:
    No. 1:  Johnson, Larrea, Friend, O'Donnell
    No. 14: Wharton, O'Connor, Thornburg, O. Johnson, Ruscito
    No. 16: White, Dover, Pace, Glick"

Bob's diary contains a September 18 post-mission report from one of the glider crews written by Flight Officer Norbert Henderson:

*"On the first day (Monday, September 18), we climbed in our gliders and started on the invasion of Holland. We left here [Cottesmore] at 1230 and when we were given the release signal, it was bout 1545. I was flying the number two position and just before I released, [F/O Robert] Schaeffer went down with his ailerons shot out. I released at 250 feet and was in a crossfire all the way to the ground. I landed in a large field but it was plowed the wrong way for us and I smashed up the glider. We piled out of the glider and the only cover was a farm house about 100 yards away. It took us almost an hour and a half to make those 100 yards.*

*"The Germans were firing at us all the time, so we whipped out a 105mm howitzer and drove them off their gun positions. A few minutes later some German ME-109 fighters came over to strafe us, but some P-51s ran them off. About 2000 we decided to try to make it back to friendly territory. By this time we had figured out our position and we were six miles from our intended drop zone, and a mile and a quarter from the German border. When we started out, we had four jeeps, two 105mm howitzers¶ and two trailers. Also we had about 75 men and officers. We sent out skirmishers and patrols in front and to the sides.*

*"On the second day (September 19), at about 0200 in the morning we stopped at a cross roads and I fell asleep. The rest of the outfit turned to the right, but I woke up and took off to the left. I had gone about a quarter of a mile when a fellow stepped out of the shadows. I asked him where the rest of the outfit was and he grabbed for his rifle. I figured he*

§ T. Moffatt Burriss, *Strike and Hold: A Memoir of the 82nd Airborne in WWII*. Washington D.C.: Brassey's, 2000. p. 106.
¶ This is possibly a case of mistaken identity. Parachute and glider field artillery units were normally equipped with the lighter 75mm pack howitzer which could be broken down into components for air transport.

*must be a damn German so I whipped out my pistol and stuck it in his belly. He gave a grunt and I had captured my first German. He said he had four comrades in a field to my left so I looked and there were four of the bastards.*

"*I yelled 'Surrender!' and they took off like ruptured ducks. I ran this Jerry until I caught up with the rest of the outfit. We got to our Command Post at 0700 the next morning. I was feeling pretty tough by then so I went with the paratroops hunting for snipers until I got scared. So I quit that and worked on a supply detail instead.*\*\*"

"*The third day, (September 20) while on this supply detail, we were strafed by German Focke-Wulf 190s. On the forth day (September 21) I ran into Larrea and O'Donnell. I gave them my last few cigarettes and put them on the road for Brussels. The fifth day (September 22), three of us headed for Brussels where they gave us a ship to come back to the outfit. We were damn lucky in this Squadron in that we only lost two glider pilots out of that mess.*"

Bob also received a firsthand account of glider operations from Lt Hugh Farler, a glider pilot from the 36th. He wrote up his experiences for Bob in October 1944:

"*After the usual period of sweating – my contention that some day the 316th would pull gliders into combat – came true – and I had the doubtful pleasure of 'driving' the first one to become airborne.*

"*We left this field at 1120 on September 17 1944, a couple of circles to wave goodbye and set out for Holland – or so I thought. Our mission was briefed for Landing Zone T around Groesbeek, southeast of Nijmegen. The trip to the coast of Holland was uneventful. From there to the Initial Point, the periodic burst of flak and the occasional hammering of a machine gun arcing tracers into our formation gave no warning that this was to be a bit 'rough.' Shortly after reaching the [Dutch] coast, some son of a bitch shipped a string of 50 caliber bullets through my tail, but that was the last hit to my aircraft until about three minutes from the LZ.*

"*Perhaps the most helpless feeling in the world, for me at least, was to watch that conglomeration of corruption come sailing into our midst and not being able to do anything about it. It started, as I said, about three minutes from the LZ and it was about this time that I realized we were slightly north of where we should be. When the ack-ack seemed the worst, bingo, the green light went on to release from the tow plane and I remember picking up the mike and yelling 'Bonjour' and then cut an immediate turn to the left and slowing up to 70 mph. The only field that caught my eye – and you always head for the first one that looks reasonable – was one that appeared to be about 150 yards long, over a road and some trees approximately 25 feet high.*

"*As we were a bit high, and a bit warm from large quantities of ground fire, I released my arresting chute and tossed the glider into a left-hand slip. Below I could see various Jerries darting out of the tree-lined road and throwing hand grenades up in slow tumbling arcs, exploding with a bang right in front of our faces, but exploding slightly below the glider. The fragments from one grenade wounded one of my airborne passengers. At this point we were cris-crossed from three directions by machine gun fire and other small arms. I set it down in a field and had slowed down to about 40 when I spotted an irrigation ditch in front. Hauling back on the stick I got the nose clear of the opposite bank by inches and we felt the landing gear part company from the aircraft. This turned out to be a lucky break because it prevented Jerry from spotting our immediate movements after leaving the glider.*

"*We were still under heavy small arms fire, so I went through the window of the glider sans any equipment but my M3 machine gun and .45. Here I lost the clip to the M3, rendering it useless and dropped my .45 which was then retrieved by the Sgt. I crawled around the glider, opened the door and started sawing on the ropes holding the jeep and at this point Jerry took over in the extreme, ventilating our glider to no small degree. When the mortars started dropping the five of us hit the ditch which was fortunately only about ten feet away.*

"*When we reached the ditch, we immediately started moving. It was the only thing we could do for we realized we weren't in the right place. We crawled away after circling back to the west and south and east and arrived at a double irrigation ditch. After crawling to the opposite side I observed our glider from the concealment of a bush. I saw some three or four others all still under intense fire and perhaps 15 of our men being taken prisoner. Evidently Jerry thought he had all of us because they did not start an immediate search of the area. We stayed in this place until dark and then after locating ourselves on the map started out in an endeavor to infiltrate back through German lines. We found that <u>we had landed s</u>ome 600 yards behind German lines.*

\*\* Unlike British glider pilots who were trained and equipped for ground combat, American glider pilots were strictly aviators, not infantrymen, and were expected to make their way back to friendly lines as soon as possible after they had deposited their cargo. They were not expected to stay "on the line" and fight with the other glider troopers.

"Since we had with us a wounded man the only course we could take was to cross the bridge at Wyler, Germany. It took a lot of debating and was definitely a last measure to walk across the bridge, especially since there were only five of us. We had very little ammo and I for one only had my .45. Thank God Jerry hadn't posted a guard on the damn bridge and we walked across unmolested. Then we cut back to the right, skirted Wyler and headed generally southeast working back to the southwest.

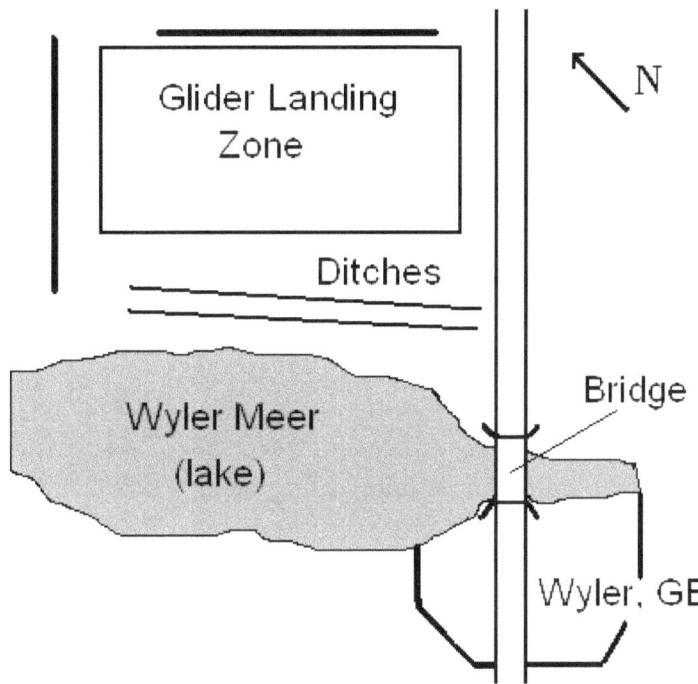

Diagram of the area described by glider pilot Lt Hugh Farler in his post-mission account.
Reproduced from the hand drawn map he provided to Bob Uhrig.

"We walked practically all night, through numerous encampments and barbed wire entanglements. Finally we entered some woods at about 0300 exhausted and ready for a rest. Here we had reached a point where we no longer gave a damn, lit cigarettes and slept three at a time while the other two watched. Early in the morning we crawled under some very thick bushes to observe the terrain and possibly wait until nightfall again. I fell asleep and was rudely awakened by a battery of heavy guns firing directly over our heads from about 75 yards away. We saw numerous Jerries burning gliders in the field directly in front of us. We quickly moved to the opposite side of the bush and moved out in a southerly direction. We shortly came to a field of gliders, thickly dispersed and assumed that we had entered Holland.

"We made a bee line for the nearest house which we surrounded and then the Major and I entered the place in search of food. The people gave us milk (which I don't like) and a few apples and told us we were in Holland. From there to the Command Post was a matter of engaging snipers occasionally (not particularly dangerous) and walking

"I reported from the 319th Glider Field Artillery Battalion Command Post to the Group Command Post and after eight days of being shelled, strafed and a tour of duty at the front lines in the Mook area prepared my group for evacuation via truck to Brussels. Unfortunately we were caught in a road block on the way out and went back to the ditches for some three hours.

"To my knowledge Jerry knocked out three trucks and three tanks in about 30 minutes at this point. Anyway, we went back to the first town where we were driven out by shelling. After a couple of days of sleeping on the ground again and some sweating, the Jerries were driven back some 500 yards. It was enough to allow us to squeeze through to Eindhoven and then to Brussels, and then on to our base. Confidentially, ain't that a hell of a way to get clusters for Air Medals?"

**Frustrating Mission Delays**

By September 20, it looked like Bob was finally going to get his chance to go along on a mission. "I am going on a mission today to the same place that we dropped the paratroopers the first day. This is a glider mission and I am going as crew chief on No. 2 with Lt Lee. It is now 1230 and they just called the mission off for 24 hours.

"All the fellows that went to the States on leave just came back yesterday. They all look just the same as when they left. Pappy Street did not come back and he was very happy about it. I will sure miss him but I am glad that he got to stay in the States. The only other person that got to stay was one of our radio operators, Berkpile."

In the few days that followed, bad weather, including dense fog, kept the Troop Carrier aircraft grounded as the bitter struggle on the ground in Holland continued. The American airborne forces were barely able to keep the single road open so that the British XXX Corps could reach their airborne comrades in Oosterbeek and Arnhem. By September 23, "They keep postponing the mission every day because of bad weather but it looks like we will make it today. The Colonel told me that I could not go as crew chief, but if they were ever short of pilots again I could go as copilot. But since all the crews are back from leave in the States it looks like my chances of going on a mission are shot. He told me that I could go on the air landings which start the day after tomorrow."

The air landings Bob spoke of were part of the plan to capture airfields in Holland and land troops on them directly, so they could go into battle almost immediately. "Engineers went in the gliders today and they are to build landing strips for us to land with freight."

There was some encouraging news concerning the crews of the aircraft lost on the first missions. "We received word that the crews on No.1 and No. 16 are safe, apart from Dover on 16. They think the crew from No. 14 bailed out over Germany and were taken prisoner. These men all returned today: Larrea and O'Donnell (ship No. 1), Pace (No. 16) and a glider pilot, Flight Officer Henderson.

"Our ships took another beating today. The corridor is so small that the Germans can shoot at them from both sides. I don't see how No. 2 made it back – it had 656 holes in it and that was the ship I was to have gone on. The airborne is taking a terrible beating in Holland. No. 8 had to land at Brussels, Belgium. He had his left engine shot out and all the controls and gas lines to the left engine were also shot out."

Amidst the work supporting the airborne operation, Bob was able to get a few letters off to Toots. "It has now been eight days since we received any mail. We have been pretty busy the past few days and have been unable to give the King's rabbits any trouble. Welter and Sullivan are now back in the room with me. That is the arrangement we made before they left. They have been telling me all about the States and the things they did. Some of the things they done are unbelievable. I wish I could have come home with them."

The final MARKET GARDEN missions flown by the 316th were on September 25. "We sent all our ships to Holland today to haul British equipment to a landing strip that the engineers had built. They all came back without any bullet holes for a change."

**The Loss of Another Comrade**

Among the many casualties of the Troop Carrier missions was Bob's longtime friend, Colonel Frank Krebs, who was reported as Missing in Action. Bob revealed his sadness over the many losses to Toots. "It is now 1205 am September 25 and we have just finished work. Do you remember the Sicily deal when we were in North Africa [referring to the parachute assault]? Well, the going now is just as bad, that is why we have so much work to do. I guess you have been reading about us. Yes, we are the First Allied Airborne Army. I wish I could tell you more of our doings. Krebs is gone. You know how much I thought of him. There are many empty beds these days but the Sicily deal hardened us. I am writing this letter with tears in my eyes. The Germans are playing music that would make anyone homesick. One of my roommates is asleep and the other one is missing, but we think he is in friendly territory. I am so dead tired and homesick. I guess these

tears will do me no good.

"Love received a letter from his wife asking him for a divorce. She said that she did not love him anymore and he could have the child. I felt so sorry for him. I told him to take the day off and get drunk. Received a letter from Pappy Street thanking me for all I did for him. He told me that he met you and he could now understand why I talked about you all the time."

The full story was that Colonel Krebs was lost on a Market Garden mission on September 17. Initially he was reported as MIA; probably captured. His airplane collided with another C-47 on the first day's mission. Both aircraft were lost and Krebs and another crewman parachuted safely. They were picked up by the Dutch Underground and evaded capture by the Germans for several weeks before returning safely to Allied lines. Bob probably delayed in writing to Toots about this since the details were sketchy and he was careful about revealing details for security purposes in case Krebs was captured. Although he would not learn the real story for many weeks, Bob feared the worst for his friend.

On September 27, Bob was finally able to see the Continent for the first time and witness the devastation wrought by the air campaign against the V-1 launching sites along the coast. "We hauled ammunition to Brussels today and I flew with Lee. I went as copilot. It is terrible how France is torn up. We flew over most of the 'flying bomb' sites and the bomb craters around them are so thick that they overlap. Every forest area has two or three sites. It is no wonder they were sending them over London every five minutes. From the air, Brussels is a beautiful city. It is one of the cleanest looking cities I have seen for a long time. We were only there for about 30 minutes and we ran into some terrible weather on the way home."

### Aftermath of the Invasion of Holland

As the men who participated in MARKET GARDEN – aircrews, wounded paratroopers and glider pilots - began to return to England, Bob was able to piece together what had happened to the units that had engaged the Germans. "We now have all but two of our glider pilots back and they had it plenty rough. I am having some of them write up their experiences for me. They had to evacuate the British Airborne today because the infantry had not been able to reach them. They had been at that bridge at Arnhem for over two weeks and they had 90% loss. The ones that escaped had to swim the lower Rhine River. Our own paratroopers suffered 75% loss. This mission was so costly. The corridor that we have is only three miles wide in places and the Germans shell the road both day and night. The Germans are able to cut the road about twice a day and then our fighters have to run them out."

The operation was formally over. As Bob noted, British XXX Corps was unable to capture Arnhem Bridge or reach the paratrooper stranded on the far side of the Rhine. Although in the days to come a few score would escape from the surrounded pocket, many hundreds of British and Polish troops were captured adding to the cost of the dead and wounded around Arnhem. The two American airborne divisions would remain in Holland for a few weeks to secure the lines of communication and prevent the Germans from infiltrating back across the Rhine. The 101st in particular was stuck in static positions along the river until the end of November. At that point, all the American paratroops (82nd and 101st) had relocated from England to new, permanent bases in France.

When the month of October arrived, the 316th had settled down once again into a quiet period. The missions flown were primarily resupply and casualty evacuation. The fighting on the Continent had moved on to the border of Germany and the threat of air attacks on England or fighter interception was greatly diminished. The Luftwaffe had pulled back their fighters to defend the German homeland or to keep the Russians at bay on the Eastern Front.

### Some Good News

There were happy events during those days, including a celebration with their British Allies. "Went to a wedding today. One of our radio operators (SSgt Paul Neading) married an English girl. I took the best man

to town with me and we went to the wrong church, so we were 15 minutes late to the wedding. When we arrived they were already being married. They picked another best man and borrowed the bride's mother's wedding ring. It was a beautiful church wedding and a very beautiful bride.

"They gave us a bad time at the reception for being late. Neading gave a speech and said, 'Ladies and Gentlemen, and Captain Uhrig.' He called me 'Captain, he came up the hard way, Uhrig.' I had a wonderful time at the reception. I thought there would be a bunch of stuffed shirt English people but I was very mistaken."

Further into October, information on the missing Troop Carrier crews was still being received. "F/O Johnson and TSgt Gooding left for the States today. It is a permanent transfer. I hate to see them go. The original bunch of enlisted men is now down to 14 and the officers are down to 9. We have been hauling gasoline and ammunition to Brussels every day now. We send out every ship that is flyable.

"All of the crews [from MARKET GARDEN missions] have been accounted for but the crew on No. 14. There were all seen bailing out but as yet there is no word on them, so they must have landed in German territory. The bad part of that is that the Germans are so sore at the Air Corps for bombing and strafing them that they have been hanging our flyers up by their feet and slitting their throats.

"We learned that Lt Dover, the copilot on No. 16, was killed. He did not have a parachute on and got hit. He made it to the back door of the airplane before he died. The ship was on fire and they were only at 400 feet altitude so everyone else jumped. The radio operator checked on Dover but seeing a large pool of blood and no signs of life, he jumped himself. The pilot, Lt White, got burned in the face and is still in the hospital."

On October 7, Bob received some encouraging news about his friend, Colonel Frank Krebs, and passed it on to Toots. "I have been told that Krebs is a prisoner of war. I hope he is safe. I will give him a bad time when I see him again." Although technically still an "evader" since he was never actually captured, Krebs would not make his way back to Allied lines for several weeks. Bob believed that coincidently it was Krebs's C-47 that struck the back of the 36th TCS airplane flow by Lt Fulton and caused it to crash.

**A Significant Decoration for Captain Uhrig**

A troop review and awards ceremony was held on October 9. It was a big day for Bob, as his work in North Africa was finally and formally recognized. "Today I was awarded the Legion of Merit. The ceremony was held at the 61st TCG field and each of the five Groups in the 52nd Troop Carrier Wing had a detail present. There were 44 awards given out. The Legion of Merit was the highest one awarded and there were only two of them given out. Gen Williams pinned on the medals and then the troops passed in review. It was quite an affair." It was indeed. The Legion of Merit decoration had only been around for a little over two years at that time, so Bob was one of the first recipients. The other soldier decorated with Bob was 1st Sgt Roy Johnson, also of the 36th TCS.

Bob immediately shared his excitement as well as his continued thoughts of home in a letter to Toots. "Yesterday I was awarded the Legion of Merit. It was a big affair with band and all. There were only two Legions of Merit given out and they were the highest awards given. General Williams made the awards and believe it or not, I was nervous. Then all the troops 'passed in review' and it was a very beautiful sight. In fact you might say we got the works. It almost felt as though you were there beside me. This is still a miserable country to have to live in. It is a wonder to me that they live as long as they do here. I have been getting in a lot of flying time and I am getting pretty hot at landing these old birds. I want to get home where we can fly together again. I have been thinking about us buying an airplane instead of a car because cars will not be any good for a couple of years."

His next letter a few days later expressed his disappointment that he would be missing yet another holiday at home. "Received the envelopes with the soaps. We don't have cars and filling stations over here so they don't do me as much good as if I were in the States. They are about the handiest things I have seen for a long time. I keep wondering if I will be home for Christmas. If not that will make the fourth Christmas that I have been away from my darling wife."

*Left:* Bob's official photo as a Captain.  *Right:* Bob is decorated with the Legion of Merit, October 9, 1944. (AFHRA)

Unfortunately, on October 14, there was another fatal mishap in the 316th. "We were ferrying gliders from Greenham Common on Saturday and one glider ran into the tow rope of another and it spun in, killing two pilots and a mechanic. It was a double glider tow but only one glider crashed. The dead are F/O Krohm, F/O Lucier and Sgt Basham. They are to be buried tomorrow (Oct 17). We were told that the glider went almost straight in, mangling the bodies beyond identification. Krohm was our new Special Service Officer. Out of the four Special Service Officers†† we have had in the past two years, three of them have been killed.

"The Germans announced today that Rommel died. His car was strafed during the Normandy invasion." The reality of this was later revealed after the war. Rommel was "allowed" to commit suicide on October 14 as punishment for his part in the July 20, 1944 bomb plot against Hitler. Because he was still a national hero, he was buried with full military honors and his involvement in the plot was covered up to maintain the illusion of his loyalty to the Nazis.

Since the excitement of the airborne operations had wound down, Bob wrote to Toots about taking his first extended leave in Scotland. "We make coffee and toast here at Engineering every morning. We have a little electric stove here and it makes the best toaster. Then we chisel butter from the mess hall and we have a complete breakfast. The squadron commander and four other fellows want me to go to Scotland deer and pheasant hunting. I don't know if I should or not because of the cost. In fact I wish that we were out of England because of the expense. You know how I hate someone that does not pay his share. One item alone is whiskey which costs $12 to $14 a bottle. It costs about $40 a month just for that. That is the hell about being an officer. I used to think it was a bunch of crap but I am finding out different. Every time the officers go to London it costs them about $100 a day. So as yet I have not seen London."

---

†† Special Service Officers were responsible for morale, welfare and recreation for the troops. They organized sporting events, musical entertainment and the like for their units. It was not normally a hazardous job, unlike the members of the First Special Service Force, who were specially trained commandos.

## Low Morale in the European Theater

The fall of 1944 was not a pleasant time for most Americans in the ETO. The two airborne divisions were still recovering from their losses in Holland and the other American armies were still trying to break through the border defenses into Germany. A significant number of American divisions were fighting a battle of attrition in a region known as the Hurtgen Forest. The best contribution the Troop Carrier forces could make was to keep up the resupply channel from the UK to the front line.

Bob spoke in general about this tough time in his letters home. "I am here in the Engineering office. I am Officer of the Day and will be until 11 tomorrow. Lt Welter and Lt Phillips are here with me and we are all writing letters. We brought the radio down here with us so we could have some music. I have to remain close to a phone so that is why I am here. I talked the other two into coming down here with me. People always have the battles won before they are started and you know how I hate saying anything until it is a sure thing. We still have a long road ahead; at least that is what I think. I remember how they fought in the desert and we had them on the run all the time but it still took nine months to run them out of Africa and now they are close to their own back yard. As you know, the old Desert Fox (Rommel) is dead. We all had great respect for him. Even the English had praise for him. Of course he made believers out of a lot of us."

Morale among Allied troops was sagging. Memories of the heady days of the summer advance through France had faded, and the war on the ground had returned to a brutal slugging match as the Germans, with their backs to the Rhine, fought for every yard of ground. The low morale led to a number of men going Absent Without Leave (AWOL). "We are going to be restricted to base for the next two days (Oct 24-25) so that the Military Police can catch most of the men that are AWOL. This is being done all over the ETO.

"We are still hauling freight to France, Holland and Belgium every day, weather permitting. The weather here is terrible. This time of the year it is almost impossible to get off the ground before 1100 because of the fog. On top of the weather there is very little daylight this time of year. Just as soon as it gets dark, the fog and haze starts moving in.

"I was sitting here at about 1900 when I heard a crash so we rushed outside and sure enough it was one of our ships, but no one was hurt. The pilot, Lt Cook, had taken off from Paris with his parking brakes on and when he landed the ship went up on her nose. It damaged the nose and both propellers and both engines. The right prop broke loose and was about 50 yards from the airplane. The airfields in France are so terrible since the rains have started and the mud is so slick that the brakes do no do any good. That is how Cook was able to take off with his brakes on.

"When I returned to Engineering, Operations called and told me that No. 4 (Lt Wheeler) had crashed in France and No. 9 was stuck in a bomb crater in Paris. Lt Wheeler washed out the landing gear and both propellers when he could not get the ship to stop in the mud. It began sliding sideways and hit a bomb crater. No one was hurt and No. 9 should be back tomorrow. This day has been one large headache.

"Lt Sullivan and his flight just landed from Cherbourg and they almost lost the whole flight. Sullivan said that only by the Grace of God did they not lose all three ships. The all landed and slid down the runway sideways. Sullivan himself slid between the windsock and a stack of gasoline barrels and the only damage that was done was to his wing which tore off the windsock. You can see we are going to have a lot of trouble this winter. In fact, we have had a small stack already. It is a wonder we have not lost more ships flying through this weather. It makes old men out of young ones. Two ships from the 37th Squadron got lost and flew over Dunkirk. The Germans threw up every thing they had but did not hit either one of them. Some luck."

The following day, Bob made plans to begin recovery of the damaged C-47s. "The Colonel wanted Maj McMenamin and I to go over to France and look at No. 4, but the weather remained bad most of the day, so the trip was cancelled."

**No Assignments**

Although Bob continued to write to Toots almost daily, there was only so much he was allowed to say in his letters. To preserve operational security, he could not give details of missions, operations or units. He could not even say much about the men in his own unit. There was always the chance that the mail would get intercepted and used by the Germans for intelligence-gathering. He "talked around" this in one letter home in late October. "If I am not home this Christmas it will make four in a row that we have not been together on. I hope and pray that I can see you this year but I don't have a very bright outlook on the subject. At times it is hard to find things to write about. I can't write things like I want to. You can write everything you know but I can't do that. Received a letter from Koczan and he is a TSgt in a Bomb Group over here so I am going to try to look him up. You know that will be one large session when I do see him. I am very anxious to see the big lug."

Through the end of October, the Squadron was still unable to retrieve all their aircraft that had been wrecked on other bases. Unlike the bomber and fighter groups which took off and landed at their home base on nearly every mission, the transport crews had to fly into a number of new airfields on every mission. Many of these were unimproved strips left over from the advance through Europe and others had better runways but few navigational aids to assist the crews. When the aircraft went down or were damaged, it was the responsibility of the Engineering teams, such as those led by Bob Uhrig, to go out and do what they could to get the aircraft back to flying again.

While waiting to repair his damaged ships, Bob had a surprise visitor on October 30. "Who should walk in this morning, but Koczan, my old assistant crew chief on No. 7. He is on a three-day leave and he found out that I was here at Cottesmore so he caught a train from London up here. He is a flight engineer and waist gunner on a B-24 bomber and has been on 16 missions already.

"Together we visited with Maj McMenamin, Stracke and Scheft. What a session we had. About 10 of us tied one on in the Officers' Club and there was no one left but officers from the 36th Squadron. Then about 0030 we decided to get everyone in the Squadron out of bed. Then we had a rough-and-tumble in our room and now none of our beds are straight. Koczan said he has not been on one like that for a long time. One fellow had his door locked and would not unlock it, so we broke it down and now we have a door repair to pay for."

The group then planned a short trip for the next day but it too was cancelled due to continued bad weather. "We were going to take Koczan to Paris with us today, but the weather was bad so we could not go. About 1600 we flew him back to his home base which is about 60 miles from here."

All-girl band at Cottesmore Officer's Club. They were part of a USO show that visited the base in October, 1944. (AFHRA)

**City of Lights**

Finally on the morning of November 1 the weather had cleared enough for Bob and his men to make the long-awaited trip to Paris to check on the damaged aircraft. But when they arrived, "We found that the Service Group from Le Mans had already sent a crew up to work on No. 4, so we had to bring all the men and equipment home again. Everyone was in low spirits because they planned on seeing Paris tonight. All we saw of it was from the air.

"The field in Paris is known to us as A-54 but its given name is LeBourget and it is where Lindbergh landed after he had flown the Atlantic. The bombing of the airfield and the nearby section is terrible. There is nothing left standing. The hangars and administration buildings must have been beautiful. Even after all the bombings you can tell what they used to look like. There are German mines on the paved runways but the Americans are clearing them as fast as possible. Everyone is landing on the dirt strips. The mud is a foot deep everywhere and even deeper in spots.

"There are bomb craters everywhere and they have only been filled in with dirt. Now every time a ship taxies over one it goes in all the way up the landing gear. There are ships stuck everywhere. We got stuck too, but not bad. We had the airplane pulled out backwards and then we made it out OK. The destruction there is terrible. It is unbelievable how they can completely destroy a factory by bombs.

"Every place I go I find someone I know. We have a large time comparing notes. I ran into Captain Miller who used to be a private in the old 1st Transport Squadron. He is Assistant Engineering Officer in a Service Group that is moving into Paris. He has been in Russia for the past nine months."

**Time off for Leave in Scotland**

Bob was able to take his first long leave to Scotland in early November. It was his first real time off since the short break in Gibraltar. On his return he wrote a long letter to Toots with all the details. "We just returned from our five day leave in Scotland. The other fellows on the trip were Major Wright, Lt Whattley, and Lt Phillips. We took the train from here to Edinburgh and then another train to Callander from where we caught a bus for the last seven miles to the Trossachs Hotel. It was an eight hour trip and the last two hours were through mountain country. The country there is beautiful with plenty of snow-capped mountains, roaring mountain streams and lakes (known locally as lochs.) I never knew that water could travel so fast.

"When we got there I could hardly believe my eyes. It is built like an old castle but is as modern as any hotel in the States. The rooms are very comfortable and the beds are wonderful. Each one has a down-filled comforter. All the flooring in the hotel is carpeted. There are lounges of every size and each one is very homey and each one has a fireplace with a roaring fire in it. The lounge next to the bar where drinks and tea are served is very large and built in a circle which his half windows. The windows are very large and you can sit here and look at the lake and mountains.

"We hunted the first two days and about killed ourselves from climbing mountains. There are deer, pheasants, mountain goats, duck and geese to hunt. We saw about five deer and two mountain goats but we were unable to get a shot at any of them. We took along plenty of shotgun shells but we could not get a shotgun, so we were unable to hunt pheasants, duck or geese. The hotel is so comfortable and home-like that you don't even want to go outside. The mountain streams run past the hotel. They make you think the wind is blowing all the time. In the spring each stream is full of salmon. There are many rapids and waterfalls. The whole country reminds you of the western mountain region of the United States.

"There was no train from Callander to Edinburgh on Sunday (12th), so we had to leave on Saturday and spend the night in Edinburgh. We ate at two different places so we could get enough. Then we had a few drinks and retired. We caught a train home on Sunday morning at 1115. All the time I was there I wished that you were there with me."

There was a little surprise duty awaiting Bob when he returned, but after a relaxing leave he took it in good spirits. "I am O.D. [Officer of the Day] today. It is the last one that I have to pull. Engineering Officers

don't have to pull O.D. anymore. There is not much for me to say because this is only the third time that I have had to pull it."

## Milestones

November 15 was Bob's 30th birthday. Since most of the original 36th TCS men had been transferred or returned to the States, Bob was now older than most of the men in the Squadron; many of whom had enlisted or been drafted after the war started. He was likely significantly older than most of the other junior officers he worked with. As a result he spent a good amount of his off-duty time with the enlisted men, particularly the senior NCOs, since they were closer to his age and had been through North Africa and Sicily together. "Today is my birthday and your daddy is now 30 years old. But I do not feel any older than ever. Last night I took some of my Granddad's popcorn over to the Enlisted Men's barracks and we popped it in butter. Then we sat around and shot the bull about the things we used to do in the desert.

"Tomorrow morning I will have completed two years overseas. It is about the longest two years I have ever spent in my whole life. I miss Toots so terribly that sometimes I think I cannot stand it any longer."

The next day, November 16, was the Group's two year anniversary of overseas service. In his letter to Toots that day he reflected on all that he had seen and done in that time. "This morning makes us two years overseas. I am glad I did not know that I would be over here two years when I left the states or I should have died. That is a long time to be away from you. The officers are having a celebration Saturday night and the enlisted men are having one next month so they can celebrate both the Air and Ground Echelons' birthday.

"Last night I went over to the enlisted men's barracks and then we had a session. There are very few of the original bunch left. We had a lot of laughs about the things we've done in the desert. It is a wonder that we are still alive. We were more of a menace to ourselves than the enemy was to us. How we used to go out in the mine fields to look for motorcycles and guns and other equipment gives me the shudders. On top of everything else someone was always setting an old ammunition dump on fire and then we would really have a show. When the aircraft were flying we were always shooting. Almost all of us had a German rifle and you could get ammunition almost anyplace in the desert. We had trench mortars, machine guns and 20mm cannons. In fact we looked like an army ourselves. You should have seen all the inventions. We could build most anything because we always had about 50 German or Italian planes to pick parts from."

Also heavy on Bob's mind was the approach of another winter season in England. "We had our first snow yesterday but it did not lay on the ground long before it melted. Today it is pretty cold and damp. There is very little flying these days because of the weather which is bad two-thirds of the time. It is a wonder to me that we have not lost more ships due to the bad weather."

## Hunting Parties

On Saturday, November 25, Bob went for a little game hunting, or "poaching" in the local area. "Last night Captain Scott, the 37th Engineering Officer, got two shotguns from the Wing Supply and we went hunting. You have to pay $2 for a license to carry a gun and with this you are permitted to hunt here on the field, but Scott and I strayed about a mile off base. If you want to hunt off base you have to have the landowner's permission, which we did not have.

"We got one pheasant and one partridge. At one time I saw Scott running for all he was worth out of the field next to me and as he ran through the gate he fell flat on his face in the mud. I thought all the game wardens in the country were after him so I ran to a hedgerow and hid under it. A little while later two fellows rode by on bicycles.

"This morning we still had the shotguns so we decided to go hunting again. As before, we could not stay on the base and started going through the hedgerows. Scott and I each got a pheasant. Later Scott jumped one of those big hares and missed him twice. I took a crack at him and stopped him. I tell you I am half afraid they might take my gun away from me because they are so large. We did not miss a thing this time."

In early December, the 316th began to receive some new aircraft, different from their normal complement of C-47 transports. These were Consolidated C-87s and C-109s which were transport versions of the four-engine B-24 bomber. These aircraft were intended to augment the cargo carrying capability of the Troop Carrier Groups. The first airplane, a C-109 received by the 36th Squadron, was a tanker version and as Bob noted, was "equipped with nothing but gasoline tanks. It is a new one and that is some help. Each Squadron in Troop Carrier Command gets one for training all their pilots."

On December 4, the 36th Squadron received a second C-109, prompting the squadron members to speculate that they were also being given a new mission. "We think that after we are finished here that they will send us to India to haul gasoline to China. If I have to go over there before I get to see Toots I am going to go mad. More headaches, they just gave us another B-24. It is all we can do to keep up with the maintenance of our ships without inheriting these two new crates."

But no new mission was to come and the 316th continued its normal routine of resupply and casualty evacuation missions. But they took time off on December 12 to celebrate another milestone. "We had our dance last night for two years overseas. They waited this long so that both the Air and Ground Echelon could celebrate together. It was as nice a dance as I have ever been to."

**Trouble in the Ardennes**

In mid December 1944, most of the Allied troops were looking forward to Christmas, thinking that it might be the last "wartime" holiday they would be seeing. There was complacency in the Allied camp. The Germans seemed all but beaten, and it would only be a matter of time before they would throw in the towel. All that changed on December 16, when a massive German offensive began in the Ardennes Forest region of Belgium and Luxembourg. In what would come to be known as the Battle of the Bulge, German armored spearheads knifed through thinly-held American lines with the objective of reaching Antwerp and dividing the Allied armies in two.

SHAEF reacted as quickly as possible by ordering its only spare reserve divisions, the 82nd and 101st Airborne, to move to the battle area and set up defensive positions. It was hoped that they could delay the Germans until Patton's forces could drive up from the south and Montgomery's forces could attack from the north and stop the German offensive. Because of time and weather, the two airborne divisions would be moved from around Mourmelon, France to Belgium by truck instead of dropping in by C-47.

Once the airborne divisions were in place in Belgium, Troop Carrier units were used to provide much needed aerial resupply to the besieged town of Bastogne, where the 101st was eventually surrounded. Although the 316th was not one of the groups called on to execute these missions, other Troop Carrier units performed heroically – parachuting in supplies and towing gliders full of ammunition and medical units directly into the Bastogne perimeter. Four Troop Carrier Wings including the 52nd were assigned to move the 17th Airborne Division from England to airfields near Reims, France so that they could be rushed to the front to plug gaps in the American lines. The 316th was one of several groups that moved the 13,000 men of the division along with their equipment. This movement operation began on December 24 and lasted for five days, eventually involving over 500 airplanes from the four Wings. As soon as the paratroopers arrived in France, they were packed off to the front in Belgium via truck convoys.[60]

On December 23, Bob had heard enough about the battle raging in Belgium and Luxembourg to give Toots some hints about how serious it was. "Major McMenamin just came back from Scotland and he brought me four quarts of gin. It sure came at the right time. We did not know what we were going to do for Xmas whiskey but this was the answer to our prayers. The cheapest you can buy it for is $14.00 a quart so you see what it cost to get your friends a drink. It cost $.50 a shot at the bar in our club, well a single shot you can almost see so we always buy double shots which cost $1.00. Now then take five fellows sitting around drinking double shots and chasing them with beer which is $.20 for a very small bottle runs into money. About the stars on our ribbon [ETO ribbon]. We wear five now but in a couple of weeks we shall wear seven. That is if the Germans don't run everyone back to this island again. They seem to be doing pretty good for

themselves. I hope this is their last go and after it is stopped that we can whip them. I tried to get you some Scotch plaid when I was in Scotland but you just cannot get it without coupons and I could not find anyone that would sell any."

For Christmas, Bob and his comrades (Love, Tucker and Fenimore) spent the day in town with a local English family, the Thompsons. "They have a home where we can get rooms to spend the night when we have 24 or 48 hour passes. They are a lovely family but it is still not home. This Christmas we did not get in the bag like all the other years. In fact it was all very quiet. This makes four Christmases that I have been away from Toots and I hope and pray that I can spend the next one with her."

The next day, Bob and his mates went hunting again, but this time there was practical-joking involved and Bob displayed his puckish streak again. "Scott, the 37th Engineering Officer and I went hunting. We were very poor shots today. We also went with Herman the bartender. Scott and I took the shot out of three shotgun shells but did not tell Herman. While we were out hunting, Scott called to me, 'Let's shoot that cow' and I said 'I will if you've got guts enough' so we just up and shot straight at the cow. Herman just about died because he thought that we had real shot in the shells. When he found out what happened he wanted to shoot the two of us."

**The Soldier's View of the War**

Bob's Christmas Day letter to Toots described his day in town and continued to speak of the serious implications of the Battle of the Bulge. "About five of us spent Xmas day at some people's house in town. These people rent out rooms to us so they wanted us to come for Xmas. They showed us a wonderful time and sure made us feel at home. I hope and pray that you and I can be together next year. One Xmas out of five now, that is not asking too much. It has now been 10 days since we received any mail. This set back [in the Ardennes] should at long last wake the people up at home that the war is not over until the last German has surrendered. The people in the States finished the war on the day after the D-Day invasion and now I hope that they wake up."

About this time a paratrooper in the 101st Airborne, David Webster, shared the same sentiments as Bob Uhrig about the differences between life in the States and at the front lines. "Back in America, the standard of living continued to rise. Back in America the race tracks were booming, the night clubs were making their greatest profits in history, Miami Beach was so crowded you couldn't get a hotel room anywhere. Few people seemed to care. Hell, this was a boom, this was prosperity, this was the way to fight a war. We read of black-market restaurants, of a manufacturer's plea for gradual reconversion to peace time goods beginning immediately and we wondered if the people would ever know what it cost the soldiers in terror, bloodshed, and hideous agonizing deaths to win the war."‡‡

As 1944 drew to a close, the Allies had stabilized the front in Belgium and Luxembourg and began to slowly push the Germans back again into Germany. It would still be many weeks before the "bulge" in the lines was erased and the front was back to where it was in mid-December. The Troop Carrier groups were still supporting the Allied ground forces with supplies as they attempted to break through the German border and make their final push towards Berlin. Although the noose was tightening around the Germans, there were still many months of hard fighting ahead and the combat missions of the 316th TCG had not yet come to an end. There was one more significant airborne operation still to come in 1945.

---

‡‡ David Kenyon Webster, as quoted in Stephen E. Ambrose, *Band of Brothers*, p. 239.

An fine informal portrait of Captain Bob Uhrig taken in England. He has the 316th Group insignia painted on his A-2 jacket.

The "Senior Staff" of the 36th Engineering around Bob's desk at RAF Cottesmore. Bob depended on his senior non-commissioned officers to keep the troops in line and the airplanes in the air.

# Chapter 12

# Victory

*My dear general, the German is whipped. We've got him. He is all through.*
Prime Minister Winston Churchill to General Eisenhower,
watching the aerial armada for Operation VARSITY fly overhead,
March 24, 1945

The New Year 1945 started very quietly for Bob Uhrig and the rest of the 316th. He made no diary entries at all until the third week in January. The weather in Europe that winter was especially harsh and it restricted flying operations throughout the region. Mike Ingrisano of the 37th TCS noted in his Group history that, "In January, for all practical purposes, the aircraft were snowbound. Snow fell, disappeared, fell again and lay over the fields. The personnel at Cottesmore were restricted to base because the roads were ice-covered." Besides the snow there was fog, "so liquid and dampish that it melted on the trees, fences and wires… frequently limiting visibility, further hindering our air and ground movements."[61]

Just after New Years, Bob had a laugh at Toots' expense when he got a newspaper from back home. "I have just returned from Major Mac's and he has a very interesting clipping from a certain paper and a certain Ivea Nelle Uhrig seems to have been stopped for speeding. It goes on to tell about a car load of girls in a convertible. I just know it must have been a misprint because I just know that my Toots would never get caught for speeding because she drives too fast to be caught! There are eight of us planning a hunting trip to Scotland this month. That many of us together should be able to take over that section of Scotland."

Unfortunately, Bob's trip to Scotland was postponed. "I have been a little too busy to do any hunting this week. We had a little snow here yesterday and it made me very homesick because I can still remember how our winters used to look back home. We go to the mess hall and get a thermos bottle (1 gal) full of coffee and a box of bread and a pound of butter then we bring it down here to Engineering and then we have toast and coffee. It is not quite as good as going to the PX at Patterson Field but it will do. There are more and more pilots coming in every morning so we will have to start running them out so we can get some work done."

On January 23, Bob noted that, "There are all sorts of rumors about our Group going home and I hope they all come true. We have stopped the training program and have gone back to hauling freight. Sullivan just came back from a six-day trip to France. He had a flight of five ships and they were hauling pigeons. So you might know that the pigeons did not make a safe journey.

"Last night I was writing to Toots and telling her about Sullivan eating pigeons and Sullivan said 'Here are their dog tags.' And he threw a couple of leg bands on the table. I thought I would die laughing."

The following day, an unexpected visitor showed up at Cottesmore, Colonel Frank Krebs. "The snow is still on the ground and it is about 18 degrees. I received the biggest and most pleasant surprise this afternoon that I have had for a long time. Col Krebs walked in on me. We never did know for sure what had happened to him. He is now a full Colonel and has not changed a bit.

"He was leading a flight of 45 ships home from the Holland invasion when a transport came up from below and collided with him, knocking off one engine and propeller and tearing away a large portion of the fuselage. He had to give the order to bail out. When he hit the ground he broke his foot. He was with the Underground for 42 days. The first two weeks he could not be moved and he said he received very good care. His outfit is now stationed in France and he left this afternoon to go home. I wish that he could have stayed overnight."

He also wrote to Toots with the happy news, a little less concerned about security at this late period in the war. "If you ever wanted to see your husband surprised and glad you should have been here today because this afternoon Col Krebs walked in on me. I almost passed out. I could not find my tongue for a long time. He looks just as good as ever. We laughed about old times. On the invasion of Holland another ship few into his and he had to jump."

Colonel Krebs, his copilot Howard Cannon (who was later elected to the U.S. Senate from Nevada) and the other four crew members had managed to parachute from their stricken aircraft after the left wing and engine were heavily damaged in the mid-air collision. They were hidden by Dutch civilians in the town of Breda along with another group of downed Allied flyers. Eventually they walked back to Allied lines with assistance from another group of Dutch Underground members. They were taken first to Antwerp and then made their way to Brussels. They finally arrived at their base in France still wearing the civilian farmers' clothing they had been given by the Dutch. After the town of Breda was finally liberated, Krebs and Cannon flew a planeload of donated food and clothes to the town to repay the kindness of the Dutch residents.[62]

New missions with the new aircraft were also beginning. "Tomorrow (January 25) we start hauling gasoline to the 'far shore' [France] in our C-109s. We should have some fireworks around here with 5600 gallons of gasoline on each ship."

Unfortunately these cargo versions of the B-24 did not hold up too well. By early February Bob's men were having difficulty keeping them in commission. "We have a B-24 [C-109] with a broken nose gear and for the past month have been trying to get spare parts for it, but have been unable to do so. We heard about a B-24 being salvaged near Brussels at airfield A-83, so Major McMenamin said that I could go over and check on it. I went with Captain Welter and Lt Lee was flying the other ship. We had a load of cargo for airfield A-42 which is just outside Paris. But the time we landed there it was too late to fly to A-83, so we all went to town."

**Gay Paree**

It was of course Bob's first and long-awaited visit to "liberated" Paris. It was definitely not the same austere, wartime environment he was used to in Britain. "We caught a French bus to Ponte de Sevres. Then we went on what they call the Metro and it is fairly fast. Some of the stations are still closed because the Germans used a lot of underground stations for aircraft factories. After arriving downtown we tried to get billets but were unable to do so. We finally found rooms in a small hotel. Then we started out to see the nightlife of 'Gay Paree.'

"We used the Metro to go every place. We ate at the Officers' Mess which used to be a German Officers' Club. We had a couple of bottles of champagne and a couple of French 75s*. Then we went to see the Follies, which is 'The Show.' These scenes were the best that I have ever seen in my life. We only were able to see the last half of the show. It was better than any stage show that I have ever seen in the States. The girls in the show were, well you know how the French women look, out of this world. The girls in the chorus wore nothing from the waist up and very little from the waist down.

"These French women can dress the best of any race of people that I know of. They can take a couple of yards of cloth and just wrap it around then and they look like a million dollars. The hats and hairdos also are all their own. If you were to see some of their wild hats in the States you would go batty but to them they are just ordinary hats and the same goes for their hairdos. They are not short of any clothes except hosiery. Food is the item that there is a terrible shortage of.

"All the women have perfume on and the minute you step inside a building you can smell it, also in all the Metro stations. You can even smell it when you pass them on the streets. They all use beautiful makeup and nail polish. Their complexions are the best of any race. The only thing that stopped me was seeing so many French women with blonde and red hair because I always thought that all French women had black hair.

"Most of the good clubs where the world's best floor shows are put on were closed due to the shortage of electricity, so we were unable to visit the 'world renowned' clubs. So after the one show, we had a few drinks and retired."

The following morning, February 8, Bob and the crew had to get on to Brussels, but not before some brief sightseeing in Paris. He was impressed by what he saw. "We were up at 0800 and dressed. We caught the Metro to go the field but stopped off at the Eiffel Tower. I took some pictures of both the palace and the tower.

\* The French 75 was a cocktail made from gin, champagne, lemon juice and sugar. It was supposed to be as powerful as a French 75mm cannon, hence the name.

The tower is just across from the palace and is an enormous thing. The French go in for beauty and they carry it out from their buildings to their gardens. Everything they have is a thing of beauty. Surrounding the tower is a formal garden that must cover 100 acres. The Seine River flows just in front of it. After taking a couple of pictures we had to rush on.

"We caught the next Metro but arrived too late to catch the bus for the flying field. So like any GI, we starting thumbing a ride. The first to pick us up was a jeep which took us part of the way and the next was a 20-ton tractor pulling a howitzer.

"The airfield A-42† is nothing but a mud hole and once you are off the taxi strip you are finished. But that is like all the fields in France just now because the Germans, along with our bombers, have done a pretty fair job of tearing them up. The destruction is terrible. At about 1030 we took off for A-83, arriving at about 1115. Low and behold, after walking about two miles, I find that the 8th Air Force has taken away the wrecked B-24 so the trip to A-83 was useless.

"The weather in England was very bad so they would not clear us for takeoff. So in the meantime we had used two small gasoline stoves to cook up a meal of canned ham and eggs, vegetable soup and coffee. About 1430 they told us we could go. When we reached southern England we ran into terrible weather and had to land just outside London. We called Cottesmore and they told us to remain where we were overnight. We were at an RAF base, so they put us up for the night. We all retired very early because our heads were still pretty large from the night in Paris. Welter, of course, caught a train to see his girlfriend."

Once safely back at Cottesmore, Bob's barracks-mates overindulged a few nights later. "Last night Welter and Crow Lee got in the bag. Crow cut the hose off his gas mask and was sucking beer up through it. Everyone that saw him just about died laughing. They said the beer got clear up to the eye pieces and you could see it running around inside the mask. He and Welter went to chow and they sat at the Field Grade Officers table (reserved for Majors and above). Everyone says they sat there like there was no one else in the Mess Hall but them. I guess they were a scream. They also had a fight with seltzer bottles.

"I just came back from giving a pint of blood and I cannot tell any change. A lot of the fellows passed out and a lot were very sick and had to lie down. We were given a shot of whiskey afterwards. I told Doc that the whiskey was pretty expensive when it costs a pint of blood."

On February 20, Bob and a small group went to a nearby base to see a sporting match. It was a bit of culture shock for both the Americans and the British. "We went to the RAF field last night to see the fights but our boy Sullivan did not win. He put up a good fight but the Polish boy out boxed him a little and won the decision. Sully says that he is going to fight him again. The next time it should be a good fight. When a boxer lands a good one the English just clap a little and that is all there is too it. Well they just about died when us Yankees started hollering, 'Kill him, what do you think he is, your brother?' and, 'We want blood.' I guess their eyes are still out on their cheeks. I guess we changed things for them."

At the end of February, the Squadron got some good news on one of the crew members lost on the MARKET GARDEN missions. "Eustice and Lee received a card from [Lt Maurice] O'Connor. He is a POW in Germany. He was shot down during the Holland invasion (aircraft No. 14.) Everyone saw the whole crew bail out so we all thought they were still alive if the Germans did not shoot them. He said they got burned pretty badly but they were OK now.

"Now O'Connor, Eustice and Lee have always roomed together and it has always been a big fight over clothes. The first one to the room is always the best dressed and every time we move they all three pack at the same time. They would never think of leaving one another to pack alone because they steal from each other so much.

"Well after the other two saw O'Connor bail out they thought he would be OK so when they got back they split up his clothes. On the card he sent them he said, 'You so and sos, better have my clothes sent home.' Well now Eustice and Lee are laughing their heads off. I told them I hope O'Connor comes back to this outfit so he could kick the hell out of them. He may be back soon because last week the Russians captured the prison camp where O'Connor was held prisoner."

---

† A-42 was also known as Villacoublay Air Base and was about 8 miles from Paris.

## Spring Comes to Europe

Although the winter of 1944-45 had been severe, spring came early to Britain. With the better weather fresh thoughts of going home came to Bob Uhrig and the rest of the 316th at the end of February. "Today is payday and everyone is ready for it. I never saw so many people broke in my life. The weather here the past week has been wonderful. It is just like spring out. Everything is turning a nice fresh green. It puts everyone in a better mood when we have weather like this.

"We are back on a training program again with glider towing. We fly day and night formations and all the rest that goes with the training program. Of course the new men can use a little training. It begins to look like another mission coming up. Most of the groups are moving to France. As yet they have not ordered us to move and we think that our chances of going home are better. We are hoping that they just use us as a spare Group."

Bob's suspicions about further airborne operations were well founded. The U.S. now had four full airborne divisions in Europe – the 82nd and 101st, which were still recovering from the winter battles; the 17th and newly arrived 13th Airborne Divisions. Allied planners continued to look for opportunities to use these divisions to seize key points in Germany, cut off the retreat of German forces or otherwise hasten the end of the war.

Further confirming suspicions about another operation was the arrival in early March of additional new aircraft for the 316th and other Troop Carrier Groups. Curtiss twin-engine C-46 *Commando* aircraft were distributed in small numbers to all of the Groups. These aircraft had already been in use in the China Burma India Theater to haul supplies over "The Hump" from India to China. They were also thought to be better aircraft for paratrooper operations because they could carry twice as many paratroopers and had cargo doors on both sides of the fuselage. This permitted the sticks of paratroopers to all exit the aircraft much faster than they could from the C-47, which had only one paratroop door.

The 316th received its first C-46 on March 3. Bob's information was that "All Troop Carriers are going to change to C-46s. They are a much larger airplane than the C-47. We also received another Cub again [Piper L-4]."

A brand new Curtiss C-46 transport showing the dual cargo doors on either side of the fuselage. Although with greater carrying capacity than the C-47, they were more vulnerable to ground fire. (US Air Force)

Although Allied troops were now in Germany in force and closing in on the Rhine River, the war was certainly not yet over by any means. A stark reminder of this was an air raid at Cottesmore on the night of March 3-4, 1945. "Well we had some excitement here last night. We had the first air raid for a long time. A flight of German fighters followed a group of British Lancaster bombers home from a mission. The bombers are stationed about three miles from here. There was plenty of shooting and fires for everyone. We don't know as yet how many, if any, German ships were lost but we do know that seven Lancasters were shot down. Two of them were shot down just over our field. We have so much work to do today that I have not been able to go out and see them yet.

"Last night made one feel as if he were back in the desert again because you could hear the German engines and we have heard enough of them to tell when they are overhead. When we left the desert we were so terribly glad to be away from it all but I guess it gets in your blood, because after a short time you want to see some action or something that puts you back on your toes."

Bob was able to get to the crash site on the field later in the day. Seeing the wreckage reinforced the knowledge that the air war was still dangerous and the Germans had not yet 'quit.' "Just returned from seeing the Lancaster and it is a hell of a mess. It fell straight down and even as large as they are, most of it is underground and what little bit was left above ground was blown to bits. I got there just as they were finishing picking up the crew and all they had was two litters full and the biggest piece was about the size of your hand.‡ While walking around looking at the debris we found many more small pieces of remains. We just found out that the Germans dropped seventeen 25 pound bombs with delayed action fuses. The engineers have removed the fuses and are going to blow them up soon. The 45th Squadron had two ships hit when they strafed the field last night."

The tension continued the next night with another air raid. "They sure are after the Troop Carrier fields. Our troops are on the banks of the Rhine, and the Germans know that we will have to use airborne troops to get a hold on the far side, so they are trying to knock out as many Troop Carrier Groups as possible. We expect German parachute troops over here any night now.

"We had another air raid last night. They did not bother us, but we could see them working over another Troop Carrier field close by. They shot up quite a few ships on the ground and shot down one C-46 that was in the air. We hear that around 60 troops were killed by strafing at the other field. It is 2030 now and we have not had any air raid yet. We have lots of protection on the field tonight in case they did drop any paratroopers. We are working every night now, getting ready for the mission. I hope after this one that we can all go home.

"Today Major McMenamin found out he can go home on a 30 day leave. After his leave is up, he is going to be put on Detached Service at Troop Carrier Command until the Group comes back to the States. It sure is a break for him. He has been trying every way to get home for the past year and a half. I hope his wife and kids will feel better when he gets home. I hate to see him go, because he is about the best officer that a man could ask for. He is so thrilled about the whole thing that he can hardly talk."

**Across the Rhine**

March 7 was a significant day for the Allied forces. American troops of the 9th Armored Division seized a bridge across the Rhine at Remagen, Germany. It was the first breach of this formidable obstacle and caused considerable excitement for the Allies since it was assumed that all the bridges over the Rhine would either be destroyed by bombing or blown up by the Germans. The news reached Cottesmore quickly. "Our troops crossed the Rhine this afternoon at 1600. The Germans were blowing up all the bridges as the Yanks reached the river. The infantry captured one railroad bridge before it could be blown up and the engineers removed the demolition charges before they were set off.

"Everyone is in high spirits because it was though that the Rhine crossing would be the bloodiest of the war. We hope that they won't need airborne troops now. But one bridgehead is not enough for the front that we have on the Rhine."

Bob was completely correct that one bridgehead over the Rhine was not enough. Although the U.S. First Army pushed hard to get troops over the river in the Remagen area, General Patton's Third Army and General Montgomery's 21st Army Group had river crossing plans of their own. Montgomery's plan envisioned the use of airborne troops from the FAAA, and as the Americans drove out from Remagen, this plan was solidified.

Bob wrote to Toots on March 9 about the departure of his longtime friend Frank McMenamin and told her to be on the lookout for him when he returned to the States. "Major Mac received his orders to come home today and he is so happy that he is about to bust. We talked over an hour this afternoon about what we are going to do when we get home. The squadron commander gave me another job two weeks ago. He put me in charge of the enlisted men's barracks because he was not satisfied with the inspection reports. I kept the

---

‡ The normal crew of the Lancaster was seven airmen.

men in a couple of nights and I have had two men working on the lawn and making clothes racks. I have only had it for two weeks and last week we took first place in the group. So in the morning I have to get up at six and go over and snap the men out of it and see if we can't take first place again. You see, between getting Engineering and the barracks ready for Saturday morning inspection, my Friday nights are plenty busy and my Saturday mornings are clear out of this world."

Bob's next letter on March 11 told of his barracks mates continuing adventures, and of finally flying the 36th Squadron's L-4. "Crow Lee and Sullivan are here in the room and they are both in the bag. I tell you if you ever see these two you will laugh until tears come into your eyes. I only hope that we all come home together so you can see what I have put up with all of the time. I got in eight landings and eight takeoffs in our [Piper] Cub today. They named it *Der Little Fuhrer* and the number on it is 000. Very rare, don't you think? We have it all fixed up with a shark's mouth painted on the nose and all the trimmings. Each squadron has one now so I get a little time in ours now. Our lovely weather still continues and we hope it lasts until we can leave this country."

On March 14, Bob heard of a new weapon employed by the British Lancasters for the first time – the *Grand Slam* 22,000 pound bomb. The weapon was intended to destroy railroad viaducts, submarine shelters or large bridges. "Today the British dropped a new 11 ton bomb on Germany. The largest up till then had been six tons. The pilots say the destruction was terrific. Even the 6 ton bombs pierced 15 foot thick roofs of U-boat pens, so one might imagine the force of the new bomb. It has been designed to destroy underground structures."

A few days later, amidst the preparations for another airborne operation, the hijinx in the 36th Squadron Officer's Quarters continued. "Last night, about midnight, Crow Lee came running into our room and he was all wet and all he was wearing was a pair of shorts and one shoe. I don't see how he was even able to walk as drunk as he was. He had been swimming in the emergency water storage just outside the barracks.

"About a dozen fellows were watching him and I guess he was a sight. It was plenty cold and would have killed a sober person. Every time he got out he would run in and tell Welter and me all about it. He also ran everyone out of the Officer's Club with a fire extinguisher. He was down here this morning and all the buttons were gone from his uniform blouse and every pocket including his pants pockets were filled with bacon."

The Remagen Bridge, weakened by bombing, artillery concussions, a failed demolition attempt and heavy vehicle traffic, finally fell into the Rhine on March 17. It had been closed to traffic a few days before and pontoon treadway bridges were constructed in the area to absorb the vehicle traffic crossing the river. Bob learned the news on March 20 and made an important observation. "Bad luck for the infantry because the railroad bridge has collapsed. It is the bridge that our infantry used to cross the Rhine. It has been bombed and shelled every day by the Germans. Our bridgehead there is 11 miles long and 2 miles deep. One bridgehead is not nearly enough, so something will have to be done, and done soon."

**Last Airborne Operation in Europe**

Once again, Bob was completely correct about the need for another bridgehead. And the "something" that had to be done was going to be done soon, and by the 316th TCG along with the rest of the First Allied Airborne Army. A jump just across the Rhine was being planned for the area near Wesel, Germany, in the British zone of advance.

The British 21st Army Group had fought its way to the Rhine in early March. Once assembled on the west bank of the Rhine, the British forces would need to stage an assault across the waterway to the German-defended east bank. Prior to this main assault crossing of the river, two airborne divisions would be dropped on the far side to secure the east bank of the river and delay any German reinforcements.

The two divisions selected to make the drop were the British 6th Airborne and the U.S. 17th Airborne. The XI Troop Carrier Command would supply all the aircraft for the paratroopers and the British Troop Carriers would be towing the glider element. This would be the first time that paratroops and gliders would land

at the same time. It would also be a daylight drop like MARKET GARDEN, and also mark the first combat missions for the C-46 aircraft in the European Theater. The planners felt a daylight drop would be more successful since at this late date in the war, the threat of Luftwaffe interference was thought to be less of a risk than night formation flying. There was also the hope that Allied fighters would be able to suppress any flak batteries or other air defense installations along the route used by the Troop Carriers.

The operation, known as VARSITY, was scheduled to take place on March 24. To shorten the distance that the 316th had to fly to reach the drop zones in Germany, on March 21 the Group was temporarily deployed to RAF Wethersfield which was closer to the coast. It was here that the 316th would pick up the British airborne troops it was assigned to drop. Many of the other Troop Carrier Groups had relocated to France to be closer to the U.S. airborne divisions. That made the 316th the logical choice to carry the British troops. "Yesterday they had us install all the British paratrooper equipment and then they sent us down here to Wethersfield which is a little north east of London, and is in the center of Buzz Bomb Alley. The only personnel that I could bring with me were Sgts Tucker and Milliman and the electrician, Sgt Bruns. We brought just enough equipment to last for three or four days."

Unfortunately the vast preparations for the Rhine crossing did not escape the Germans. They sensed something was "up" and began an intensive air attack on airfields and installations in Britain in France to spoil the attack. "Last night and this morning there have been over 20 buzz bombs and V-2s landing in the area. Two that landed during the night were just a little too close because the concussion blew open some of the doors on the Nissen huts. All during the night you could hear one after the other going off. It makes one a little nervous because the V-2s give no warning at all.

"We have a hut fixed up here on the line to use for Engineering. We stole six chairs, two tables and a blanket from the RAF Club this morning and we are now all set up nice and comfortable like. This is one time that we don't have a lot of work to do. Welter, Sullivan, Lee and I are all in one hut so you can imagine what a place that is to live in. We fixed it all up last night and got a good fire going. We came out on the line and gathered up enough wire to hook up the radio. I went to the farms that are close to the flying field and bought what few eggs they had. We now have four dozen and each farmer promised me more tomorrow or the next day. You have to do a lot of talking before they will sell you any because the British government will fine them for not selling them to the government. Sure is a free country if you have holes in your head and a flexible imagination. Well, anyway, we now have fresh eggs.

"The 36th Squadron starts operating as soon as they hit a place. The British paratroopers will be here to load the ships tomorrow morning at 1000. We took a walk this afternoon and stopped at a farm where they raise thoroughbred Holstein cattle. They have two bulls that are worth $8000 each. They have vacuum milking machines. The owner asked if we wanted to ride the horse and we told him that we would be back tomorrow to ride him.

"In the evening we got the shotgun from Cottesmore so Scott, Sullivan and I went hunting. Scott and Sullivan got pheasants and I got feathers out of a partridge, but no meat. Later we fried eggs, played cards and drank the rest of our quart of whiskey. We retired about midnight."

## Operation VARSITY

On Friday, March 23, Bob and his maintenance crews had the aircraft prepared for the Operation VARSITY mission. His diary for the day includes a concise and accurate account of the plans for the operation. "This morning Lee, Scott and Milliman went over to the farmer's house and rode the horse. I had to stay here because the paratroopers are loading our ships. The pilots were briefed this morning and we are going to drop our paratroopers about seven miles east of the Rhine River. Our ships should only be over enemy territory about 10 minutes, but at 800 feet that is long enough. This will be the largest airborne operation since the war started.§ Other American and English Groups are going to take in gliders and troops.

"The B-24s are going to be used for re-supply and are going to start dropping supplies 15 minutes after
§ This point is debatable. Both the D-Day drop and MARKET GARDEN involved dropping at least three full airborne divisions and VARSITY involved only two, but the VARSITY drops were completed on one day.

the paratroopers are dropped. British Lancasters and Halifaxes are going to re-supply the glider troops. The entire area is being bombed and shelled today. All the crossings to be made across the river will start tonight at midnight.

"All the missions to be made from England will start from the airfields in this area and Jerry has sure been trying to stop some of us from leaving by the use of his rockets. All night last night and most of this morning we have been able to hear them going off and then a few seconds later feel the concussion. As yet he has not laid any on this field. Most of the Troop Carriers are in France now and that is where most of the glider missions will be flown from."

A formation of 36th TCS aircraft late in the war. By this time, most of the invasion stripes applied for the Normandy invasion had been painted over. (NARA)

The aircraft for the VARSITY mission departed on schedule on March 24 for the last airborne operation of the European war. Bob watched them all go off and then the ground crews "sweated" their return as they had for so many missions before. "We were all up this morning at 0400 and had breakfast at 0500. We were all laughing and joking but behind it everyone was tense. We were all on the line by 0545 and the takeoff was at 0722.¶ I went around before takeoff and wished the crew chiefs the best of luck.

"Just as our ships were taking off, a buzz bomb came across the field at 500 feet headed for London. It looked funny because the sky was full of all types of airplanes and here comes this one little German ship through the whole mess. I think there were more ships in the sky than there were on D-Day. After the ships took off we loaded all our equipment and came back to Cottesmore, damn glad to be out of Buzz Bomb Alley."

One British airborne officer recalled the scene at his field and the steadiness of the Troop Carrier crews. "The noise and acceleration and pace of any take-off is always exciting but that American mass take-off with something like eighteen to twenty-seven Dakotas [C-47s] moving down the runway together in successive waves was a revelation of skill and nerve."[63] The airborne serials were now on their way to Germany. They flew in three columns, each some 150 miles long, taking more than two hours to pass.[64]

Once back at their home base, Bob and his men awaited the return of the Group. "At 1215 the ships started coming in and right away we could tell that all was not well because they were scattered all over the sky. Our Squadron Commander, Col Wright, Lt Lee and Lt Sullivan were the only ones that held their formation through the entire mission. This mission was almost as bad as the one in Sicily. We did not have as many shot down but we only have 3 ships out of 20 that did not get hit. The crews say everything went along swell until after the paratroopers were out. It was when they started making their turn to come home that all hell broke loose. Lee says they threw everything at them except the kitchen stove."

Although German fighters had not been much of a problem, German anti-aircraft had done a significant amount of damage to the Troop Carrier aircraft. What the Germans lacked in fighters they partially made up for in air defense capability. After enduring several years of Allied bombing, their anti-aircraft crews were very experienced and accurate. Some of these Germans who weren't "whipped" by the Allied fighters took a horrible toll on the C-47s and in particular, the C-46s.

---

¶ The official USAF history of the operation noted, however, that the loading the American aircraft at Boreham was briefly delayed, "while the British finished their inevitable tea."

Once the men paratroopers landed, they watched in horror as the Troop Carrier aircraft were picked off by German gunners. They remembered the sight of aircrews standing in the doors of burning planes unable to jump because of the low altitude and having to go back inside and take their chances in a crash landing. Other aircrew did manage to jump if their pilots managed to gain enough altitude, but many forgot to remove their flak vests and were unable to open their parachutes before they plunged to their deaths.[65]

**VARSITY Damage Assessment**

Bob carefully noted the status of each aircraft in the 36th Squadron as well as what needed to be done to repair them, if possible.

| Ship | Tail Number | Crew Chief | Outcome |
|---|---|---|---|
| 1 | 42-93780 | Sgt Garcia | Minor damage, made it back |
| 2 | 43-15258 | TSgt Babbitt | Major hit, landed at B-86 |
| 3 | 42-15265 | TSgt Kesner | Minor damage, made it back |
| 4 | 42-100973 | TSgt McAllister | Major hit but made it back |
| 5 | 43-15205 | TSgt Keating | Major hit, crew bailed out |
| 6 | 42-108909 | TSgt Przbylski | Minor damage, made it back |
| 7 | 43-14179 | TSgt Rucker | Minor damage, made it back |
| 8 | 42-68758 | Sgt Little | Major hit, landed at B-90 |
| 9 | 42-68765 | TSgt Saunders | Minor damage, made it back |
| 10 | 42-24389 | TSgt Nordness | Minor damage, made it back |
| 11 | 42-23503 | TSgt Turner | Major hit, landed at B-100 |
| 12 | 42-108902 | TSgt Hearn | Major hit but made it back |
| 13 | 43-30652 | TSgt McPherson | Minor damage, made it back |
| 14 | 43-30721 | TSgt Erwin | Major hit but made it back |
| 15 | 42-100872 | TSgt Moreno | Major hit but made it back |
| 17 | 42-23931 | TSgt Shanta | Major hit, went down in flames |
| 18 | 43-15227 | Sgt Griffith | Minor damage, made it back |
| 19 | 42-24392 | TSgt Richey | Major hit, landed at B-100 |
| 00 | 42-68769 | Cpl Renken | Minor damage, made it back |
| 01 | 42-100678 | TSgt Campbell | Minor damage, made it back |
| 02 | 43-15634 | TSgt Brown | Minor damage, made it back |

(According to the VARSITY flight schedule provided in Appendix A, ship No. 16 was not on the schedule and ships No. 00 and 03 were listed as spares in case of an abort.)

"Aircraft that did not return:

No. 2: One engine shot out and about 200 other holes. They landed at B-86 (Eindhoven, Netherlands).**

No. 5: The entire crew jumped and the British hauled two of them back tonight. As yet we do not know how the rest of them fared. There is a full report from Lt Robert E. Pace.

No. 8: Shot up and had to land at B-90. The crew chief (Saunders) is not expected to live.

No. 11: Shot up bad and had to land at B-90.
** B numbered airfields in the ETO were British operated fields. B-86 was at Helmond (Eindhoven), NL, B-90 was at Kleine Brogel, BE and B-100 was at Goch, GE.

No. 17: Went down in flames and no one got out.

No. 19: Shot up and had to land at B-100. Full report from George T. Greenstein.

"That makes a total of 6 that were unable to make it home. Of the aircraft that did return, this is the status.

No. 4: Took a direct hit in the wing center section which tore up just about everything in the belly including one gas tank. The ship burned for a short time and then the fire went out. What put the fire out is beyond me. It also had two direct hits in the tail end which made the empennage section look like someone had used a shotgun on it. Captain Miller was flying it and I cannot see how he was able to fly it back here.

No. 12: One wing shot up beyond repair and the fuselage damaged in many places from flak.

No. 14: Hydraulic system shot out and numerous hits from small arms fire.

No. 15: One engine shot up, hydraulic system shot out and numerous small holes.

No. 00: Captain Etter was flying and he only got one hit with small arms fire but that one small bullet cleaned all the pulleys off the throttle quadrant and made the throttles useless but he still flew it home.†† He was unable to change the throttle settings, because when the quadrant was hit, it reduced the setting so he had to fly it home on reduced power then turn the switches off to make his landing here.

"All the rest of the ships (1, 3, 6, 7, 9, 10, 13, 18, 01, 02) were full of small holes but we were able to make temporary repairs."

The day following the VARSITY mission (March 25), Bob was able to capture some post mission reports from two of the pilots and record them in his diary.

"Report on Mission by Lt Robert E. Pace, Pilot of Aircraft No. 5:

"Everything went off smooth and we were doing fine until about six or seven minutes to the Drop Zone. You could see planes coming out one or two at a time, flying low and fast. It was plain then that the Jerries were ready for us. As we approached the DZ about five or six planes could be seen burning on the ground and Col Lewis [Mars Lewis, CO of the 45th Squadron] was going down with flames all over his left side. I don't believe anyone in the 36th Squadron was hit going in, but when we turned to the left we caught merry hell.
"I was leading B Flight in the last serial and the turn threw me right in the middle of a flak barrage. My left wingman went down and it is reported that two parachutes came out but he wasn't high enough for them to help any. I got a burst of .50 caliber which came in through the right wing into the companionway, down the center of the cabin floor and out the cargo door. The holes were eight to ten inches apart.
"I thought we were hit pretty bad so I started getting enough altitude to jump if necessary. When I reached about 500 feet I got about eight 40mm bursts and five or six of those hit and exploded in the plane itself. The others were too close for comfort. The 50 caliber shells had knocked out my rudder and aileron controls but I could still climb. The 40mm had set the right engine and one gas tank on fire but my only thought was to get back across the Rhine before we had to jump and I told the crew to get out of their flak suits and put their parachutes on.
"About a mile this side of the river I called Pussycat 'H' and told him I was going to an emergency field but Sullivan was the only one who heard me. Just as I finished calling the gas tank blew, making a hell of a noise, so I told the crew to bail out. I knew I was hit in the leg but I knew it wasn't bad. It felt like someone had hit me directly in the back of the

---

†† In modern day Air Force parlance, that one lucky shot is known as a "Golden BB."

leg with a baseball bat. I thought, 'Brother this is going to be embarrassing.'

"I was busy trying to keep the plane level because my elevator control went out when the gas tank blew up. I set the fire extinguisher on the right engine and jumped over to the right seat to see if it did any good, but of course, it didn't. I got back in the left seat and started working the trim tabs because I didn't think the copilot had gone out yet. I can't remember whether they did any good or not. It was getting pretty hot by this time so I looked back and the whole cabin was one big mess of flames. It seemed as if I had all the time in the world and I didn't have to hurry a bit.

"I thought perhaps the copilot had burned up trying to get out but I had to try to make it, thinking I might be able to find him on the way out and carry him out too. I put my left hand over my eyes and started back through the cabin. The floor was burned so I kept stepping through it, and having to pull myself out again. I thought I would be pretty badly burned but only my hair was burned.

"I will never forget how cool the air felt when I jumped. I had a chest parachute which has the rip cord on the right end of it. But I started clawing for the rip cord on the left side and of course I couldn't find it, so I tried to tear the flap open but that wouldn't work either. I was getting pretty mad by this time and the first thing that popped into my mind was that those bastards back in the States had made a parachute without a rip cord. I finally found the thing and then pulled it.

"The pilot chute jumped out and waved back and forth in front of my face for what seemed to be 30 minutes. I said to myself, 'That little bastard isn't slowing me down a bit, let's get the big one out here.' I remember thinking that my wife is going to have a lot of husband if I hit going fast because I'm sure going to splatter. I helped the big chute out and I was surprised at how close I was to the ground. I made one swing in the risers before I hit. I landed on top of a tree and made an unbelievably soft landing.

"Men from the 15*th* Scottish Division were waiting when I got down and I thought they were Germans at first. They said I was cursing and yelling when I came over their camp at about 100 feet in the air. I jumped about 1030 and by 1200 I was so drunk I didn't know what time it was. They sure fed me some whiskey.

"I was evacuated pretty fast and got back here the next afternoon (March 25), none the worse for my experiences and a happy man. Robert E. Pace, Teshomingo, Oklahoma."

The second mission report was from Lt George T. Greenstein, pilot of aircraft No. 19:

"We took off as scheduled at 0731, formed up and started out on course. It was smooth flying, very good weather and ideal for a mission. Lt Rewinkle, my copilot and I split the flying time going over. I didn't want him to feel as though he was just along for the ride. Everything went off smoothly. I didn't realize it at the time, but we were about six or seven minutes ahead of schedule, but I was quick to realize it as soon as we approached the Rhine.

"Ships were coming out in all directions, low and fast, off course, high. Something was going on over there, I thought. Well, something was going on. The closer we got the more plainly I could see the flak and burning 'geese' (aircraft). I saw Col Lewis' ship – what a mess it was, the flames must have been 100 feet long. As we approached Drop Zone A, I saw a ship go into a very steep dive, exploding when it hit the ground – right on the Drop Zone. We were almost to the DZ now. 'Rip' (Lt Rewinkle) told me we had a hole in our right wing.

"Over the DZ now, and our paratroopers went out. No sooner had they all jumped when I felt the ship lurch a couple of times, rather violently. We were hit in all four gas tanks. I started to turn to get out. Rip told me that three of our tanks were dry. We switched both engines to the left main which had about 90 gallons in it, and going down. The plane had a strong odor of burning gas about it and we could smell gas in the cockpit.

"We went into a screaming dive with both engines pulling 60 inches of mercury. The airplane leveled off on the deck doing 240 miles per hour. We approached a Jerry gun battery and he hit my left engine starting it afire. TSgt Richey, the crew chief, told us we were on fire. I didn't know what would be best to do but I decided I would try to get it to airfield B-100, about 20 miles away.

"We got to B-100, passed over the strip and I wiggled my wings to indicate that I wanted to land. I made a short pattern but when I approached on my final, they shot a red flare at me. I saw a peashooter just under me so I went around. Just about then Richey ran up and told me that our cabin was on fire. He tried to put it out but couldn't. We made another pattern, landed, and when we stopped we all ran like hell.

*"Rip jumped from the top escape hatch and damn near killed himself. A couple of minutes afterwards 'Old T for Tare' burned violently in spite of the fire trucks that attempted to stop her from burning. They all stood clear once she got started and let her go – finis No. 19. – George T. Greenstein, Chicago, Illinois."*

## The Reckoning

Overall, the airborne drops and glider landings had been enormously successful. Lessons learned from MARKET GARDEN had been taken to heart. Once again a daylight drop was executed but this time all the airborne forces were taken in via a single lift as opposed to several lifts over several days. This accentuated the shock and surprise advantage of an airborne assault.

Although anticipated by them, the actual landings did take the Germans by surprise. The British, Canadian and American paratroopers and glidermen quickly secured all their VARSITY objectives. "By 1400 on the afternoon of D-Day (March 24), the airborne part had finished as a complete success. The U.S. 17th Airborne Division and the British 6th Airborne Division had paved a wide path for Montgomery's ground forces to drive deep into the German homeland...[the] Red Devils had cleared the northern part of the drop and landing zones, captured thousands of prisoners and made contact with Montgomery's ground forces on the eastern shore of the Rhine."[66]

Although VARSITY was the best executed and most successful airborne operation in Europe, it was the least ambitious. The two divisions landed in a six mile by five mile area, well within the reach of Allied artillery on the other side of the Rhine. By the end of the day on March 24 they were relieved by the advancing ground forces.

The VARSITY losses though, were significant. Among the airborne troops, the British lost 347 men and another 731 wounded, the Americans lost 393 men, 934 wounded and 164 missing in action. Among the Troop Carriers, the losses were just as severe – 41 killed, 153 wounded and 163 missing in action.[67]

The decision to drop in daylight also meant that the aircraft were subjected to more accurate anti-aircraft fire. The IX Troop Carrier Command lost 20 of the 74 new C-46 aircraft used on the mission. Because of their fuel system design, the C-46 aircraft were less resistant to battle damage than the older, more durable C-47s. Even a minor hit to a C-46 could cause a fuel system fire that would consume the aircraft. Of the 1100 C-47s on the mission, 38 of them were lost. Included in that number are the aircraft losses from the 316th – two aircraft from the 45th Squadron and the two from the 36th Squadron.‡‡ The blow was especially hard to the 45th as they lost their Squadron Commander, Lt Col Mars Lewis. Fifteen of the B-24s dropping supplies were also shot down. Several of the 300-odd aircraft that were damaged made it to an emergency field, but would never fly again. Most of the gliders used in the mission were unrecoverable afterwards. It is still debatable whether the VARSITY airborne assault was really necessary, given that crossings were made in other places on the Rhine without such preparations and with far fewer losses.

Bob dashed off a letter to Toots also on March 25 to let her know what he was up to and why he had been so busy; still only able to hint at the big operation. "Tell Maj. McMenamin that the little deal we were getting ready for when he left came off yesterday. You can tell him it was rougher than all the rest. Tell him I had to repair everything but two so he can guess for himself. Tell him all my old buddies came back so we can still all go home together. This same little thing is why I have not written to you for the past four days, but you will forgive me, won't you?"

Bob noted in his diary his early misgivings about the VARSITY operation. "When I heard the pre-mission briefing I felt like it looked too easy and told TSgt Brown that I did not trust it. Little did I know. However, the mission was a success and all the generals are now across the Rhine and have established good bridgeheads. This is the first time that they have had to operate over 100 miles from the open sea. They took assault boats overland and used them for the Rhine crossing.

"We were to move to France today (March 26) and pull a mission for Patton but we hear that Patton is

---

‡‡ Several sources were consulted to get an accurate figure for VARSITY losses. All of them stated conflicting numbers. The statistics presented are from Air University, USAF Historical Study No. 97, *Airborne Operations in World War II, European Theater* by Dr. John C. Warren, Maxwell AFB AL, 1956.

across the Rhine and is going hog wild again. We are set up to leave tomorrow now but if he keeps on there will be no need for us."

Bob was correct about Patton. He had stolen a march on Montgomery by sneaking across some troops from the U.S. 5th infantry division on the night of March 22, two days before the massive VARSITY operation. Patton reported with glee that he was able to get a bridgehead across the Rhine by using a handful of troops and no massive artillery or airborne preparation. By the next day he had the full 5th Infantry Division across the Rhine, completely stealing the thunder from Montgomery's crossing the next day. This fact was behind Bob's statement of March 27: "I guess Patton does not need us because we are still sitting here waiting."

The VARSITY mission had been successful, but at a terrible cost. Bob noted that, "The 313th Group lost 25 C-46s on the crossing. Now none of us believe that the new C-46 is going to be very good in combat." He was full of praise for the old Gooney Bird. "These C-47s are the only airplane for the job because they come back so full of holes and with so many parts missing that no one can believe they can make it back, even after seeing them with your own eyes."

As the damaged aircraft were being repaired, life quieted down at Cottesmore. Details about the Rhine jump had now made it to the U.S. Bob wanted to make sure Toots knew it was his unit that was involved. "What do you think of your Troop Carrier husband? They are sure playing us up for a change in the papers and on the radio. Welter and Sullivan went to town tonight and I am here alone. I just got back from a flight tonight at seven and the cooks in the mess hall cooked us up a big steak dinner and also a batch of onions, and were they ever good. We went to a depot to get a few parts for these new type ships that we have a couple of. Crook went along to try to see his brother-in-law. He never found him but he found a neighbor of his that lives in the same house there in Dayton."

On March 30, Bob and his crews gathered up their tools for a trip to the Continent to recover two of their ships damaged in the VARSITY mission. "We took two ships and crews and equipment to Holland and Belgium today. We are going to repair Nos. 2 and 8 ourselves. No. 2 is at B-86 which is close to Eindhoven and No. 8 which is at B-90 which is about 30 miles away.

"At B-90 [Kleine Brogel] the base is run by Canadians and they had us stay for lunch. They have all Spitfire fighters.§§ It only takes them 10 minutes to reach the front. They carry one 100lb bomb under each wing and a full load of ammunition. They go out and look for anything at all to shoot up.

"One of the fellows had just lost his buddy that morning. They were diving on three trucks and he got hit with a 20mm cannon shell. His microphone button was pressed in and his buddy could hear every word he said when he went down. He said that his right arm was shot off and he was also hit in the side and then could hear him moaning just before he hit the ground.

"At B-86 they have all Typhoon fighters that carry six 4-inch rockets under each wing. The Typhoon is the fastest fighter in the air. All the fighter pilots tell us the war is going so fast since the Rhine crossing that there is no bomb line anywhere so they just go out looking for enemy equipment. I saw my first pair of wooden shoes being worn today. They sure look funny. Holland still has as much water as ever and also has all her windmills."

When he was home again at Cottesmore, Bob apologized to Toots for being away from writing for so long. "I just returned from Belgium, France and Holland. I had to do a little inspection after the Rhine crossing and also to take some crews over to do some work caused by old Jerry. That is the reason I have not written to you for the past few days."

**First Visit to Germany**

Bob's first chance to get to Germany came on April 1. Along the way he saw the results of the battles for the German frontier. "Today is Easter Sunday and we still hauled a load of gasoline to Germany. First of all I fixed the trip up for Sullivan, then he goes to Nottingham last night and got in the bag and did not show up until about noon today and we took off at 1230. I had to take more equipment to the crews at B-90 and B-86,

---
§§ The Canadian unit was likely No. 414 Squadron, RCAF, flying Spitfires.

so I fixed it that I could go on one of the ships hauling gasoline to Germany and then stop with the equipment on the way back. We went in ship No. 2 and were a one-ship flight.

"The weather was terrible all the way. I got some pictures of the Ruhr River, Siegfried Line and Rhine River. The Siegfried Line is a sight to see. You can look for miles and see the rows of tank traps. We were flying very low and got a good look at the pill boxes. Some of them looked to be at least 12 feet thick. All along the route we could see where tank battles had been fought. They sure did cut up the fields, and in small towns where the Germans tried to hold out the buildings were all leveled to the ground.

"Then we flew over the Ardennes Salient where the Germans had made their breakthrough and killed all the American prisoners.¶¶  The Ardennes Forest has large amounts of treetops shot out. The country is torn up like they had fought over it for years. We saw lots of pontoon bridges and I did not see one original bridge across any river still standing. The weather was terrible and by this time we were just clearing the treetops on some of the hills because the ceiling was so low.

"When we reached the Rhine we all kept our fingers crossed because General Patton's salient is only 20 miles wide on the east side of the Rhine and it is about 60 miles long. As we reached the river we could see Cologne burning. It was to the north of where we were flying and the Germans still have the city on the east side of the Rhine. We flew to Giessen which is north east of Frankfurt and only 10 miles from the front. So that put the Germans within 10 miles on three sides of us. Although Patton is going hog wild there is a lot of cleaning up to do and his corridor is very narrow, but they are widening it every day.

"After landing, the first thing we did was to go exploring in all the buildings. The post was a permanent German air base. We found the supplies still in the bins and gaskets were still hanging on the hooks. Hershberg would have gone wild if he could have seen their supply rooms. By the looks of things, the Germans are still not short of parts for their airplanes. The propeller shop was still intact with all the tools and equipment. There were about 25 German ships on the field waiting for repairs and they had to leave them all behind.

"They left in such a hurry that they did not have time to mine or booby trap anything so we had the time of our life going through the equipment. I took a lot of instruments out of the German aircraft. Ever since I have been overseas I have been trying to get German instruments but they have always had time to remove them before now. We picked up numerous odds and ends.

"After giving the hangars an inspection we started on the quarters and Officer's Club. In the bar we found all the glasses and pictures still hanging on the walls. The cash register was still in working order but there was no money. We each took a few glasses to send home and we each took a reproduction picture. I took a light fixture from the wall in the club. We always did want to tear up a bar and now we had the chance and took it.

"In the Mess, all the tables and chairs were still there. They must have been having a movie when they had to leave because we found film about half run through the projector and the other projector was full of film. We brought all the film back with us but cannot tell if it is a newsreel or actual battle pictures. They did not destroy the film projectors. All they did was take the lens with them. I don't see why they did not have time to destroy more. All the equipment in the club was just as it was when they left. Even the pianos were in the club and they had not touched them.

"The field was beautiful. The flying field runway is grass and is located in a sort of a valley. The hangar line is built in a half moon shape. The administration building is in the middle of the hangar area and there are six hangars on each side of it. In the rear of the hangar line are the quarters, built on and around small hills overlooking the hangar line. The barracks and hangar area is filled with pine trees and they only cut away enough trees for the hangars and barracks. It is the prettiest location of any flying field I have ever seen. Ours was the first ship to land on the field and the only one to remain overnight.

"As soon as we landed the tank boys came in trucks and started hauling the gasoline away. By night they were not quite finished so they told us that they would have to wait until morning to finish. There are infantry patrols in the hills around the field and they just caught a sniper on the field the night before. They told us not to sleep near the ship and we did not argue with them. We slept near a creek about 100 yards

---

¶¶ The most famous incident of this during the Battle of the Bulge was the Malmedy Massacre.

from the ship and each one of us had our guns close enough to reach. Of course we saw Germans all night long in our sleep."

Bob and the rest of the aircrew felt brave enough to venture out in the surrounding area the next day. "I always wanted to go to a German town, so in the morning we said, snipers or not, we are going to town. It felt funny because we did not know if they were going to take a shot at us or not. We looked the town over and never got shot at once, but everyone was looking out the window at us. As yet the Military Police have not reached the town so there is no one to tell you what you can and cannot do. Some of the infantry boys told us there was a cemetery in back of the town and there were a lot of bodies of Jews just lying on top of the ground. The Germans had starved them to death. We went back to the airfield and then flew to B-83 and B-90 to take more food to the crews that are working on our ships and also check to see when they are going to be finished."

After the trip to the front, Bob wrote a long letter to Toots describing what he had seen. He included some details he did not put in his diary. "Yesterday Sullivan, crew and I flew to Germany. We went to a German air base 60 miles east (sic) of the front. We were the first to land at the base. All the rest of the ships had to return but we remained all night. It was a field day for us. The only other soldiers there were some medics that had just moved in to set up a field hospital and a few infantrymen left behind to clean out the snipers.

"First we went through all the hangars which were almost destroyed by our bombers. About the only thing they had time to take was their tools and personal belongings. It was wonderful to be able to see just how they kept things. They had plenty of everything. I have always wanted some German instruments and the ships they left behind still had instruments in them so we removed what we wanted.

"Next we went to the Officer's Club and found cheese and a pitcher of beer and bread still sitting on the tables. The club was just like they had used it except they took all the drinks and money. There were all types of glasses and dishes and also pictures still hanging on the walls. It was not safe around that place for a short time. They must have left in the early evening because we found a movie just half run in the projection machine.

"The base was a permanent one and I must say it was a beauty. The quarters were beautiful with the bottom cement and the top half wooden like all the German houses. The quarters were built on these small hills which are covered with pine trees. The infantry fellows stay in town. They just walk up to a house and give the people 15 minutes to get out. They told us not to sleep in the ships because of snipers so we slept on a creek bank away from the ship and each one of us had our gun very handy. All during the night we could hear shooting in the woods close by and heavy artillery in the distance.

"This morning we decided to go to town, snipers or no snipers. Every house, and I do mean every house, had a white flag hanging out the window. I have never seen such a sight in my life. We are not permitted to speak to anyone but I did take a few pictures of the kids. There were some of the cutest little girls that I have ever seen. We were the first airmen in the town and everyone was hanging from the windows looking at us and of course we thought we were going to be shot at any minute. I hope someday that we can go over together and travel all over that part of the country."

*Left:* A German postcard of Giessen airfield. (Author's collection)

*Right:* Bob's photo of white flags hanging from German houses.

**For You, the War is Over**

By early April, Bob Uhrig's long war was winding down. Other than supply runs and returning liberated POWs, there were fewer and fewer missions for the Troop Carrier units. After over a year in Britain, Bob finally got to London on April 4. It was his first trip to the city and he went with his friends Sullivan and Fenimore. "In the evening we had dinner at the Trocadero, and then we went to a bottle club where I got in the bag." As a result of an overdose of fun he was sick most of the next day.

Bob had fully recovered by April 6 and he and his fellows went out to witness the devastation felt by London in six years of war. "We hired a cab and he took us sightseeing, showing us all the points of interest. One has to see London to know how terrific the bombing has been. There are areas where as many as five city blocks are missing.

"Then came the buzz bombs which have landed all over the town. Next came the V-2s which are rockets and go to 60 miles high and travel 3000 miles an hour. The V-2s can destroy an entire block because of the terrific charge and the depth they go below ground before exploding. Their destruction is terrible because it kills so many people in the vicinity. We saw where two of them had hit the week before, killing over 300 people. They were still removing the debris."

"Fenimore is on leave and is going to spend the rest of it in Leicester. Scott and I made him miss his train because we made him take a ride with us on the Underground. We got on and then did not know how to get back. He was moaning and groaning all the time. We had a pretty good time and it is a wonder we did not kill one another by some of the tricks we played.

"When we arrived back at Oakham, Scott and I went to the movies, then we came on back to the field. When we arrived everyone asked us what had happened to Whatley. He rode to London on the train with us and got a little too much to drink. When he left us in London he got something else to drink, and from then on he does not know what happened. All the Squadron knows is that the Military Police called from London and asked what to do with him. They sobered him up the next day and sent him back to the Squadron."

Bob took one more mission to Germany on April 8, probably knowing it would be his last chance. He went with some of the men he had worked and served with for many months in the 36th TCS. "The most motley crew that ever took the air took off today for Germany. We had to fly supplies to Frankfurt and then take parts and food to the crews working on No. 8 at B-90 in Belgium. Pilot (Crow Lee), copilot (Eustace), engineer (Fenimore) and passengers (Scott and I). Everyone was willing to bet odds that some of a crew of this type would not return. We fought all the way over.

"When we landed at Frankfurt it was too late to fly to B-90 because you cannot fly through the corridor at night. So we went to Brussels and spent the night there. We had to see the Town Mayor to get a bed. All the hotels are military controlled. I wanted to buy Toots some perfume but we had to leave before the stores opened."

By April 10, the handwriting was on the wall for the 316th. They would either be disbanded or sent home in the near future. "Things have begun to happen. All the new personnel are being transferred out, and we are transferring 2 x C-47s, 1 x C-46 and 3 x C-109s. And the next day; "The personnel have started coming in from the other Groups. All old men in the other Groups are being transferred into our Group. The Group that has arrived is going to replace us. We are turning over all our supplies and equipment to them."

A war-weary C-47 in the markings of the 36th TCS (4C). This particular aircraft (43-30652) survived the war and 65+ years after; a tribute to the men and women who built her and the mechanics and engineers like Bob Uhrig who "kept her flying." (Robert Shawn collection)

# Home

The last entry in Bob Uhrig's wartime diary was on April 12, 1945. It was short and to the point, and the only political comment he made throughout the entire war. "President Roosevelt died today. May America always have a leader as capable as he."

Bob's last letter to Toots was dated April 16. It was not much different from all the others, yet Bob knew it would be the last he would write from overseas. The 316th was coming home. "Did you get the letter where I told you about staying in Germany one night? Our ship was the only one that remained over night. Well we knew it was close to the front but we did not know it was as bad as it was. They have now declared it a mission. They say ignorance is bliss. We did know that there were still plenty of snipers in the area because we could hear the shots. I have some perfume here on the desk for you. It is Paris perfume and I paid $18.00 for it."

The long months of combat, deployments and stress were over. Bob was going home for the first time in two and a half years. It would give him little comfort, but the Air Force would eventually recognize the impact that long overseas service had on non-flying personnel who did not have as a goal a number of missions or flying hours before they could return to the U.S.

> "The lack of a rotation policy for personnel other than combat crews resulted in unnecessarily long overseas service under arduous conditions for such personnel. By VE Day, non-flying personnel assigned to Ninth Air Force usually served overseas from 18 months to three years. This service involved intensive pre-invasion training, long hours of staff planning and, frequently, actual combat. It is recommended that a rotation policy for non combat personnel be established so that such personnel can be returned to the Z.I. (Continental U.S.) for rest and recuperation after two years overseas service."[68]

A fitting finish to Bob Uhrig's WWII story was the April 20, 1945 letter he received from Toots, just before he left for the U.S. at the end of his tour. Of the hundreds she had written, it was the last one. It was also the only letter of hers that he was able to keep, but the most important – the two of them would finally be together. "All the girls in Dayton got a letter telling them not to write anymore because there is going to be a change in address – which of course must be you are coming home. I just am afraid to believe it, I want it so badly. Ain't I crazy? It just has to be true – you're coming home, because I just will it so."

## "That Big Boat Home"

The USNS *Henry Gibbins* - a sight every G.I. dreamed about - the transport back to the United States. This ship carried Bob Uhrig home along with many other members of the 316th Troop Carrier Group.
(US Navy)

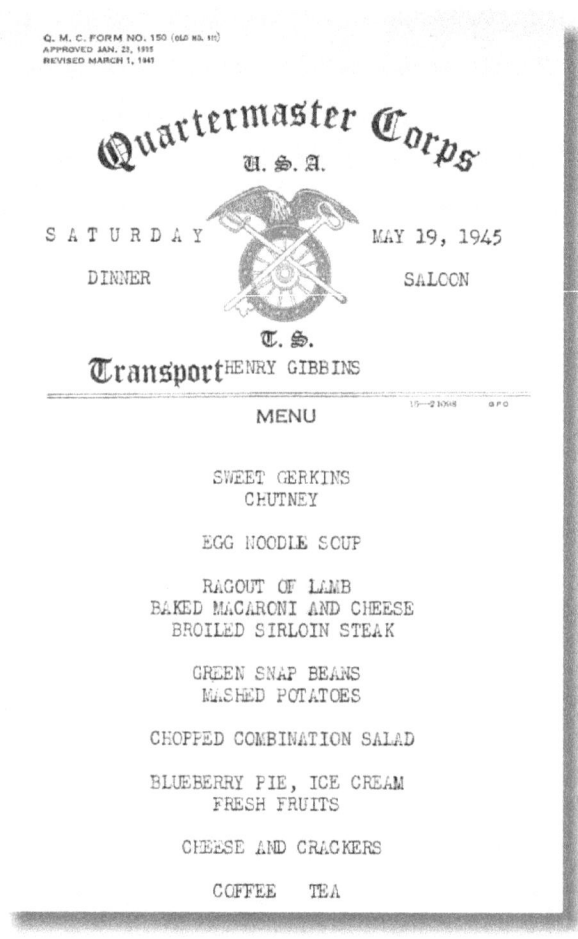

# Epilogue

## The Years that Followed

**B**ob Uhrig, along with other members of the 316th TCG, returned to the U.S. in May 1945. Bob himself traveled home on the troop transport ship the *Henry Gibbins* in mid May. He kept one of the menus from the voyage, dated May 19, where dinner included egg noodle soup, sirloin steak, blueberry pie and ice cream.

Once Bob arrived in the States, he stayed just briefly at Pope Field in North Carolina where the 316th TCG was now based. He decided to stay in service, and having been overseas for the entire European war, he received his choice of duty for his next assignment. He asked to return to the Dayton area so he could be near his home and family. The request was granted and he was reassigned to the Flight Test Division of Air Materiel Command at Wright Field. Bob was finally leaving Troop Carriers after almost 10 years. Once reunited with Toots, they spent a few years there at Wright Field.

After the Air Force became a separate service in September 1947, Wright and Patterson Fields were combined into one base and renamed Wright-Patterson Air Force Base. Bob decided to stay with in the Air Force and so, with Toots and family in tow, moved to Edwards Air Force Base, California in 1952.

At Edwards, Bob was assigned as the Chief of Aircraft Maintenance in the Directorate of Flight Test and Development. He soon became close friends with a number of men who would soon write their names in the pages of Air Force and aviation history: Pete Everest, Jack Ridley, Chuck Yeager and Gus Julian.

After two years at Edwards, Bob went back overseas again. This time it was with his family and this time to Germany itself. The Uhrigs were stationed in Wiesbaden from 1954 to 1958. They returned to their "home station" of Wright-Patterson for another tour from 1958 to 1962. At that time the conflict in Southeast Asia was escalating. Bob was transferred to Laos in early 1962.

Bob's last Air Force tour was as the Chief of the Air Force Section of the Joint U.S. Military Assistance Group for Laos. He spent considerable time in Vietnam where most of the heavy maintenance and repair of Air Force aircraft was performed. In late 1962 the Geneva Accords were signed and all countries had to pull their representatives out of Laos. The Uhrig family moved with the JUSMAG organization to Bangkok, Thailand and continued there until Bob retired from the Air Force as a Lieutenant Colonel in 1966 with 30 years of active duty service. He then spent one year with "Air America" from 1967-1968.

Bob decided to stay on in Thailand and was hired by Northrop Aircraft to run their Foreign Military Sales operations in Southeast Asia. He lost his beloved Toots in 1982 while they were living in Bangkok. Bob continued to work for Northrop in Thailand and Malaysia until the mid 1980s when he returned to a quiet life in the U.S., spending time with his family in Castile, New York and New Carlisle, Ohio. He passed away in 2003 and is buried next to Toots in Medway, Ohio, right off the flight path to the runway at Wright-Patterson AFB.

Bob and Toots in Thailand, 1966.

The technology has changed, but the mission remains the same:
"Anything, Anywhere, Anytime."

*Above:* USAAF C-47 (US Air Force)   *Below:* USAF C-17 (US Air Force)

Appendix A
Flight Schedule for Operation VARSITY

36TH TROOP CARRIER SQUADRON
OFFICE OF THE SQUADRON OPERATIONS OFFICER

APO 133, U.S. Army,
24 March 1945

## FLIGHT SCHEDULE

OPERATION "VARSITY"

| | | | | |
|---|---|---|---|---|
| (141) | 43-15179 | H | 7 | Wright - Hance - Schechter - Rucker - Maltais |
| (142) | 43-15634 | V | 02 | Sullivan - Rhoden - Phillips - Brown - Flaherty |
| (143) | 42-24389 | K | 10 | Lee - Mittwoch - Nordness - Friedman |
| (144) | 43-15256 | D | 3 | Welter - Pepple - Schwind - Kesner - Kunz |
| (145) | 42-24392 | T | 19 | Greenstein - Rewinkle - Richey - Reed |
| (146) | 42-68758 | I | 8 | Eustice - Rose - Little - Horne |
| 147) | 42-100973 | E | 4 | Miller - O'Daniel - Ballash - McAllister - Murphy |
| (148) | 42-93780 | B | 1 | Cooke - Minotti - Garcia - Higgins |
| (149) | 42-23503 | L | 11 | Konrad - Murphy - Turner - Fowler |
| (150) | 43-30652 | N | 13 | McCullough - Dronen - Nagle - McPherson - Ouellette |
| (151) | 42-100678 | U | 01 | Wheeler - Spence - Campbell - Hill |
| (152) | 43-30721 | O | 14 | White - Stringer - Erwin - Balco |
| (153) | 43-15205 | F | 5 | Pace - Ransford - Piazza - Keating - O'Donnell |
| (154) | 42-108902 | M | 12 | Jackson - Wilde - Hearn - Glick |
| (155) | 42-23901 | R | 17 | Smith,C.T. - Rudolph - Shanta - Swartz |
| (156) | 42-100872 | P | 15 | Baxter - Chambers - Yurgelun - Moreno - Pranke |
| (157) | 43-15258 | C | 2 | Ash - Mayer - Babbitt - Davis |
| (158) | 42-68765 | J | 9 | Shank - Tazelaar - Sanders - Trevarton |
| 159) | 43-15227 | S | 18 | Etter - McBrayer - Cunningham - Griffith - Robbins |
| (160) | 42-108909 | G | 6 | Whatley - Ernst - Przybylski - Szego |

### SPARES

| | | | | |
|---|---|---|---|---|
| | 42-68769 | A | 00 | Bynum - Thornton - Rneken - Skolnik |
| | 42-100502 | W | 03 | Smith,F.A. - Cornelius - Uselman - Smith |

By order of Lt Colonel WRIGHT:

JAMES L. ROBERTS,
Major, Air Corps,
Operations Officer.

Appendix B
36th Troop Carrier Squadron
Airmen in Overseas American Cemeteries

| Name | Rank | Ser No | State | Date of Death | Cemetery |
|---|---|---|---|---|---|
| ALLEY WILBUR K | 2LT | O1703993 | AR | 11-Jul-43 | Florence, IT |
| BEST ROSCOE M | TSGT | 14040414 | TN | 11-Jul-43 | Sicily-Rome, IT |
| CHURCHILL RICHARD | CAPT | O-790516 | ME | 11-Jul-43 | Sicily-Rome, IT |
| CULL DALTON W | TSGT | 15065400 | OH | 11-Jul-43 | Normandy, FR |
| DOVER ROLAND V JR | 1LT | O-481608 | TX | 18-Sep-44 | Nettuno, IT |
| FARRIS JAMES R | MAJ | O-789698 | TX | 12-May-44 | Cambridge, UK |
| FLANNIGAN HOWARD T | SGT | 32234566 | NY | 11-Jul-43 | Sicily-Rome, IT |
| JUNO FRANK | CPL | 36195078 | MI | 11-Jul-43 | Sicily-Rome, IT |
| LOFFRFDO JAMES A | FLT O | T-190980 | IA | 11-Jul-43 | Sicily-Rome, IT |
| MAC GREGOR RODNEY J | 2LT | O-666766 | CT | 11-Jul-43 | North Africa |
| MOBLEY LUCIEN D | FLT O | T-190987 | KY | 11-Jul-43 | Sicily-Rome, IT |
| MORGAN WILLIAM H | 1LT | O-662077 | OH | 12-Jul-43 | Sicily-Rome, IT |
| POPPLETON ALBERT R | TSGT | 39827062 | ID | 5-Mar-44 | Cambridge, UK |
| SINGER GEORGE A | TSGT | 15200766 | OH | 11-Jul-43 | Sicily-Rome, IT |
| STILL GLENN R | 1LT | O-744871 | NE | 12-May-44 | Cambridge, UK |
| TOMCHEK THADDEUS T | SSGT | 33131463 | PA | 12-May-44 | Cambridge, UK |
| TYSON HILDREN | 1LT | O-680121 | PA | 23-Dec-44 | Luxembourg |

Information on these burials is from the American Battle Monuments Commission (ABMC), who maintain America's overseas commemorative cemeteries and memorials.

# Appendix C
## 36th Troop Carrier Squadron Officers - 1944

36th TCS officers outside their quarters at Cottesmore. Photo taken in mid-1944. Bob Uhrig, then a captain, is in the bottom row, fifth from the left. (courtesy David Nagle collection)

# Notes

## Chapter 2  Join The Air Corps

1 Some of the material in this section is adapted from "Early History of Patterson Field," unpublished manuscript by Leith Levi Mancum, 1997; and from Lois E. Walker and Shelby Wickam, *From Huffman Prairie to the Moon: The History of Wright-Patterson Air Force Base*. (Washington D.C.:  U.S. Government Printing Office, 1986.)

## Chapter 3  War is Declared

2  Charles Young, *Into the Valley: The Untold Story of USAAF Troop Carrier in WWII*. (Dallas:  PrintComm, Inc, 1995.), pp. 55-56.

## Chapter 4  Airborne All the Way

3  John C. Andrews, *Airborne Album, Volume 1: Parachute Test Platoon to Normandy* (Williamstown, NJ: Phillips Publications, 1982), p.9.
4  Richard Ladd, 502$^{nd}$ PIR, 101$^{st}$ Airborne, interview with the author, October 2010.
5  Military Airlift Command Office of History, *Anything, Anywhere, Anytime: An Illustrated History of the Military Airlift Command, 1941-1991*. (Scott Air Force Base IL: Headquarters Military Airlift Command, 1991.), p. 44.
6  Michael N. Ingrisano, Jr., *Valor Without Arms: The History of the 316$^{th}$ Troop Carrier Group, 1942-1945* (Bennington, VT: Merriam, 2001), p. 16.
7  Ingrisano, pp. 16-17.

## Chapter 6  Chasing the Afrika Korps

8  Kenn C. Rust, *Ninth Air Force Story in World War II* (Temple City, CA: Historical Aviation Album, 1992), p.7.
9  Bernard Law Montgomery, *The Memoirs of Field Marshall The Viscount Montgomery of Alamein, K.G.* (London: Collins, 1958), p. 143.
10  Rust, p. 9.
11  David Isby, *C-47/R4D Units of the ETO and MTO* (Oxford, UK:  Osprey Publishing Ltd, 2005), p.14.
12  Lewis H. Brereton, *The Brereton Diaries: The War in the Air in the Pacific, Middle East and Europe, 3 October 1941-8 May 1945* (New York: W. Morrow and Company, 1946.), p. 167.
13  Information on the P-40 units at Marble Arch and the incident with the wounded soldiers was taken from the web page "A Medic's Memory Lane" by Stuart Morris, from the website of Number 3 Squadron, Royal Australian Air Force, accessed February 2012. URL: http://www.3squadron.org.au/subpages/stu.htm. It describes the exact incident Bob Uhrig recorded in his diary.
14  Carl Molesworth, *57th Fighter Group: First in the Blue* (London: Osprey Publishing Ltd., 2011) p. 31.
15  Information on the Robeson crew from Ingrisano, p. 30-31.
16  William G. F. Jackson, *The Battle for North Africa 1940-1943* (New York: Mason/Charter, 1975), p.419.
17  Rick Atkinson, *An Army at Dawn: The War in North Africa, 1942-1943* (New York: Henry Holt and Company, 2002), p. 529.
18  Jackson, p. 410.
19  *316$^{th}$ Troop Carrier Group* (Pope Field, NC: privately printed, 1945), no page numbers.
20  *316$^{th}$ Troop Carrier Group*.

## Chapter 7  Training and Tragedy

21  Some of the material in this section is taken from Phil Nordyke, *All American All The Way: The Combat History of the 82$^{nd}$ Airborne Division in World War II* (St Paul, MN: Zenith Press, 2005), pp 35-39.
22  Deryk Wills, *Put On Your Boots and Parachutes: Personal Stories of the Veterans of the United States 82$^{nd}$ Airborne Division* (Oadby, Leicestershire, UK: Deryk Wills, 1982), p. 19.
23  Nordyke, *All American All The Way*, p. 36.
24  Some of the material in this section is adapted from Nordyke, *All American All The Way*, pp. 36-38.
25  Wills, p. 28.
26  Wills, p. 29.
27  Phil Nordyke, *More Than Courage: The Combat History of the 504$^{th}$ Parachute Infantry Regiment in World War II* (St Paul, MN: Zenith Press, 2008), p. 58.
28  Isby, p. 4.
29  Albert N. Garland and Howard Smyth, *The Mediterranean Theater of Operations:  Sicily and the Surrender of Italy* (Washington DC: Center for Military History, United States Army, 1965), p.425.
30  Rex Shama, *Pulse and Repulse: Troop Carrier and Airborne Teams in Europe During World War II* (Austin, TX: Eakin Press, 1995), p. 22.
31  Army Air Forces Historical Office, *Army Air Forces Historical Study No. 37: Participation of the Ninth and Twelfth Air Forces in the

*Sicilian Campaign* (Headquarters U.S. Army Air Forces, 1945) p.145.
32  Some of the material in this section is adapted from Nordyke, *All American All The Way*, pp. 97-98.
33  To the extent possible, the author attempted to determine the final resting place of a number of 316th men killed in action or who died overseas. (See Appendix B.) Some remains may have been repatriated. Records of those men still buried in overseas cemeteries are available from the American Battle Monuments Commission, URL: www.abmc.gov.
34  Wills, p. 14.
35  Wills, pp. 14-15.
36  Adapted from Nordyke, *All American All The Way*, p. 99.

## Chapter 8 On to Sicily

37  Gerald M. Devlin, *Paratrooper! The Saga of U.S. Army and Marine Parachute and Glider Combat Troops during World War II* (New York: St Martin's Press, 1979), p. 300.
38  Devlin, p. 300.
39  Atkinson, p. 297.
40  Ingrisano, p. 341.

## Chapter 9 Crossroads

41  Office of Air Force History, *Condensed Analysis of the Ninth Air Force in the European Theater of Operations* (Washington DC: Office of Air Force History 1946 (1984 reprint)), p. 74.
42  *Condensed Analysis of the Ninth Air Force*, pp. 74-75.
43  George Quisenberry's assignment to the Pathfinders from Young, p. 27.
44  *Condensed Analysis of the Ninth Air Force*, p. 75.
45  Ladd, Interview with the author, October 2010.

## Chapter 10 Invasion

46  Wills, p. 202.
47  Wills, p. 62.
48  Some of the material in this section is adapted from Ingrisano, pp. 103-109.
49  Wills, p. 78.
50  Wills, p. 66.
51  Napier Crookenden, *Dropzone Normandy: The Story of the American and British Airborne Assault on D-Day 1944* (Abingdon, Oxon: Purnell Book Services Ltd., 1976.), p. 113-117.
52  Wills, p. 69.
53  John C. Warren, U.S. Air Force Historical Study No. 97, *Airborne Operations in World War II, European Theater* (Maxwell Air Force Base, AL: U.S. Air Force Air University,1956), p. 50.
54  Hugh M. Cole, *The Lorraine Campaign* (Washington DC: United States Army Historical Division, 1950), p. 23.

## Chapter 11 Liberation

55  Nordyke, *All American All The Way*, p. 413.
56  Nordyke, *All American All The Way*, p. 414.
57  Wills, p. 138.
58  Wills, p. 138.
59  Story of the lost aircraft from Ingrisano, pp. 161-162.
60  Ingrisano, p.224.

## Chapter 12 Victory

61  Ingrisano, p.224.
62  Story of Col Krebs' escape from Young, pp. 291-296.
63  Peter Allen, *One More River: The Rhine Crossings of 1945* (New York: Barnes & Noble 1994, reprint of Simon and Schuster 1980 ed.), p. 254.
64  Allen, p. 261.
65  Allen, p. 271.
66  Edward M. Flanagan, *Airborne: A Combat History of American Airborne Forces* (New York: Ballantine Books, 2002), p. 296.
67  Flanagan, p. 297.
68  *Condensed Analysis of the Ninth Air Force*, p. 117.

# Bibliography/Sources

## Books and Reports

Allen, Peter. *One More River: The Rhine Crossings of 1945*. New York: Barnes and Noble, 1994 (reprint of Simon and Schuster 1980 edition.)

Andrews, John C. *Airborne Album Volume 1: Parachute Test Platoon to Normandy*. Williamstown, NJ: Phillips Publications, 1982.

Army Air Forces Historical Office. *Army Air Forces Historical Study No. 37: Participation of the Ninth and Twelfth Air Forces in the Sicilian Campaign*. Headquarters, U.S. Army Air Forces November 1945.

Atkinson, Rick. *An Army at Dawn: The War in North Africa, 1942-1943*. New York, Henry Holt and Company, 2002.

Bowers, Peter. *The DC-3: 50 Years of Legendary Flight*. Blue Ridge Summit, PA: TAB Aero Publishers, 1986.

Brereton, Lewis H. *The Brereton Diaries: The War in the Air in the Pacific, Middle East and Europe, 3 October 1941-8 May 1945*. New York: W. Morrow and Company, 1946.

Center for Air Force History. *Airborne Assault on Holland: An Interim Report*. Wings at War Series, No. 4. Reprint of document originally published by Headquarters, Army Air Forces. Washington DC: Center for Air Force History, 1982.

Cloe, John H. *Top Cover and Global Engagement: A History of the 11th Air Force*. Anchorage: Alaska Quality Publishing, 2001.

Cole, Hugh M. *The Lorraine Campaign*. The United States Army in WWII Series: European Theater of Operations. Washington, DC: United States Army Historical Division, 1950.

Crookenden, Napier. *Dropzone Normandy: The Story of the American and British Airborne Assault on D-Day 1944*. Abingdon, Oxon: Purnell Book Services Ltd., 1976.

Davis, Larry. *C-47 Skytrain in Action*. Squadron/Signal Aircraft Number 149. Carrollton, TX: Squadron/Signal Publications, Inc., 1995.

Devlin, Gerard M. *Paratrooper! The Saga of U.S. Army and Marine Parachute and Glider Combat Troops During World War II*. New York: St. Martin's Press, 1979.

Flanagan, Edward. M. *Airborne: A Combat History of American Airborne Forces*. New York: Ballantine Books, 2002.

Garland, Albert N. and Howard Smyth, *The Mediterranean Theater of Operations: Sicily and the Surrender of Italy*. The United States Army in World War II Series. Washington D.C.: Center for Military History, United States Army, 1965.

Glines, Carroll V. *The Amazing Gooney Bird: The Saga of the Legendary DC-3/C-47*. Altgelt PA: Schiffer Publishing, 2000.

Ingrisano, Michael N. *Valor Without Arms: The History of the 316th Troop Carrier Group, 1942-1945*. Bennington, VT: Merriam, 2001.

Isby, David. *C-47/R4D Units of the ETO and MTO*. Osprey Combat Aircraft Series No. 54. Oxford, UK: Osprey Publishing Ltd., 2005.

Jackson, William G.F. *The Battle for North Africa 1940-1943*. New York: Mason/Charter, 1975.

Military Airlift Command Office of History. *Anything, Anywhere, Anytime: An Illustrated History of the Military Airlift Command, 1941-1991*. Scott AFB IL: Headquarters Military Airlift Command, May 1991.

Molesworth, Carl. *57th Fighter Group: First in the Blue*. Osprey Aviation Elite Units No. 39. London: Osprey

Publishing, Ltd., 2011.

Montgomery, Bernard Law. *The Memoirs of Field Marshal The Viscount Montgomery of Alamein, K..G.* London: Collins, 1958.

Nordyke, Phil. *All American All The Way: The Combat History of the 82nd Airborne Division in World War II.* St Paul, MN: Zenith Press, 2005.

------------------. *More Than Courage: The Combat History of the 504th Parachute Infantry Regiment in World War II.* St Paul, MN: Zenith Press, 2008.

Office of Air Force History. *Condensed Analysis of the Ninth Air Force in the European Theater of Operations.* Washington, DC: Office of Air Force History, 1946 (1984 reprint.)

Pearcy, Arthur. *A Celebration of the DC-3.* Leicester, UK: Promotional Reprint Company, Ltd., 1985.

Rottman, Gordon. *US Army Airborne 1940-90: The First Fifty Years.* Osprey Elite Series No. 31. London: Osprey Publishing Ltd., 1990.

Rust, Kenn C. *Ninth Air Force Story in World War II.* Temple City, CA: Historical Aviation Album, 1982.

Shama, H. Rex. *Pulse and Repulse: Troop Carrier and Airborne Teams in Europe During World War II.* Austin, TX: Eakin Press, 1995.

Shamburger, Page and Joe Christy. *The Curtiss Hawk Fighters.* New York: Sports Car Press, 1971.

*316th Troop Carrier Group.* Unit history published by group members. Pope Field, NC: Privately printed, ca. 1945.

Walker, Lois E. and Shelby Wickam. *From Huffman Prairie to the Moon: The History of Wright-Patterson Air Force Base.* Washington D.C.: U.S. Government Printing Office, 1986.

Warren, John C. *Airborne Operations in World War II, European Theater*, U.S. Air Force Historical Study No. 97. Maxwell Air Force Base AL: U.S. Air Force Air University, 1956.

Weeks, John. *Airborne Equipment: A History of its Development.* London: David & Charles, 1976.

Wills, Deryk. *Put On Your Boots and Parachutes: Personal Stories of the Veterans of the United States 82nd Airborne Division.* Oadby, Leicester, UK: Deryk Wills, 1992.

Young, Charles H. *Into the Valley: The Untold Story of USAAF Troop Carrier in WWII.* Dallas: PrintComm, Inc, 1995.

Zaloga, Steven. *US Airborne Divisions in the ETO 1944-1945.* Osprey Battle Orders Series No. 25. Oxford, UK: Osprey Publishers Ltd., 2007.

## Archival Materials

Archives of the 316th Troop Carrier Group 1942-1945, Air Force Historical Research Agency (AFHRA), Maxwell Air Force Base, Alabama.

Stuart Morris, "A Medic's Memory Lane." Webpage from the website of the Number 3 Squadron, Royal Australian Air Force. URL: http://www.3squadron.org.au/subpages/stu.htm. Accessed February 2012.

Unpublished manuscript of the History of Patterson Field, Leith Levi Mancum, 1997.

## Interviews

Richard Ladd, 502nd Parachute Infatnry Regiment, 101st Airborne Division, October 2010.

Mahlon Hamilton, Troop Carrier Pilot, US Army Air Forces, China-Burma-India Theater, August 2010.

## About the Author

Brian J. Duddy attended Embry Riddle Aeronautical University, earning a BS in Aeronautical Engineering and was commissioned through the Air Force ROTC program. He has a MA in History from Wright State University; an MS in Management from St Mary's University and is a graduate of the Air Force Air War College. He served 24 years in the United States Air Force, retiring in 2007 as a Lieutenant Colonel. He has previously published articles in *Flying Safety*, *Defense Acquisition, Technology & Logistics*, *Defense Acquisition Research Journal*, *Unmanned Systems*, *Finescale Modeler* and *Military Modeling* magazines and been a finalist in the International Imitation Hemingway Competition. His first book: *Wings Over LeRoy: the History of the Donald Woodward Airport, LeRoy, NY*, was published in 2008. He lives in western New York.

www.ingramcontent.com/pod-product-compliance
Lightning Source LLC
Chambersburg PA
CBHW082115230426
43671CB00015B/2708